Study Guide for

Reteaching and Practice

by Kay Thompson

Structure and Method Book 2

Brown
Dolciani
Sorgenfrey
Kane

McDougal Littell
A HOUGHTON MIFFLIN COMPANY
EVANSTON, ILLINOIS
BOSTON • DALLAS • PHOENIX

About the Study Guide for Reteaching and Practice

The Study Guide for Reteaching and Practice is a set of worksheets designed for reteaching, extra practice, and other make-up work.

- **Two pages** of material are provided for each lesson in *ALGEBRA AND TRIGONOMETRY, Structure and Method, Book 2.*
- **Vocabulary review, worked-out examples,** and **cautions** about common errors prepare students for the practice exercises.
- **Practice exercises,** mainly at "A" level, follow the worked-out examples and build gradually in difficulty.
- Short **Mixed Reviews** in every other lesson boost students' skill retention.
- Answers are provided in a separate **Answer Key.**

2004 Impression

ISBN 13: 978-0-395-47064-0 ISBN 10: 0-395-47064-1

26 27 28 29 30 – VEI – 12 11 10 09 08

13 Trigonometric Graphs; Identities

14 Trigonometric Applications

15 Statistics and Probability

16 Matrices and Determinants

Symbols

Symbol	Meaning	Page
$\{\ \}$	set	1
$\in$	is an element of	
$=$	equals *or* is equal to	1
$>$	is greater than	1
$<$	is less than	1
$\lvert a \rvert$	absolute value of a	1
$-a$	additive inverse of a *or* opposite of a	1
$\stackrel{?}{=}$	is equal to?	13
$\neq$	does not equal	13
$\therefore$	therefore	13
$\emptyset$	empty set *or* null set	13
$\geq$	is greater than or equal to	21
$\leq$	is less than or equal to	21
$a<x<b$	x is greater than a and less than b *or* x is between a and b	21
$P(a, b)$	point P with coordinates (a, b)	35
x_1	x sub 1	37
$f(x)$	f of x *or* the value of f at x	47
a^n	nth power of a	55
$\approx$	is approximately equal to	75
$\pm$	plus-or-minus sign	89
$\sqrt[n]{b}$	nth root of b	89
i	imaginary unit ($i^2 = -1$)	101
$b^{1/n}$	nth root of b	155
$b^{p/q}$	qth root of the pth power of b	155

Symbol	Meaning	Page
f^{-1}	inverse function of f	159
$\log_b N$	logarithm base b of N	161
$\ln x$	natural logarithm of x *or* logarithm base e of x	169
t_n	nth term of a sequence	171
Σ	summation sign	177
∞	infinity	177
S_n	sum of the first n terms of a series	179
$!$	factorial	185
$^\circ$	degree	187
$'$	minute	187
$''$	second	187
$\overrightarrow{AB}$	vector AB	223
$\lVert \mathbf{v} \rVert$	norm of vector $\mathbf{v}$	223
Cos^{-1}	inverse cosine *or* Arc cosine	233
σ	standard deviation	241
r	correlation coefficient	245
${}_nP_r$	number of permutations of n elements taken r at a time	249
${}_nC_r$	number of combinations of n elements taken r at a time	251
$P(E)$	probability of event E	255
$\cap$	intersection	257
$\cup$	union	257
$\overline{A}$	complement of event A	257
$A_{m \times n}$	matrix A with m rows and n columns	259
$\det A$	determinant of matrix A	267
A^{-1}	inverse of matrix A	269

Greek letters: α, β, γ, θ, π, σ, ϕ, ω *alpha, beta, gamma, theta, pi, sigma, phi, omega*

Table of Measures

	Metric Units	
Length	10 millimeters (mm) =	1 centimeter (cm)
	100 centimeters, 1000 millimeters =	1 meter (m)
	1000 meters =	1 kilometer (km)
Area	100 square millimeters (mm^2) =	1 square centimeter (cm^2)
	10,000 square centimeters =	1 square meter (m^2)
Volume	1000 cubic millimeters (mm^3) =	1 cubic centimeter (cm^3)
	1,000,000 cubic centimeters =	1 cubic meter (m^3)
Liquid Capacity	1000 milliliters (mL) =	1 liter (L)
	1000 cubic centimeters =	1 liter
Mass	1000 milligrams (mg) =	1 gram (g)
	1000 grams =	1 kilogram (kg)
Temperature in degrees Celsius (°C)	0°C =	freezing point of water
	100°C =	boiling point of water

	United States Customary Units	
Length	12 inches (in.) =	1 foot (ft)
	36 inches, 3 feet =	1 yard (yd)
	5280 feet, 1760 yards =	1 mile (mi)
Area	144 square inches (in.^{2}) =	1 square foot (ft^2)
	9 square feet =	1 square yard (yd^2)
Volume	1728 cubic inches (in.^{3}) =	1 cubic foot (ft^3)
	27 cubic feet =	1 cubic yard (yd^3)
Liquid Capacity	16 fluid ounces (fl oz) =	1 pint (pt)
	2 pints =	1 quard (qt)
	4 quarts =	1 gallon (gal)
Weight	16 ounces (oz) =	1 pound (lb)
Temperature in degrees Fahrenheit (°F)	32°F =	freezing point of water
	212°F =	boiling point of water

Time	
60 seconds (s) =	1 minute (min)
60 minutes =	1 hour (h)

NAME ______________________ DATE ______________

1 Basic Concepts of Algebra

1–1 *Real Numbers and Their Graphs*

Objective: To graph real numbers on a number line, to compare numbers, and to find their absolute values.

Vocabulary

Real numbers The set consisting of all the rational and irrational numbers. A rational number is the result of dividing an integer by a nonzero integer. An irrational number is one that is not rational.

Examples of rational numbers: $5 \quad 0 \quad -2 \quad 6.3 \quad \frac{4}{7} \quad 1.33\ldots$

Examples of irrational numbers: $\pi \quad \sqrt{2} \quad \sqrt{5}$

Coordinate of a point The real number paired with that point on a number line. Example: The coordinate of point *M* is 3.

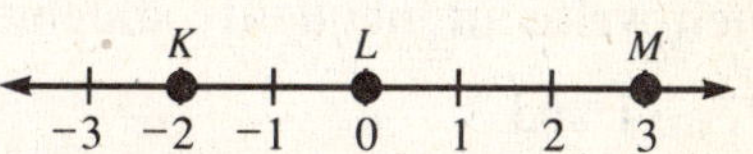

Origin The graph of zero on a number line.

Opposites On a number line, numbers that are the same distance from zero but on opposite sides of it. Example: -2 and 2

Absolute value of a number On a number line, the distance between the graph of the number and zero. Examples: The absolute value of 3 is 3 (write $|3| = 3$). The absolute value of -3 is also 3 (write $|-3| = 3$).

Symbols $-$ (opposite of) $\quad | \; |$ (absolute value) $\quad >$ (is greater than) $\quad <$ (is less than)

Example 1 Find the coordinate of the point one fourth of the way from *S* to *W* on the number line below.

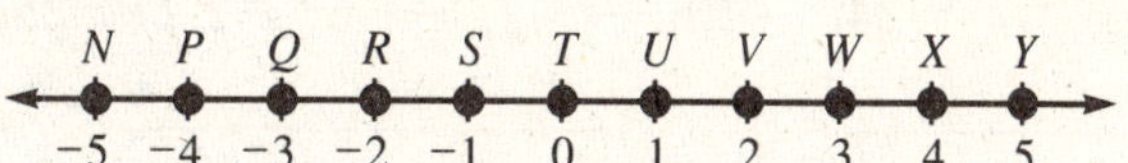

Solution The distance from *S* to *W* is 4 units. One fourth of 4 is 1. Move 1 unit to the right of *S* to find the desired point. The coordinate is 0.

Find the coordinate of each point using the number line in Example 1.

1. *Y* **2.** *T* **3.** *U* **4.** *P* **5.** *X* **6.** *N*

7. The point halfway between *Q* and *Y*

8. The point one fourth of the way from *R* to *V*

9. The point two thirds of the way from *P* to *V*

10. The point three fourths of the way from *Q* to *W*

Example 2 Write each statement using symbols.

a. Five is greater than negative two. **b.** Negative ten is less than zero.

Solution **a.** $5 > -2$ **b.** $-10 < 0$

NAME ______________________ DATE ______________

1-1 *Real Numbers and Their Graphs* *(continued)*

Write each statement using symbols.

11. Four is less than six.

12. Three is greater than negative three.

13. Negative two is less than one.

14. Negative one is greater than negative six.

15. Negative one half is less than zero.

16. Five is greater than negative eleven.

Example 3 Graph the numbers 1.5 and -2.5 on a number line. Then write an inequality statement comparing them. Remember: the numbers increase from left to right.

Solution

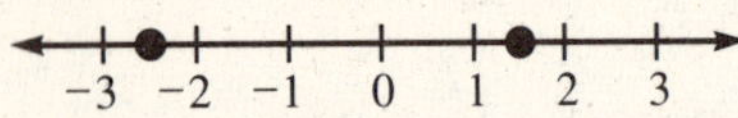

$-2.5 < 1.5$, or $1.5 > -2.5$

Graph each pair of numbers on a separate number line. Then write an inequality statement comparing the numbers.

17. -1 and 3

18. 0 and -3

19. $\frac{1}{3}$ and $-\frac{1}{3}$

20. $-\frac{2}{3}$ and $-\frac{4}{3}$

21. -1.25 and 0.75

22. -0.75 and -1.5

Example 4 Arrange -3, 0, 1.5, and -2.5 in order from least to greatest.

Solution First graph the numbers on a number line. Then write the numbers as they appear from left to right.

$-3, -2.5, 0, 1.5$

Arrange each list of numbers in order from least to greatest.

23. $4, -2, 0, -3, 2$

24. $-2.6, -3.2, -1.6, -2.2$

25. $-\frac{2}{3}, \frac{1}{6}, -1, \frac{1}{3}, -\frac{4}{3}$

Example 5 Find the value of each expression.

a. $-|-14|$ **b.** $|-8| - |-3|$

Solution First find the absolute values. Then simplify.

a. $-|-14| = -(14) = -14$

b. $|-8| - |-3| = 8 - 3 = 5$

Find the value of each expression.

26. $|-12|$

27. $-|12|$

28. $|4| - |-1|$

29. $|-7| \cdot 3$

Mixed Review Exercises

Tell whether each statement is true or false.

1. $2\frac{2}{3} - \frac{1}{2} > 2$

2. $\frac{5}{4} \div \frac{1}{2} = \frac{5}{2}$

3. $\frac{1}{4} + \frac{3}{5} = \frac{4}{9}$

4. $34.6 + 5.23 = 86.9$

5. $0.3 \times 0.2 < 0.6$

6. $10 \times 0.01 < 1$

NAME ______________________________ DATE ______________

1–2 Simplifying Expressions

Objective: To review the methods used to simplify numerical expressions and to evaluate algebraic expressions.

Vocabulary

Variable A symbol, usually a letter, used to represent one or more numbers.

Algebraic expression A numerical expression; a variable; or a sum, difference, product, or quotient that contains one or more variables.
Examples: $6 + 11$ $\quad x \quad$ $x^2 - 5y + z$

Simplify To simplify an expression you replace it by the simplest or most common symbol having the same value.

Evaluate an expression To evaluate an algebraic expression, or find its value, replace each variable in the expression by a given value and simplify the result.

Power A product of equal factors. The repeated factor is the *base*. A positive *exponent* tells the number of times the base occurs as a factor. Example: $5 \times 5 \times 5 = 5^3$ is a power in which 5 is the base and 3 is the exponent.

Absolute value If x is positive or zero, $|x| = x$.
If x is negative, $|x| = -x$ (reads "the opposite of x").

Symbols Grouping: ( ) (parentheses) [ ] (brackets) —— (fraction bar)

CAUTION Remember to use the correct order of operations. Expressions inside grouping symbols are simplified first, and then powers. Next, multiplication and division are done *in order from left to right.* Finally, addition and subtraction are done in order from left to right.

Example 1 Use one of the symbols <, =, or > to make a true statement.

a. $5^2 + 3^2 \ \underline{?}\ (5 + 3)^2$ **b.** $(7 + 2) + 8 \ \underline{?}\ 7 + (2 + 8)$

Solution Find the value of each side. Then compare the results.

a. $5^2 + 3^2 = 25 + 9 = 34$
and $(5 + 3)^2 = 8^2 = 64$
Since 34 is less than 64,
$5^2 + 3^2 < (5 + 3)^2$.

b. $(7 + 2) + 8 = 9 + 8 = 17$
and $7 + (2 + 8) = 7 + 10 = 17$,
so $(7 + 2) + 8 = 7 + (2 + 8)$.

Use one of the symbols <, =, or > to make a true statement.

1. $1 \cdot 4 \ \underline{?}\ 1 \div 4$

2. $4 \cdot 1 \ \underline{?}\ 4 \div 1$

3. $4^2 \cdot 5^2 \ \underline{?}\ (4 \cdot 5)^2$

4. $4^2 + 5^2 \ \underline{?}\ (4 + 5)^2$

5. $\frac{5 + 3}{5 - 3} \ \underline{?}\ \frac{7 + 5}{7 - 5}$

6. $\frac{5 + 4}{5 - 4} \ \underline{?}\ \frac{8 + 4}{8 - 4}$

7. $(8 + 5) + 1 \ \underline{?}\ 8 + (5 + 1)$

8. $(8 - 5) - 1 \ \underline{?}\ 8 - (5 - 1)$

9. $(8 \cdot 5) \cdot 2 \ \underline{?}\ 8 \cdot (5 \cdot 2)$

10. $(16 \div 4) \div 2 \ \underline{?}\ 16 \div (4 \div 2)$

NAME ________________________ DATE ____________

1–2 Simplifying Expressions *(continued)*

Example 2 Simplify $9^2 - 6 \div 3 + 4$.

Solution

$$9^2 - 6 \div 3 + 4 = 81 - \underbrace{6 \div 3} + 4 \qquad \text{Simplify the power: } 9^2 = 9 \times 9.$$
$$= \underbrace{81 - 2} + 4 \qquad \text{Divide.}$$
$$= 79 + 4 \qquad \text{Subtract.}$$
$$= 83 \qquad \text{Add.}$$

Example 3 Simplify $11 + 5[8 - 3(6 - 4)]$.

Solution

$$11 + 5[8 - 3(6 - 4)] = 11 + 5 \cdot [8 - 3 \cdot 2] \qquad \text{Subtract inside parentheses.}$$
$$= 11 + 5 \cdot [8 - 6] \qquad \text{Multiply inside brackets.}$$
$$= 11 + 5 \cdot 2 \qquad \text{Subtract inside brackets.}$$
$$= 11 + 10 \qquad \text{Multiply.}$$
$$= 21 \qquad \text{Add.}$$

Simplify.

11. a. $18 - 7 + 3 - 1$ **b.** $18 - (7 + 3) - 1$ **c.** $18 - (7 + 3 - 1)$

12. a. $6 \cdot 4 + 5 \cdot 2$ **b.** $6 \cdot (4 + 5) \cdot 2$ **c.** $6 \cdot (4 + 5 \cdot 2)$

13. a. $6^2 - 8 \div 2 + 5$ **b.** $(6^2 - 8) \div 2 + 5$ **c.** $(6^2 - 8) \div (2 + 5)$

14. $3 \cdot 2^3 - (7^2 - 5^2)$ **15.** $16 - 3[9 - 2(5 - 3)]$ **16.** $[4(5 - 2) + 2^3] \div 2$

17. $\dfrac{3^3}{4 - (4 - 1)}$ **18.** $\dfrac{1}{2}\left|\dfrac{1 + 9^2}{5^2}\right|$ **19.** $\dfrac{4^3 + 6}{4^2 - 6}$

Example 4 Evaluate each expression if $r = 2$, $s = 5$, $t = 6$, and $u = -9$.

a. $3r^2 + s - 6$ **b.** $\dfrac{t^2 - 3rs}{t + r^2}$ **c.** $3|u| - |s|$

Solution Substitute the given values for the variables. Then simplify.

a. $3r^2 + s - 6 = 3(2)^2 + 5 - 6 = 3 \cdot 4 + 5 - 6 = 12 + 5 - 6 = 11$

b. $\dfrac{t^2 - 3rs}{t + r^2} = \dfrac{6^2 - 3 \cdot 2 \cdot 5}{6 + 2^2} = \dfrac{36 - 30}{6 + 4} = \dfrac{6}{10} = \dfrac{3}{5}$

c. $3|u| - |s| = 3|-9| - |5| = 3 \cdot 9 - 5 = 27 - 5 = 22$

Evaluate each expression if $w = -8$, $x = 2$, $y = 3$, and $z = 7$.

20. $y^2 - y + 1$ **21.** $3x^2 + x - 8$ **22.** $(xy - x)^3$ **23.** $\dfrac{2xy}{z^2 - y^3}$

24. $\left(\dfrac{xyz}{x + y + z}\right)^2$ **25.** $|w| - |x|$ **26.** $|w| + 3|y|$ **27.** $|w|^2 - |z^2|$

NAME ______________________ DATE ____________

1–3 Basic Properties of Real Numbers

Objective: To review properties of equality of real numbers and properties for adding and multiplying real numbers.

Vocabulary

Field Properties For all real numbers:

Closure property The sum or product of any two real numbers is a real number.

Commutative property Order doesn't matter with addition or multiplication.
Examples: $7 + 4 = 4 + 7$ $7 \cdot 4 = 4 \cdot 7$

Associative property Grouping doesn't matter with addition or multiplication.
Examples: $(8 + 3) + 7 = 8 + (3 + 7)$ $(8 \cdot 3) \cdot 7 = 8 \cdot (3 \cdot 7)$

Distributive property Multiplication is distributive over addition.
Examples: $3(5 + 2) = 3 \cdot 5 + 3 \cdot 2$ $(6 + 4)8 = 6 \cdot 8 + 4 \cdot 8$

Identity properties There is a unique (one and only one) real number 0, and a unique real number 1 such that $a + 0 = a = 0 + a$, and $a \cdot 1 = a = 1 \cdot a$.

Inverse Properties

Property of opposites The sum of a number and its *opposite* (or *additive inverse*) is 0. Every real number has a unique opposite.
Example: $8 + (-8) = 0$ and $-8 + 8 = 0$

Property of reciprocals The product of a number and its *reciprocal* (or *multiplicative inverse*) is 1. Every real number except zero has a unique reciprocal.
Example: $5 \cdot \frac{1}{5} = 1$ and $\frac{1}{5} \cdot 5 = 1$

Example 1 Simplify.

a. $27 + 3x + 23$ **b.** $2(a + 3) + (-6)$ **c.** $\left(\frac{2}{3}x\right)\left(\frac{3}{2}y\right)$

Solution

a. $27 + 3x + 23 = (27 + 23) + 3x$ Reorder and regroup.
$= 50 + 3x$

b. $2(a + 3) + (-6) = 2a + 6 + (-6)$ Use the distributive property.
$= 2a + 0$ Add opposites.
$= 2a$ Add zero.

c. $\left(\frac{2}{3}x\right)\left(\frac{3}{2}y\right) = \left(\frac{2}{3} \cdot \frac{3}{2}\right)(xy)$ Reorder and regroup.
$= 1(xy)$ Multiply reciprocals.
$= xy$ Multiply by 1.

Simplify.

1. $24 + 32 + 6 + 8$

2. $\frac{1}{5}(5) + (-2 + 2)$

3. $8 + 2a + (-8)$

4. $3(-x + x)$

5. $13(2 + a)$

6. $6(2x + 3) + (-18)$

7. $\left(\frac{5}{7}a\right)\left(\frac{7}{5}c\right)$

8. $(-1 + m) + \left(\frac{1}{4} \cdot 4\right)$

9. $\left(\frac{1}{2} \cdot 2\right)[x + (-2)] + 2$

NAME ______________________ DATE ______________

1–3 Basic Properties of Real Numbers *(continued)*

Vocabulary

Properties of Equality For all real numbers a, b, and c:

Transitive property If $a = b$, and $b = c$, then $a = c$.
Example: If $x = 2 + 7$, and $2 + 7 = 9$, then $x = 9$.

Addition property Adding the same number to both sides of an equation maintains equality. If $a = b$, then $a + c = b + c$.
Example: If $x = 8$, then $x + 3 = 8 + 3$.

Multiplication property Multiplying both sides of an equation by the same amount maintains equality. If $a = b$, then $ac = bc$.
Example: If $x = 5$, then $x \cdot 6 = 5 \cdot 6$.

Example 2 Name the property used in each step.

Solution

a. $\frac{1}{3}(1 + 3x) = \frac{1}{3} \cdot 1 + \frac{1}{3} \cdot (3x)$ — a. Distributive property

b. $= \frac{1}{3} \cdot 1 + \left(\frac{1}{3} \cdot 3\right)x$ — b. Associative property of multiplication

c. $= \frac{1}{3} \cdot 1 + (1)x$ — c. Property of reciprocals

d. $= \frac{1}{3} + x$ — d. Identity property of multiplication

Name the property used in each step of the simplification.

10. $\frac{1}{4}(y + 4)$

$= \frac{1}{4}y + \frac{1}{4} \cdot 4$ **a.** ?

$= \frac{1}{4}y + 1$ **b.** ?

11. $x + (x + 5)$

$= (x + x) + 5$ **a.** ?

$= (1 \cdot x + 1 \cdot x) + 5$ **b.** ?

$= (1 + 1)x + 5$ **c.** ?

$= 2x + 5$ Substitution

12. $5 + 2(x + 1)$

$= 5 + (2x + 2 \cdot 1)$ **a.** ?

$= 5 + (2x + 2)$ **b.** ?

$= 5 + (2 + 2x)$ **c.** ?

$= (5 + 2) + 2x$ **d.** ?

$= 7 + 2x$ Substitution

13. $a(b + 1) + (-1)a$

$= a(b + 1) + a(-1)$ **a.** ?

$= a[(b + 1) + (-1)]$ **b.** ?

$= a[b + (1 + (-1))]$ **c.** ?

$= a[b + 0]$ **d.** ?

$= ab$ **e.** ?

Mixed Review Exercises

Use <, =, or > to make a true statement.

1. $0.105 \ \underline{?}\ 0.1025$

2. $4(9 - 7) \ \underline{?}\ 4 \cdot 9 - 4 \cdot 7$

3. $(24 \div 6) \cdot 2 \ \underline{?}\ 24 \div (6 \cdot 2)$

4. $2^2 + 5^2 \ \underline{?}\ (2 + 5)^2$

5. $\frac{8 + 4}{7 - 1} \ \underline{?}\ \frac{7 + 1}{8 - 4}$

6. $\frac{2}{3} + \frac{2}{3} \ \underline{?}\ \frac{2 + 2}{3 + 3}$

NAME ______________________ DATE ____________

1–4 Sums and Differences

Objective: To review the rules for adding and subtracting real numbers.

Vocabulary

Rules for addition

1. When adding two numbers with the *same sign,* you add the absolute values of the numbers and keep the sign.
   Example: $-6 + (-3) = -(|-6| + |-3|) = -(6 + 3) = -9$
2. When adding two numbers with *opposite signs,* you subtract the *lesser* absolute value from the *greater* absolute value and keep the sign of the greater absolute value. Example: $4 + -9 = -(|-9| - |4|) = -(9 - 4) = -5$

Rule for subtraction To subtract a number, add its opposite: $a - b = a + (-b)$.
Examples: $4 - (-7) = 4 + 7 = 11$ $\quad -6 - 11 = -6 + (-11) = -17$

Distributive property (of multiplication over subtraction)
Example: $4(5 - 2) = 4 \cdot 5 - 4 \cdot 2 = 12$

Similar terms (or **like terms**) Terms with the same variables and exponents.
Examples: $5xy^2$ and $9xy^2$ are similar terms, but $6ab$ and $4ab^2$ are not.

Example 1 Simplify: **a.** $-13 + (-40)$ **b.** $-3.4 + 7.2$ **c.** $-14 - 28$

Solution

a. Use the *same sign* rule for addition. The answer will be negative.
$-13 + (-40) = -(|-13| + |-40|) = -(13 + 40) = -53$

b. Use the *opposite signs* rule for addition. The answer will be positive since 7.2 has the greater absolute value.
$-3.4 + 7.2 = |7.2| - |-3.4| = 7.2 - 3.4 = 3.8$

c. Use the rule for subtraction; add the opposite of 28.
$-14 - 28 = -14 + (-28) = -(14 + 28) = -42$

Simplify.

1. $-52 + 17$ **2.** $-27 - 14$ **3.** $12 - (-33)$ **4.** $-16 + (-36)$

5. $-96 - (-28)$ **6.** $0 - (-23.1)$ **7.** $-22.7 - (-22.7)$ **8.** $-16.5 - 12.5$

Example 2 Simplify: **a.** $-17 + 15 - 19 + 31$ **b.** $(3 - 5) - (7 - 2)$

Solution

a. *Method 1:* Add left to right.

$$\underbrace{-17 + 15} - 19 + 31$$
$$\underbrace{-2 - 19} + 31$$
$$\underbrace{-21 + 31}$$
$$10$$

Method 2: Group the negative terms and the positive terms.

$$-17 + 15 - 19 + 31$$
$$\underbrace{(-17 - 19)} + \underbrace{(15 + 31)}$$
$$\underbrace{-36 + 46}$$
$$10$$

b. Be sure to perform the operations inside parentheses first.
$(3 - 5) - (7 - 2) = -2 - 5 = -2 + (-5) = -7$

NAME ______________________ DATE ______________

1-4 Sums and Differences *(continued)*

Simplify.

9. $-18 - 19 + 12$

10. $-2 + 8 - 6 + 3$

11. $-12 + 3 + (-4) + 7$

12. $1 - (3 - 4) + (7 - 8)$

13. $3 - (5 + 1) - (1 - 4)$

14. $-6 - (12 - 20) + (6 - 10)$

15. $(-15 - 20) + [8 + (-3) - (-5)]$

16. $[-5 + (-17)] - (3 - 7 + 2)$

17. $|2 - 6| + |8 - 20|$

18. $|-21 - 3| - |8 + (-12)|$

Example 3 Multiply: **a.** $3(7x - 2y)$

b. $\frac{3}{4}(8x - 4y + 20)$

Solution Apply the distributive property.

a. $3(7x - 2y) = 3(7x) - 3(2y)$
$= 21x - 6y$

b. $\frac{3}{4}(8x - 4y + 20) = \frac{3}{4}(8x) - \frac{3}{4}(4y) + \frac{3}{4}(20)$
$= 6x - 3y + 15$

Multiply.

19. $5(2x + 1)$

20. $5(6 - y)$

21. $2(x - 8y)$

22. $\frac{2}{3}(6x - 3y)$

23. $\frac{1}{2}(10x - 4y - 2)$

24. $\frac{3}{5}(5x + 10y - 20)$

Example 4 Simplify by combining similar terms.

a. $3(2x + 3y) - 5x$ **b.** $(2r + 5s - 3) + 3(r - 2s - 4)$

Solution

a. $3(2x + 3y) - 5x = 6x + 9y - 5x$ Use the distributive property.
$= 9y + [6x - 5x]$ Group similar terms.
$= 9y + (6 - 5)x$ Use the distributive property.
$= 9y + x$ Simplify what is in parentheses.

b. $(2r + 5s - 3) + 3(r - 2s - 4)$
$= 2r + 5s - 3 + 3r - 6s - 12$ Use the distributive property.
$= [2r + 3r] + [5s - 6s] - 3 - 12$ Group similar terms.
$= (2 + 3)r + (5 - 6)s - 15$ Use the distributive property.
$= 5r - s - 15$ Simplify what is in parentheses.

Simplify by combining similar terms.

25. $19t + 3(8 + t)$

26. $2(3y - 5) - 18y$

27. $7(g - 2) + 8(g + 3)$

28. $3(9 - y) + 5(1 - y)$

29. $5d - 8 - 2d - 8$

30. $7x - 3y - 12x - (-5)y$

31. $(-6m - 3n + 2) + 5(m - 2n + 1)$

32. $4(3m - 6n) + 2(2m + 5n - 3)$

NAME ______________________ DATE ______________

1–5 Products

Objective: To review the rules for multiplying real numbers.

Vocabulary

Rules for multiplication

1. The product of two real numbers with *like signs* is a *positive* real number.
   Examples: $(3)(8) = 24$ $(-6)(-9) = 54$
2. The product of two reals with *opposite signs* is a *negative* real number.
   Example: $(5)(-14) = -70$
3. A product of nonzero numbers is *positive* if the number of negative factors is *even.*
   Example: $(-8)(-3)(-5)(-4) = 480$ (4 negative factors)
4. A product of nonzero numbers is *negative* if the number of negative factors is *odd.*
   Example: $(-6)(5)(-1)(-9) = -270$ (3 negative factors)
5. The absolute value of the product of two or more numbers is the product of their absolute values. Example: $|(3)(-8)(4)| = |3| \cdot |-8| \cdot |4| = 96$

Multiplicative property of 0 The product of any number and zero is zero.

Multiplicative property of −1 The product of any number and negative one is the opposite of that number. Examples: $(5)(-1) = -5$ $(-1)(-18) = 18$

Property of the opposite of a product For all real numbers a and b, $-ab = (-a)(b) = (a)(-b)$.
Example: $-(3)(5) = (-3)(5) = (3)(-5) = -15$

Property of the opposite of a sum For all real numbers a and b, $-(a + b) = (-a) + (-b)$.
Example: $-[3 + (-19)] = -3 + 19 = 16$

Example 1 Simplify.

a. $\left(\frac{1}{2}\right)(-8)(-6)\left(-\frac{1}{6}\right)$ **b.** $(4x)(-3y)(2)$ **c.** $(5 - 7)(-8 + 3)$

Solution

a. Multiply, beginning with the reciprocals.

$$\left(\frac{1}{2}\right)(-8)(-6)\left(-\frac{1}{6}\right)$$
$$= \left(\frac{1}{2}\right)(-8)(1)$$
$$= (-4)(1)$$
$$= -4$$

b. Reorder and regroup factors.

$$(4x)(-3y)(2)$$
$$= 4(-3)(2)xy$$
$$= -24xy$$

c. Simplify expressions in parentheses first.

$$(5 - 7)(-8 + 3)$$
$$= (-2)(-5)$$
$$= 10$$

Simplify.

1. $4(-2)(-3)(-5)$ **2.** $(1.4)(-3)(-0.2)$ **3.** $(-0.6)(-4)(-3)(-5.2)$

4. $\frac{1}{2}(-6)\left(-\frac{1}{12}\right)(-12)$ **5.** $(4)\left(-\frac{3}{8}\right)(12)$ **6.** $\left(-\frac{1}{2}\right)\left(-\frac{1}{3}\right)(0)(-2)(-3)$

7. $5(2x)(-3y)$ **8.** $\left(-\frac{1}{3}\right)(3u)(-v)$ **9.** $(4x)(-2y)(-3z)$

10. $(-a)(-b)(-c)$ **11.** $(4 - 5)(3 + 8)$ **12.** $(-12 - 3)(2 + 5)$

1–5 Products *(continued)*

Example 2 Simplify: **a.** $(-2)^3\left(-\frac{1}{4}\right)$ **b.** $3(-6) + 3(-2)$

Solution

a. Write $(-2)^3$ as a product of factors. Then multiply.

$$(-2)^3\left(-\frac{1}{4}\right)$$
$$= (-2)(-2)(-2)\left(-\frac{1}{4}\right)$$
$$= (4)(-2)\left(-\frac{1}{4}\right)$$
$$= (-8)\left(-\frac{1}{4}\right) = 2$$

b. Follow the order of operations.

$$3(-6) + 3(-2) = -18 + (-6)$$
$$= -24$$

Or, apply the distributive property.

$$3(-6) + 3(-2) = 3[-6 + (-2)]$$
$$= 3(-8)$$
$$= -24$$

Simplify.

13. $(14)(-1)^8(-3)^2$ **14.** $(-3)^2\left(-\frac{1}{9}\right)^2$ **15.** $(-2)^5(-3)(-1)$

16. $6(-8) + 6(5)$ **17.** $7(8.5) + 7(-1.5)$ **18.** $(-4)^3(-1 - 1)(-3)$

Example 3 Simplify $3(x^2 - 4) - 2(5x^2 - x)$.

Solution Distribute and simplify. Remember that a minus sign before a set of parentheses reverses each sign inside the parentheses.

$$3(x^2 - 4) - 2(5x^2 - x) = 3x^2 - 12 - 10x^2 + 2x$$
$$= -7x^2 + 2x - 12$$

Simplify.

19. $-2(x^2 - 3x - 2)$ **20.** $-4(-2x + 3y - z)$ **21.** $2\left(-6a - \frac{3}{2}\right)$

22. $(-r)(6s) + (3r)(-4s)$ **23.** $-5r - 4(2 + r)$ **24.** $5(2a + b) - 3(2a + b)$

25. $2(p^2 + 2) - 3(2p^2 - p)$ **26.** $4(2x - 3y) - 2(-2y + 3x)$ **27.** $m(n + 1) - 4(mn + 2)$

Mixed Review Exercises

Evaluate each expression if $a = -2$ and $b = 3$.

1. $2a + b + 1$ **2.** $-|a| + |b|$ **3.** $a + b^2$

4. $-|a - b|$ **5.** $a - 2(b - 5)$ **6.** $a - 4b$

Name the property illustrated in each statement.

7. $8 \cdot 4 = 4 \cdot 8$ **8.** $5 \cdot 1 = 5$ **9.** $\frac{2}{3} \cdot \frac{3}{2} = 1$

10. $-3 + 3 = 0$ **11.** $3 \cdot 6 + 3 \cdot 9 = 3(6 + 9)$ **12.** $(6 + 2) + 8 = 6 + (2 + 8)$

NAME ______________________________ DATE ______________

1–6 Quotients

Objective: To review rules for dividing real numbers.

Vocabulary

Division To divide by any *nonzero* real number, multiply by its reciprocal.

Examples: $15 \div 3 = 15 \cdot \frac{1}{3} = 5$ $\quad 12 \div \left(-\frac{4}{3}\right) = 12 \cdot \left(-\frac{3}{4}\right) = -9$

Since zero has no reciprocal, *division by zero is undefined.*

Rules for division

1. The quotient of two real numbers with *like signs* is a *positive* real number.
   Examples: $18 \div 3 = 6 \quad (-22) \div (-2) = 11$
2. The quotient of two real numbers with *opposite signs* is a *negative* real number.
   Example: $6 \div (-3) = -2$
3. For all *nonzero* real numbers *a, b,* and *c*:

$$\frac{a+b}{c} = \frac{a}{c} + \frac{b}{c} \quad \text{and} \quad \frac{a-b}{c} = \frac{a}{c} - \frac{b}{c}.$$

Example 1 Simplify $-6 \div 9 \div \frac{1}{3}$.

Solution Work from left to right.

$$-6 \div 9 \div \frac{1}{3} = \left(-6 \cdot \frac{1}{9}\right) \div \frac{1}{3}$$
$$= -\frac{2}{3} \div \frac{1}{3}$$
$$= -\frac{2}{3} \cdot 3 = -2$$

Simplify.

1. $-63 \div (-9)$
2. $-4 \div 16$
3. $-48 \div 8 \div (-2)$
4. $-18 \div \frac{2}{3}$
5. $-\frac{1}{3} \div \left(-\frac{1}{6}\right) \div (-4)$
6. $(-2)^3 \div [4(-6)]$

Example 2 Simplify: **a.** $\frac{(-6)(-8) \div (-2)}{4(-3)}$ **b.** $\frac{3\left(\frac{3}{4} - \frac{1}{4}\right)}{-\frac{1}{2} \div \frac{3}{2}}$

Solution Simplify the numerator and denominator separately. Then divide.

a. $\frac{(-6)(-8) \div (-2)}{4(-3)} = \frac{48 \div (-2)}{-12} = \frac{-24}{-12} = 2$

b. $\frac{3\left(\frac{3}{4} - \frac{1}{4}\right)}{-\frac{1}{2} \div \frac{3}{2}} = \frac{3\left(\frac{2}{4}\right)}{-\frac{1}{2} \cdot \frac{2}{3}} = \frac{\frac{3}{2}}{-\frac{1}{3}} = \frac{3}{2} \cdot (-3) = -\frac{9}{2}$

NAME ______________________ DATE ____________

1–6 Quotients *(continued)*

Simplify.

7. $\frac{(-2)(-8)(-9)}{(-12)(-3)}$ **8.** $\frac{(-4)(-12) \div (-3)}{8(-2)}$ **9.** $\frac{7(-12) \div 14}{4(-3)}$

10. $\frac{4^2 - 3^2}{4 - (-3)}$ **11.** $\frac{-8\left(-\frac{1}{2} - \frac{1}{4}\right)}{-\frac{3}{4} \div 3}$ **12.** $\frac{-27\left[12 \div \left(-\frac{3}{4}\right)\right]}{12\left(-\frac{3}{4}\right)}$

Example 3 Simplify $\frac{72 - 8x^2}{-4}$.

Solution Apply the third rule for division on page 11. Then simplify.

$$\frac{72 - 8x^2}{-4} = \frac{72}{-4} - \frac{8x^2}{-4}$$
$$= -18 - (-2x^2)$$
$$= -18 + 2x^2$$

Simplify.

13. $\frac{6x^2 - 21}{-3}$ **14.** $\frac{56 - 4x^2}{-4}$ **15.** $\frac{2 - (-x)^2}{-1}$

16. $\frac{-3x^2 - 3^2}{-3}$ **17.** $\frac{12 + 4x - 8x^2}{4}$ **18.** $\frac{-20x^3 - 15x^2 - 10x}{-5}$

Example 4 Evaluate $\frac{(x^2 + 2)(x - 3)}{x - 1}$ for $x = -2$.

Solution Substitute -2 for x. Then simplify using the order of operations.

$$\frac{(x^2 + 2)(x - 3)}{x - 1} = \frac{[(-2)^2 + 2](-2 - 3)}{-2 - 1}$$
$$= \frac{(4 + 2)(-5)}{-3}$$
$$= \frac{6(-5)}{-3}$$
$$= \frac{-30}{-3} = 10$$

Evaluate each expression for the given values of the variables.

19. $\frac{x(x + 5)}{x - 2}$ **a.** $x = 1$ **b.** $x = 0$ **c.** $x = 4$

20. $\frac{(y^2 - 2)(y + 3)}{y + 1}$ **a.** $y = -2$ **b.** $y = 1$ **c.** $y = 3$

21. $\frac{(n - 3)(n + 2)(n - 1)}{\frac{1}{4}n - 4}$ **a.** $n = -4$ **b.** $n = 0$ **c.** $n = 4$

22. $\frac{a(a + 5)(a - 2)}{(a + 1)(a - 3)}$ **a.** $a = 1$ **b.** $a = -3$ **c.** $a = \frac{1}{2}$

NAME ______________________________ DATE ______________

1–7 Solving Equations in One Variable

Objective: To solve certain equations in one variable.

Vocabulary

Open sentence An equation or inequality that contains one or more variables.

Solution set The set of all values of a variable that make an open sentence true. A solution is also called a *root*. A solution is said to *satisfy* an equation.

Transformations Changes that produce equivalent equations (ones with the same solution set). They include:

1. Simplifying either side of an equation.
2. Adding the same number to each side of an equation, or subtracting the same number from each side of an equation.
3. Multiplying (or dividing) each side of an equation by the same *nonzero* number.

Solve an equation Transform the equation into a simpler equivalent one whose solution set is easily seen.

Identity An equation that is satisfied by all values of the variable. The solution set of an identity is the set of all real numbers.

Symbols $\stackrel{?}{=}$ (Are they equal?) $\therefore$ (therefore)

$\neq$ (is not equal to) $\emptyset$ (empty or null set, the set with no members)

Example 1 Solve $3(2x - 3) = 4x + 7$. (The goal is to get x alone on one side.)

Solution

$6x - 9 = 4x + 7$	Simplify the left side.
$6x - 9 + 9 = 4x + 7 + 9$	Add 9 to each side.
$6x = 4x + 16$	
$6x - 4x = 4x + 16 - 4x$	Subtract $4x$ from each side.
$2x = 16$	
$\frac{2x}{2} = \frac{16}{2}$	Divide each side by 2.
$x = 8$	
Check: $3[2(8) - 3] \stackrel{?}{=} 4(8) + 7$	Substitute 8 for x in the *given* equation.
$3(16 - 3) \stackrel{?}{=} 32 + 7$	Simplify each side.
$3(13) \stackrel{?}{=} 39$	
$39 = 39$ ✓	

$\therefore$ the solution set is $\{8\}$.

Solve. Check your work.

1. $4x - 6 = 2$ **2.** $6 = 3x + 3$ **3.** $\frac{1}{2}x - 4 = -2$ **4.** $7 - \frac{1}{5}y = -2$

5. $48 - 6x = 2x$ **6.** $x + 2 = 3x - 6$ **7.** $4(x - 3) = 2x - 6$ **8.** $3(1 - y) = 3y$

NAME ______________________ DATE ______________

1–7 Solving Equations in One Variable *(continued)*

Example 2 Solve.

a. $3x - (x - 5) = 2(x + 4)$ **b.** $3x - (x - 8) = 2(x + 4)$

Solution

a.
$$3x - x + 5 = 2x + 8$$
$$2x + 5 = 2x + 8$$
$$5 = 8 \text{ False!}$$

No value of x will make this statement true.
$\therefore$ the solution set is $\emptyset$.

b.
$$3x - x + 8 = 2x + 8$$
$$2x + 8 = 2x + 8$$

This is an identity. It is true for all values of x.
$\therefore$ the solution set is the set of all real numbers.

Solve. Check your work when there is a single solution.

9. $5(x - 2) - 3 = -(x + 1)$ **10.** $2x - 1 = 2(x + 4)$ **11.** $2(x + 2) = 2x + 4$

12. $4a + 1 = a - 5 - 3a$ **13.** $1.2(u - 2) = 4.8$ **14.** $0.4(2r + 3) = 0.6r + 3.6$

15. $3(x - 2) = 5(x - 2)$ **16.** $\frac{1}{5}(x + 3) = x - 5$ **17.** $5z - (6 - z) = 4(1 - z)$

18. $11x - 3(4x - 2) = 2(8 + 2x)$ **19.** $\frac{5k - 3(k - 2)}{4} = -6$

Example 3 Solve the equation $T = c + ct$ for the variable t.

Solution

$$T = c + ct \quad \text{Work toward getting } t \text{ alone on one side.}$$
$$T - c = c + ct - c \quad \text{Subtract } c \text{ from each side.}$$
$$T - c = ct$$
$$\frac{T - c}{c} = \frac{ct}{c} \quad \text{Divide each side by } c. \text{ (Assume } c \neq 0.\text{)}$$
$$t = \frac{T - c}{c}$$

Solve each equation for the given variable.

20. $C = \pi d$ for d **21.** $I = prt$ for r **22.** $3x - 4y = 8$ for y

23. $P = 2l + 2w$ for l **24.** $ax + by + c = 0$ for x **25.** $A = 0.5h(a + b)$ for b

Mixed Review Exercises

Simplify.

1. $5(8 - 6 + 1)$ **2.** $(-4)^2 - 3^2$ **3.** $|5 + (-7)|$ **4.** $6\left(-\frac{1}{2}\right)(-4)\left(-\frac{1}{3}\right)$

5. $\frac{-12 \div 3}{-1 - 1}$ **6.** $9y - 2(3y - 8)$ **7.** $3(4ab) - (2a)(-7b)$ **8.** $(-c)^3(-d)^5$

9. $3(x + 2y) + 4(-2x - y)$ **10.** $-5p + \frac{2}{3}(9 - 3p) - 2$ **11.** $\frac{2m - 4}{-2}$

NAME ______________________ DATE ____________

1–8 Words into Symbols

Objective: To translate word phrases into algebra expressions and word sentences into equations.

Vocabulary

Uniform motion Motion at a constant speed. It is described by the formula

distance = rate × time, or $d = rt$.

Example 1 Represent each word phrase by an algebraic expression. Use n for the variable.

a. Four less than twice a number
b. The sum of twice a number and its square
c. The difference between a number and three

Solution

a. $2n - 4$
b. $2n + n^2$
c. $n - 3$

Represent each phrase by an algebraic expression. Use *n* for the variable.

1. Four less than the product of a number and three
2. The quotient when eight is divided by twice a number
3. The square of the sum of a number and three
4. Five more than the reciprocal of a number

Example 2 A rectangle has a width of w yards and a perimeter of 48 yards. Find the length in terms of w.

Solution Make a sketch. Call the length l. The perimeter of a rectangle is twice the length added to twice the width.

$$2l + 2w = 48$$
$$l + w = 24$$
$$l = 24 - w$$

w (width), l (length)

∴ the length of the rectangle is $(24 - w)$ yd.

Example 3 Anna and Sue rode toward each other on their bicycles. Sue's rate was three times Anna's rate of r mi/h. They met after two hours. How far apart were they before traveling? Express your answer in terms of r.

Solution The chart below shows all of the given information.

	Rate	× Time	Distance
Anna	r	2	$2r$
Sue	$3r$	2	$6r$

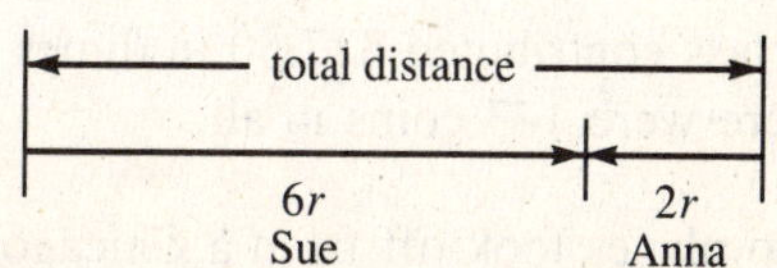

Their initial distance apart equals the sum of the distances traveled, or $8r$ mi.

NAME ______________________________ DATE ______________

1-8 Words into Symbols *(continued)*

Express each answer in simplest form in terms of the given variable.

5. Carl is a years old. His sister Jenny is six more than twice his age. What is the average of their ages?

6. The base and height of a triangle are consecutive odd integers, and the height exceeds the base. If the base is b cm, find the area of the triangle.

7. At 2:00 P.M. a train left a station traveling east at r mi/h. At 3:00 P.M. a second train headed west from the station at a rate 20 mi/h faster than the first. How far apart were the trains at 5:00 P.M.?

8. Seth sold 16 tickets to a school play. Of these, a were adults' tickets at $3.75 each. The rest were children's tickets at $2.50 each. How much money did Seth collect?

Example 4 The number of dimes and the number of quarters that Erin has earned in tips are consecutive even integers. She has fewer dimes than quarters. The total value of her coins is $7.50.

a. Choose a variable to represent the number of dimes.

b. Write an expression for the number of quarters in terms of the number of dimes.

c. Write an equation that describes the situation.

Solution

a. Let d = the number of dimes.

b. Then $d + 2$ = the number of quarters.

c. The value of the dimes is $0.10d$ dollars, and the value of the quarters is $0.25(d + 2)$ dollars.
$\therefore 0.10d + 0.25(d + 2) = 7.50$, or $0.35d + 0.5 = 7.50$.

Choose a variable to represent an unknown number, and then write an equation to describe the given situation.

9. In triangle ABC, the measure of $\angle A$ is four times that of $\angle B$. In addition, the measure of $\angle C$ is 15° less than half the measure of $\angle A$.

10. Carlos worked twice as many hours as Alan on Monday. After each boy worked nine more hours on Tuesday, the sum of their hours was three times the number of hours that Carlos worked on Monday.

11. A class contributed $27.50 in dimes and quarters to a local charity. There were 143 coins in all.

12. Two planes took off from a Chicago airport flying in opposite directions. One plane traveled 30 mi/h faster than the other. They were 1500 miles apart after 2 hours.

NAME ______________________________ DATE ______________

1–9 Problem Solving with Equations

Objective: To solve word problems by using an equation in one variable.

Plan for Solving a Word Problem

Step 1 Read the problem carefully a few times. Decide what numbers are asked for and what information is given. Making a sketch may be helpful.

Step 2 Choose a variable and use it with the given facts to represent the number(s) described in the problem. Labeling your sketch or arranging the information in a chart may help.

Step 3 Reread the problem. Then write an equation that represents relationships among the numbers in the problem.

Step 4 Solve the equation and find the required numbers.

Step 5 Check your results with the *original* statement of the problem. Give the answer.

CAUTION Problems can contain information that is unnecessary or contradictory. Sometimes there is not enough information given. Therefore, it is important to understand the given facts and their relationships before you try to solve a problem.

Example The Junior class sold shirts bearing the school insignia for $12.00 each. An extra $1.00 was charged to have a shirt monogrammed. There were 324 shirts sold, and a total of $4036.00 was collected. Of the shirts sold, 174 were bought by Juniors. How many shirts were *not* monogrammed?

Solution

Step 1 The problem asks for the number of shirts sold without a monogram.

Step 2 Let x = the number of shirts sold without a monogram.
Then $324 - x$ = number of shirts sold *with* a monogram.

	Price ×	Number =	Sales
Without a monogram	12	x	$12x$
With a monogram	13	$324 - x$	$13(324 - x)$
	Total Shirt Sales		4036

Step 3 Sales without a monogram + Sales with a monogram = Total Sales

$$12x \qquad\qquad 13(324 - x) \qquad\qquad 4036$$

Step 4

$$\begin{aligned} 12x + 13(324 - x) &= 4036 \\ 12x + 4212 - 13x &= 4036 \\ -x + 4212 &= 4036 \\ -x &= -176 \\ x &= 176 \quad \text{(shirts without a monogram)} \\ 324 - x &= 148 \quad \text{(shirts with a monogram)} \end{aligned}$$

(Solution continues on the next page.)

NAME ______________________________ DATE ______________

1–9 *Problem Solving with Equations* *(continued)*

Step 5 *Check:* Is the total number of shirts 324?

$$176 + 148 \stackrel{?}{=} 324$$
$$324 = 324 \ \checkmark$$

Do shirt sales total \$4036.00?

$$176(12) + 148(13) \stackrel{?}{=} 4036$$
$$2112 + 1924 \stackrel{?}{=} 4036$$
$$4036 = 4036 \ \checkmark$$

∴ 176 shirts were sold without a monogram.

The information about the number of shirts bought by Juniors was unnecessary.

Solve each of the following problems. If there is not enough information to solve the problem, say so. If extra information is given, identify it.

1. Cheryl's weekly allowance is \$2.00 more than Emily's. Together they get \$11.00. What is each girl's weekly allowance?
2. A child's bank contains 70 coins consisting of nickels and dimes that have a total value of \$5.55. How many of each kind of coin are there?
3. A store sold 40 baseballs and 14 softballs over a two-week period. The sales for these items totaled \$200. What was the price of one baseball?
4. The length of a rectangle is 6 cm more than the width. A square can be formed by tripling the width of the rectangle, and reducing its length by 2 cm. Find the dimensions of the rectangle.
5. The perimeter of an isosceles triangle is 36 cm, and the area is 60 cm^2. The length of the base is 3 cm less than the length of a leg. Find the length of each side.
6. The measures of the angles of a quadrilateral are consecutive odd integers. Find the measure of each angle. (*Hint*: The sum of the measures of the angles of a quadrilateral is 360°.)
7. Two trains whose rates differ by 8 mi/h leave stations that are 432 miles apart at 10 A.M. If the trains meet at 2 P.M., what is the rate of each train?
8. A hiker hikes up a mountain 1 mi/h slower than she hikes down the mountain. If it takes her 1.5 hours to hike up the mountain, and only 1 hour to hike down it, how fast does she move in each direction?
9. Sharon earned \$460 in simple interest on an investment of \$6200. Some of the money earned an interest rate of 5%, and the rest earned 8%. Her average rate of interest was 7.4%. How much did she invest at each rate?

Mixed Review Exercises

Solve.

1. $4 - 5x = 14$ **2.** $4(y - 3) = y - 15$ **3.** $7z + 4 = 4z + 16$

Evaluate each expression if $a = -5$ and $b = 10$.

4. $(a + b)^2$ **5.** $(ab)^2$ **6.** $|a - 2b|$ **7.** $\frac{b - a}{-3}$ **8.** $\frac{ab}{-2}$ **9.** $\frac{b \div a}{2}$

NAME ______________________________ DATE ______________

2 Inequalities and Proof

2–1 Solving Inequalities in One Variable

Objective: To solve simple inequalities in one variable.

Vocabulary

Properties of order (or properties of inequality) Let a, b, and c be any real numbers.

Comparison property Exactly one of the following statements is true:
$a < b$, $a = b$, or $a > b$.

Transitive property If $a < b$ and $b < c$, then $a < c$.

Addition property If $a < b$, then $a + c < b + c$.

Multiplication property

1. If $a < b$ and c is *positive*, then $ac < bc$.
2. If $a < b$ and c is *negative*, then $ac > bc$.

Equivalent inequalities Inequalities which have the same solution set.

Transformations that produce equivalent inequalities

1. Simplifying either side of an inequality.
2. Adding the same number to each side of an inequality, or subtracting the same number from each side of an inequality.
3. Multiplying (or dividing) each side of an inequality by the same *positive* number.
4. Multiplying (or dividing) each side of an inequality by the same *negative* number and *reversing* the direction of the inequality.

Symbol $\{x: x < k\}$ (the set of all real numbers x such that x is less than k)

CAUTION When each side of an inequality is multiplied or divided by a *negative* number, the inequality symbol reverses direction.

Example 1 Solve $3x + 15 < 3$ and graph its solution set.

Solution

$$3x + 15 < 3$$

$$3x + 15 - 15 < 3 - 15 \quad \text{Subtract 15 from each side.}$$

$$3x < -12$$

$$\frac{3x}{3} < \frac{-12}{3} \quad \text{Divide each side by 3.}$$

$$x < -4$$

∴ the solution set consists of all real numbers less than -4. It can be written $\{x: x < -4\}$.

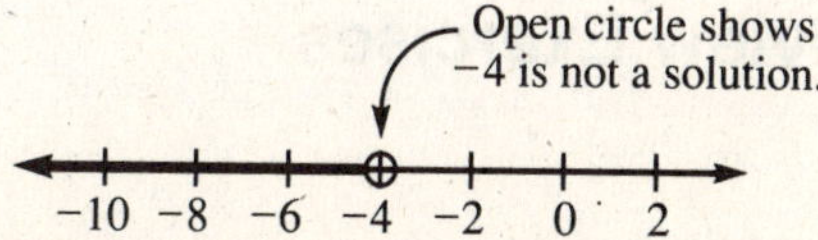

Check: Test a point from your solution set in the given inequality. Let $x = -5$.

$$3(-5) + 15 \overset{?}{<} 3$$

$$0 < 3 \ \checkmark$$

NAME ______________________ DATE ______________

2–1 *Solving Inequalities in One Variable* (continued)

Example 2 Solve $-5x < 10$ and graph its solution set.

Solution

$$-5x < 10$$

$$\frac{-5x}{-5} > \frac{10}{-5} \quad \text{Reverse!}$$

$$x > -2$$

∴ the solution set consists of all real numbers greater than -2. It can be written $\{x: x > -2\}$.

Check: Test a point from your solution set in the given inequality. Let $x = 1$.

$$-5(1) \overset{?}{<} 2(1 + 7)$$

$$-5 \overset{?}{<} 2(8) \ \checkmark$$

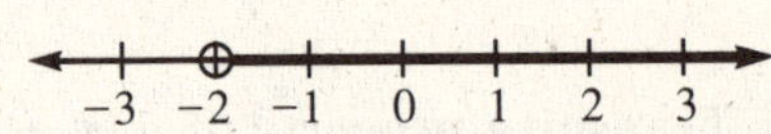

Solve each inequality and graph its solution set.

1. $x - 4 > -3$
2. $y + 5 < 7$
3. $3m > 12$
4. $4k > -20$
5. $-6r < 18$
6. $-5q > -15$
7. $-8 < -2v$
8. $-\frac{2}{5}r < 10$
9. $2w - 8 > 2$
10. $3a + 5 < 14$
11. $2 - d > 3$
12. $5 - 2c < 11$

Example 3 Solve each inequality and graph its solution set.

a. $3(x - 2) > 3x + 2$ **b.** $-10y < -2(5y - 3)$

Solution

a. $3(x - 2) > 3x + 2$

$$3x - 6 > 3x + 2$$

$$-6 > 2$$

Since the equivalent inequality $-6 > 2$ is *false*, the given inequality is false and has no solution.

∴ the solution set is $\emptyset$, and there is no graph.

b. $-10y < -2(5y - 3)$

$$-10y < -10y + 6$$

$$0 < 6$$

Since the equivalent inequality $0 < 6$ is *true*, the given inequality is true for all values of y.

∴ the solution set is {real numbers}.

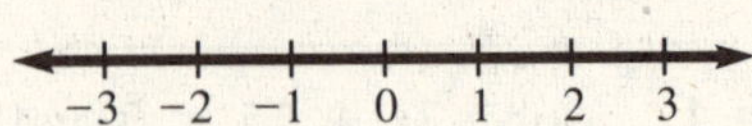

Solve each inequality and graph each solution set that is not empty.

13. $3(x - 2) > 4 - x$
14. $5b + 3 > 5(b + 1) - b$
15. $3(5 - t) - 2t > 2 + 5(3 - t)$
16. $5(1 - 3k) + 2(6k - 1) < 3(2 - k)$

Mixed Review Exercises

Simplify.

1. $(-2)^3(-x)^3$
2. $(-2)(-3)(-4)(0)$
3. $|-2| - |-3| + 2$
4. $3x^2 - 5x - (x^2 - 2x)$
5. $-4(1 - 4)^2$
6. $\frac{4ab - 8}{-2}$
7. $\frac{5(1 + 4)}{6 \cdot 4 + 1}$
8. $(6x - 2) - (5 - x)$

NAME ______________________ DATE ____________

2–2 Solving Combined Inequalities

Objective: To solve combined inequalities.

Vocabulary

Conjunction A sentence formed by joining two sentences with the word *and.*

Disjunction A sentence formed by joining two sentences with the word *or.*

Symbols $\geq$ (is greater than or equal to) $a < x < b$ (means "$x > a$ *and* $x < b$")
$\leq$ (is less than or equal to)

CAUTION A conjunction is true only when *both* sentences are true. A disjunction is true when *either* sentence is true, or when both sentences are true.

Example 1 Graph the solution set of the conjunction $x \geq -1$ *and* $x > 2$.

Solution 1 First find the values of x for which *both* sentences are true. The conjunction is only true when x is greater than 2. To graph, put an open circle at 2 to show that 2 is *not* included in the solution set. Shade to the right of 2.

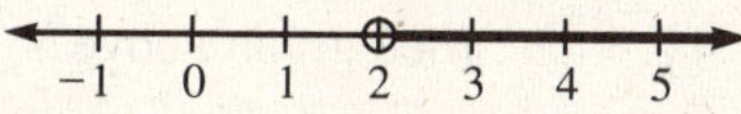

Solution 2 Begin by graphing each inequality separately, above a number line. Then make a graph of the solution set *on* the number line, including only those points that appear in *both* parts.

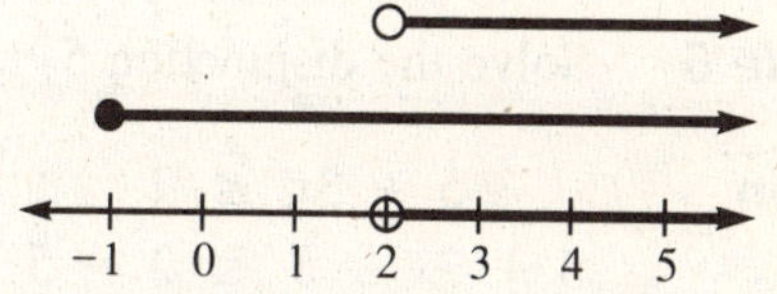

Example 2 Graph the solution set of the conjunction $x \geq -6$ *and* $x < -2$.

Solution Rewrite the conjunction as $-6 \leq x < -2$. Then draw the graph. Or, as an alternative, use the method shown in Solution 2 above.

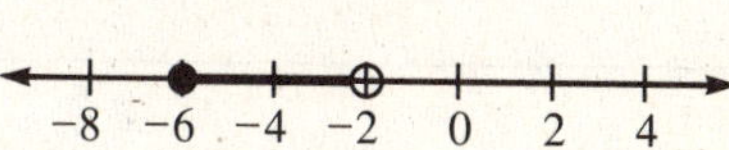

Example 3 Graph the solution set of the disjunction $x < 1$ *or* $x > 4$.

Solution Find the values of x for which *at least one* of the sentences is true. The disjunction is true for all values of x either less than 1 or greater than 4.

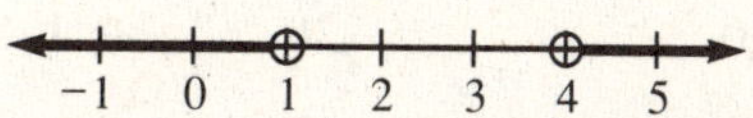

Solve each conjunction or disjunction and graph each solution set that is not empty.

1. $x > 3$ and $x < 7$ **2.** $x < 5$ and $x < 6$ **3.** $x \leq -2$ or $x > 2$ **4.** $2 \geq x > -1$

5. $x > 2$ or $x < -1$ **6.** $x < 3$ or $x > 3$ **7.** $x > 4$ and $x \leq -1$ **8.** $x < 2$ or $x > -2$

NAME ______________________ DATE ____________

2–2 Solving Combined Inequalities *(continued)*

Example 4 Solve $-1 \le 3x - 4 < 8$ and graph its solution set.

Solution 1 Rewrite the inequality using *and*. Then solve each inequality.

$$-1 \le 3x - 4 \quad \text{and} \quad 3x - 4 < 8$$
$$3 \le 3x \quad \text{and} \quad 3x < 12$$
$$1 \le x \quad \text{and} \quad x < 4$$

$\therefore$ the solution set is $\{x\colon 1 \le x < 4\}$.

−1 0 1 2 3 4 5

Solution 2 Solve both inequalities simultaneously.

$$-1 \le 3x - 4 < 8$$
$$-1 + 4 \le 3x - 4 + 4 < 8 + 4 \qquad \text{Add 4 to each expression.}$$
$$3 \le 3x < 12$$
$$\frac{3}{3} \le \frac{3x}{3} < \frac{12}{3} \qquad \text{Divide each expression by 3.}$$
$$1 \le x < 4$$

$\therefore$ the solution set is $\{x\colon 1 \le x < 4\}$. (See graph above.)

Example 5 Solve the disjunction $5 + 3x \le 2$ *or* $2x - 2 > 4 - x$ and graph its solution set.

Solution

$$5 + 3x \le 2 \quad \text{or} \quad 2x - 2 > 4 - x$$
$$5 + 3x - 5 \le 2 - 5 \quad \text{or} \quad 2x - 2 + 2 + x > 4 - x + 2 + x$$
$$3x \le -3 \quad \text{or} \quad 3x > 6$$
$$x \le -1 \quad \text{or} \quad x > 2$$

$\therefore$ the solution set is $\{x\colon x \le -1 \text{ or } x > 2\}$.

−3 −2 −1 0 1 2 3

Solve each conjunction or disjunction and graph each solution set that is not empty.

9. $0 \le x - 1 \le 4$

10. $1 > k + 3 > -2$

11. $y - 1 < 3$ and $y + 3 > 3$

12. $x + 3 < 2$ or $x - 6 > -2$

13. $w + 2 < 3$ and $2 - w < 5$

14. $-2 < 1 - t < 4$

15. $4 - 3x < -2$ or $x < 0$

16. $-1 < 2m + 1 < 3$

17. $2x + 3 \ge 9$ and $5x - 1 < 9$

18. $3x - 1 > 2$ or $2x + 6 < 2$

19. $2n + 5 < 3$ or $2(n - 1) > 0$

20. $3z + 1 > 7$ and $2(z - 1) < -4$

21. $5x - 2 \ge 8$ or $4 - x > 5$

22. $3 < 4 - 2r < 6$

23. $3p - 5 \le 1$ or $4 - 2p < 6$

24. $3 > \frac{n}{4} + 3 > 2$

NAME ______________________ DATE ____________

2–3 Problem Solving Using Inequalities

Objective: To solve word problems by using inequalities in one variable.

Symbols

$x \ge a$ (x is at least a, or x is no less than a.)

$x \le b$ (x is at most b, or x is no greater than b.)

$a < x < b$ (x is between a and b.)

$a \le x \le b$ (x is a, b, or between a and b.)

Example 1 A video store charges $19.99 for a lifetime membership. Members pay $2.00 to rent a movie, while nonmembers pay $2.25. At least how many movies would a member have to rent in order to pay less, overall, than a nonmember?

Solution

Step 1 The problem asks for the least number of movies a member would have to rent for rental costs to be less than those for a nonmember.

Step 2 Let m = the number of movies.
Then $19.99 + 2.00m$ = the amount a member would pay for m movies; and
$2.25m$ = the amount a nonmember would pay for m movies.

Step 3 Rental for a member is less than rental for a nonmember.

$$19.99 + 2.00m < 2.25m$$

Step 4 Multiply both sides of the inequality in Step 3 by 100 to clear the decimals.

$$1999 + 200m < 225m$$
$$-25m < -1999$$
$$m > 79.96$$

Interpret the result: Since the number of movies must be a whole number, $m \ge 80$.

Step 5 *Check:* Has a member who has rented 80 movies paid less than a nonmember who has rented 80 movies? Compare.

$$19.99 + 2.00(80) \overset{?}{<} 2.25(80)$$
$$179.99 < 180 \ \surd$$

Check: Is 80 the least number of movies a member must rent to pay less, overall, than a nonmember? Try 79.

$$19.99 + 2.00(79) \overset{?}{>} 2.25(79)$$
$$177.99 > 177.75 \ \surd$$

∴ a member must rent at least 80 movies to pay less, overall, than a nonmember.

Solve.

1. The owners of a skating rink sell discount cards for $15.00 that are worth $.50 off the regular admission price of $3.00. The discount cards are good for six months. At least how many times would you have to go skating in order to pay less, overall, with the discount card than without it?

2. A summer recreation department charges $45.00 for a season ticket to the town pool. Admission to the pool for one day is $1.75. How many days would you have to go swimming at the regular price in order to spend at least the cost of a season ticket?

NAME ______________________ DATE ______________

2–3 Problem Solving Using Inequalities *(continued)*

Example 2 Two sides of a triangle are consecutive *even* integers. The other side is 65 cm. If the perimeter is between 215 cm and 230 cm, what are the possible lengths for the first two sides?

Solution

Step 1 The problem asks you to find all possible lengths for the first two sides if the other side is 65 cm and the perimeter is between 215 cm and 230 cm. Draw a sketch.

Step 2 Let s = the length of the first side. Then $s + 2$ = the length of the second side.

Step 3 The perimeter is between 215 cm and 230 cm.

$$215 < s + (s + 2) + 65 < 230$$

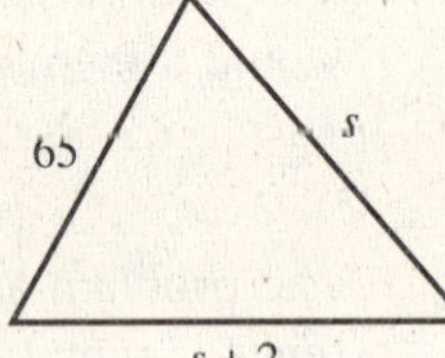

Step 4

$$215 < 2s + 67 < 230$$
$$148 < 2s < 163$$
$$74 < s < 81.5$$

Interpret the result: Since s must be an even integer, there are only three possible values for s: 76, 78, and 80. There are three pairs of consecutive even integers that meet the requirements of the problem: 76, 78; 78, 80; and 80, 82.

Step 5 To check, you must verify that the perimeter is between 215 cm and 230 cm in all three cases, and that neither the "next smaller" nor the "next larger" pair of even consecutive integers is a solution. The work is left for you.

∴ the lengths, in centimeters, of the first two sides are: 76, 78; 78, 80; or 80, 82.

Solve.

3. The length of a rectangle is 3 cm more than twice its width. Find the largest possible width if the perimeter is at most 66 cm.
4. The length of a leg of an isosceles triangle is twice the length of the base. What is the minimum length of the base if the perimeter is at least 20 cm?
5. Find all sets of four consecutive integers whose sum is between 95 and 105.
6. Ellen's first three test scores were consecutive *odd* integers. Her fourth score was 83. She had a B− average (between 80 and 82, inclusive) for the four tests. What was her lowest test score?

Mixed Review Exercises

Solve each open sentence and graph each solution set that is not empty.

1. $2x - 3 > -7$ **2.** $0.5p \le 1$ and $p + 1 \ge -2$ **3.** $3(n - 1) > 2 - 3(1 - n)$

4. $-2y < -8$ or $1 + 2y < 3$ **5.** $5 - 2d < 3$ **6.** $-2 > 1 - 2x > -6$

Evaluate if $a = -2$ and $b = 5$.

7. $|a - b|$ **8.** $|a| - |b|$ **9.** $|ab|$ **10.** $|b| - |a|$

NAME ______________________________ DATE ______________

2–4 Absolute Value in Open Sentences

Objective: To solve open sentences involving absolute value.

Vocabulary

Rules for solving open sentences involving absolute value

If a is a *positive* real number:

1. $|x| = a$ is equivalent to the disjunction $x = -a$ or $x = a$.
2. $|x| > a$ is equivalent to the disjunction $x < -a$ or $x > a$.
3. $|x| < a$ is equivalent to the conjunction $x > -a$ and $x < a$ (or $-a < x < a$).

CAUTION Watch out for statements such as $|x| > -6$ and $|x| < -6$. Since the absolute value of every real number is nonnegative, $|x|$ is *always* greater than a negative number. Similarly, $|x|$ can *never* be less than a negative number.

Example 1

a. The solution set of $|x| = 2$ is $\{x: x = -2 \text{ or } x = 2\}$.

b. The solution set of $|x| > 2$ is $\{x: x < -2 \text{ or } x > 2\}$.

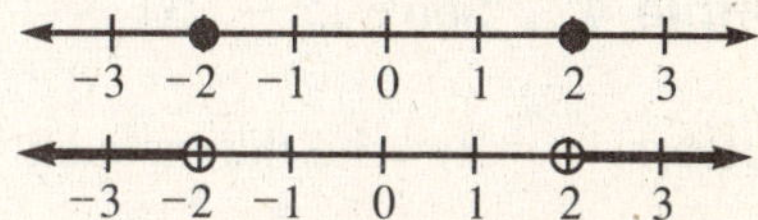

c. The solution set of $|x| < 2$ is $\{x: -2 < x < 2\}$.

d. The solution set of $|x| > -2$ is {real numbers}.

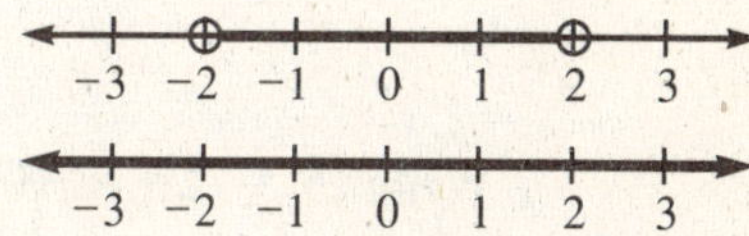

e. The solution set of $|x| < -2$ is $\emptyset$. No graph

Example 2 Solve $|2x - 6| = 10$.

Solution Use Rule 1 to write an equivalent disjunction. Then solve.

$|2x - 6| = 10$ is equivalent to this disjunction:

$$2x - 6 = -10 \quad \text{or} \quad 2x - 6 = 10$$
$$2x = -4 \quad \text{or} \quad 2x = 16$$
$$x = -2 \quad \text{or} \quad x = 8$$

Check: $|2(-2) - 6| \stackrel{?}{=} 10$ $\quad$ $|2 \cdot 8 - 6| \stackrel{?}{=} 10$

$10 = 10 \ \checkmark$ $\quad$ $10 = 10 \ \checkmark$

$\therefore$ the solution set is $\{-2, 8\}$.

Solve.

1. $|x| = 5$ **2.** $|x| = 8$ **3.** $|2 + y| = 7$ **4.** $|3 - n| = 4$

5. $|3p - 4| = 5$ **6.** $|2t + 1| = 7$ **7.** $|7k + 14| = 0$ **8.** $|5c - 20| = 5$

9. $|5 - 2a| = 9$ **10.** $|3 + 4b| = 9$ **11.** $|x - 1| = -4$ **12.** $|-t| = 0$

NAME ______________________ DATE ____________

2–4 Absolute Value in Open Sentences *(continued)*

Example 3 Solve $|4x - 5| > 9$ and graph its solution set.

Solution Write $|4x - 5| > 9$ as the equivalent disjunction:

$$4x - 5 < -9 \quad \text{or} \quad 4x - 5 > 9$$
$$4x < -4 \quad \text{or} \quad 4x > 14$$
$$x < -1 \quad \text{or} \quad x > \frac{14}{4} = \frac{7}{2}$$

$\therefore$ the solution set is $\left\{x: x < -1 \text{ or } x > \frac{7}{2}\right\}$.

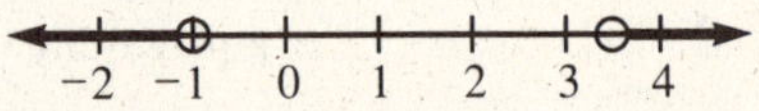

Check: Test a point in each of the shaded regions and a point between them.

Try $x = -2$: $|4(-2) - 5| = |-8 - 5| = |-13| = 13 > 9$ True √

Try $x = 4$: $|4 \cdot 4 - 5| = |16 - 5| = |11| = 11 > 9$ True √

Try $x = 0$: $|4 \cdot 0 - 5| = |0 - 5| = |-5| = 5 > 9$ False √

Example 4 Solve $|5 - 2x| - 1 < 2$ and graph its solution set.

Solution Rewrite the inequality so that the absolute value is alone on one side.

$$|5 - 2x| - 1 < 2$$
$$|5 - 2x| < 3$$

Write $|5 - 2x| < 3$ as the equivalent conjunction:

$$-3 < 5 - 2x < 3$$
$$-8 < -2x < -2$$
$$4 > x > 1$$

$\therefore$ the solution set is $\{x: 1 < x < 4\}$.

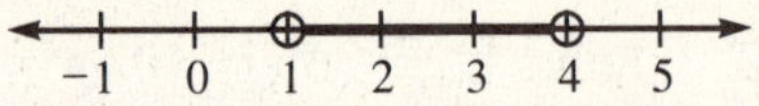

Check: Test a point in the shaded region and a point on either side of it.

Try $x = 2$: $|5 - 2 \cdot 2| = |5 - 4| = |1| = 1 < 3$ True √

Try $x = 0$: $|5 - 2 \cdot 0| = |5 - 0| = |5| = 5 < 3$ False √

Try $x = 5$: $|5 - 2 \cdot 5| = |5 - 10| = |-5| = 5 < 3$ False √

Solve and graph the solution set.

13. $|r| > 2$ **14.** $|k| \le 5$ **15.** $|-3m| > 0$ **16.** $|f + 2| > 3$

17. $|g - 4| \ge 2$ **18.** $|z - 1| < 2$ **19.** $0 > |x - 7|$ **20.** $|2y + 1| \le 5$

21. $6 < |2m - 3|$ **22.** $|3h + 1| > 5$ **23.** $|5p - 4| < 1$ **24.** $|3x - 2| \le 4$

25. $|2n + 3| > -2$ **26.** $|2 - q| \ge 5$ **27.** $|3 - a| \le 2$ **28.** $|1 - 2b| < 3$

29. $|4 - 3x| \ge 5$ **30.** $|6 - 4w| > 2$ **31.** $|6 - 7c| \le -1$ **32.** $|3 - 0.4y| \le 7$

33. $\left|2 - \frac{1}{3}k\right| > 1$ **34.** $\left|\frac{m + 1}{2}\right| \le \frac{3}{2}$ **35.** $|t| + 2 < 3$ **36.** $|r + 1| - 3 < 1$

NAME ______________________________ DATE ______________

2–5 Solving Absolute Value Sentences Graphically

Objective: To use number lines to obtain quick solutions to certain equations and inequalities involving absolute value.

Vocabulary

Distance on a number line The distance on a number line between the graphs of real numbers a and b is $|a - b|$. Example: The distance between the graphs of -2 and 3 on a number line is $|-2 - 3| = 5$.

Example 1 Find the distance between the graphs of each pair of numbers.

a. 5 and 12 **b.** -10 and -4 **c.** -6 and 3

Solution

a. $|5 - 12| = |-7| = 7$

b. $|-10 - (-4)| = |-10 + 4| = |-6| = 6$

c. $|-6 - 3| = |-9| = 9$

Note: The order in which you subtract the numbers will not affect the result.

Find the distance between the graphs of each pair of numbers.

1. 3 and 14 **2.** -5 and -11 **3.** -1 and 4 **4.** 10 and -1

5. -1 and -4 **6.** 0 and -2 **7.** 6 and -8 **8.** -12 and -9

Example 2 Solve $|x - 4| = 3$.

Solution The equation $|x - 4| = 3$ tells you that the distance between x and 4 is 3 units. To find x, start at 4 on a number line and move 3 units in each direction (left and right). You will arrive at the values 1 and 7.

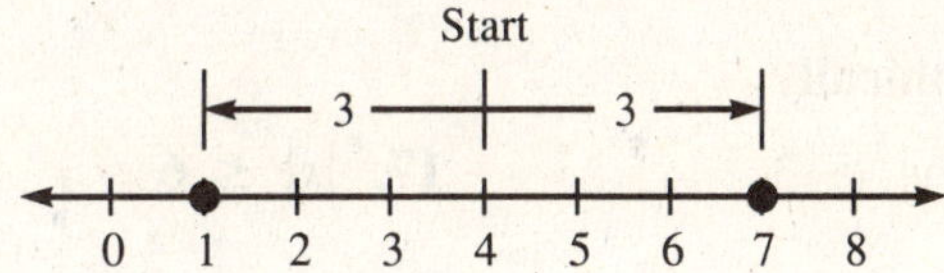

$\therefore$ the solution set is $\{1, 7\}$.

CAUTION On a horizontal number line, you can measure distance from a point either to the left or to the right. However, the distance itself will always be positive.

Solve each open sentence graphically.

9. $|q| = 2$ **10.** $|x| = 5$

11. $|m - 1| = 4$ **12.** $|a - 2| = 3$

13. $|t - 3| = 5$ **14.** $|r - 4| = 1$

2–5 Solving Absolute Value Sentences Graphically *(continued)*

Example 3 Solve $|x + 1| \le 4$.

Solution Since distance is defined in terms of subtraction, you must rewrite the inequality accordingly: $|x + 1| \le 4$ is equivalent to $|x - (-1)| \le 4$.

This inequality says that the distance between x and -1 must be 4 units or less. To find x, start at -1 on a number line and move 4 units to the left and to the right. You will arrive at the values -5 and 3. The numbers -5, 3, and all numbers *between* -5 and 3, satisfy the inequality.

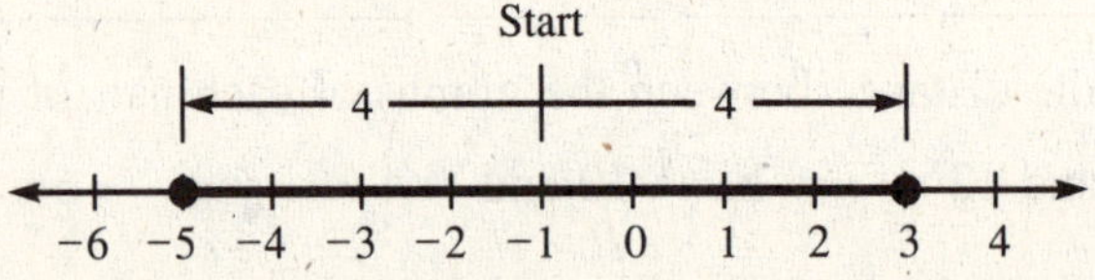

$\therefore$ the solution set is $\{x: -5 \le x \le 3\}$.

Example 4 Solve $|m - 3| > 2$.

Solution This inequality says that the distance between m and 3 must be *more than* 2 units. To find m, start at 3 on a number line and move 2 units to the left and to the right. You will arrive at 1 and 5. The numbers beyond these values are more than 2 units from 3.

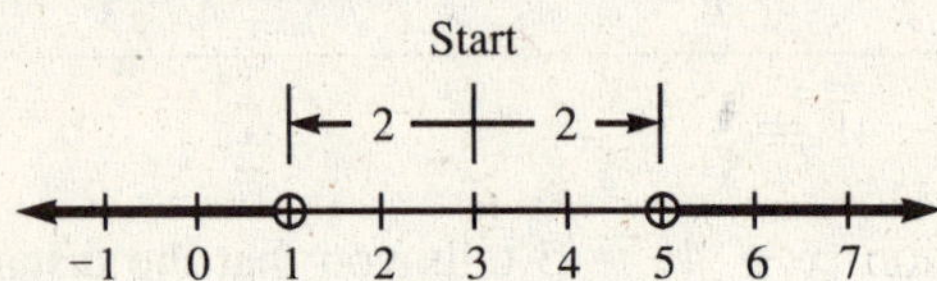

$\therefore$ the solution set is $\{m: m < 1 \text{ or } m > 5\}$.

Solve each open sentence graphically.

15. $|m| < 4$ **16.** $|v| \ge 3$ **17.** $|s| \le 6$ **18.** $|x| > \frac{3}{2}$

19. $|k + 2| < 3$ **20.** $|w - 5| \le 2$ **21.** $|d + 3| \ge 1$ **22.** $|m - 1| > 3$

23. $|h + 4| \le 1$ **24.** $5 \ge |x + 3|$ **25.** $2 < |k - 7|$ **26.** $4 < |r - 2|$

27. $1 > \left|y - \frac{1}{2}\right|$ **28.** $\left|g + \frac{1}{2}\right| < 2$ **29.** $\left|u - \frac{2}{3}\right| \ge \frac{4}{3}$ **30.** $|b + 5| > 1.5$

Mixed Review Exercises

Solve each open sentence.

1. $2 \le a - 1 \le 5$ **2.** $|3 - 2b| = 5$ **3.** $5(2x - 1) > 9x + 3$

4. $|2t - 1| \le 5$ **5.** $|y| - 1 \ge 2$ **6.** $z + 3 < 2$ or $4z > 2$

7. $3c - 8 \le 2$ **8.** $|x + 2| = 2$ **9.** $\frac{1}{2}d > 1$ and $3 - d > -1$

NAME ______________________________ DATE ______________

2–6 Theorems and Proofs

Objective: To use axioms, definitions, and theorems to prove some properties of real numbers.

Vocabulary

Axiom or postulate A statement that is assumed to be true. Examples: the substitution principle, the properties of equality, the field properties, and the properties of order.

Proof Reasoning from *hypothesis* (a given statement that you accept as true) to *conclusion* (a statement that follows logically from the hypothesis). Each step in the reasoning process must be supported by an axiom, a definition, or another statement that has already been proved.

Theorem A statement that can be proved. It is sometimes written in the form of an if-then statement, where the "if" part is the hypothesis, and the "then" part is the conclusion.

Converse The statement that results from interchanging the hypothesis and the conclusion of an if-then statement. Example: The converse of the statement "If $a = b$, then $a + c = b + c$" is "If $a + c = b + c$, then $a = b$." The converse of a true statement isn't necessarily true.

Counterexample In algebra, a single numerical example showing that a given statement is false.

If and only if statement A statement equivalent to two if-then statements that are converses of each other.

Cancellation property of addition For all real numbers a, b, and c: If $a + c = b + c$, then $a = b$; and if $c + a = c + b$, then $a = b$.

Cancellation property of multiplication For all real numbers a and b and *nonzero* real numbers c: If $ac = bc$, then $a = b$; and if $ca = cb$, then $a = b$.

Zero-product property For all real numbers a and b: $ab = 0$ if and only if $a = 0$ or $b = 0$.

CAUTION You will not be able to understand or to write algebraic proofs unless you are familiar with the substitution principle (Lesson 1-2), the field properties (Lesson 1-3), and the properties of equality (Lesson 1-3). Review these properties before you proceed.

Example 1 Give a counterexample to show that each statement is false. The domain of each variable is the set of real numbers.

a. If $x^2 = 25$, then $x = 5$. **b.** $a - b < a$

Solution

a. Let $x = -5$. Then $x^2 = 25$, but $x \neq 5$.

b. Let $a = 3$ and $b = -4$. Then $a - b = 3 - (-4) = 7$, but $7 \not< 3$.

Note: Negative numbers, fractions, and the numbers 0 and 1 are often good numbers to test for counterexamples.

NAME ______________________ DATE ____________

2–6 *Theorems and Proofs* (continued)

Give a counterexample to show that each statement is false. The domain of each variable is the set of real numbers.

1. If $|x| = 2$, then $x = 2$.

2. $|a| + |b| = |a + b|$

3. If $a > b$, then $a^2 > b^2$.

4. $x^2 > x$

Example 2 Give reasons for the steps shown in the following proof.

If $a \neq 0$ and $ab = a$, then $b = 1$.

Solution

	Statement	Reason
1.	$a \neq 0$ and $ab = a$	1. Given
2.	$a = a \cdot 1$	2. Identity property of multiplication
3.	$ab = a \cdot 1$	3. Transitive property of equality
4.	$\therefore b = 1$	4. Cancellation property of multiplication

Note: This proof shows that the real number 1 is the unique identity element for multiplication.

Give reasons for the steps shown in each proof. The domain of each variable is the set of real numbers unless otherwise stated.

5. If $x + c = x$, then $c = 0$.

Proof

1. $x + c = x$
2. $x = x + 0$
3. $x + c = x + 0$
4. $\therefore c = 0$

6. If $x = a + (-b)$, then $x + b = a$.

Proof

1. $x = a + (-b)$
2. $x + b = [a + (-b)] + b$
3. $x + b = a + [-b + b]$
4. $x + b = a + 0$
5. $\therefore x + b = a$

7. If $a \neq 0$, $b \neq 0$, and $a = b$, then $\frac{1}{a} = \frac{1}{b}$.

Proof

1. $a \neq 0$, $b \neq 0$, and $a = b$
2. $a \cdot \frac{1}{a} = 1$; $b \cdot \frac{1}{b} = 1$
3. $a \cdot \frac{1}{a} = b \cdot \frac{1}{b}$
4. $a \cdot \frac{1}{a} = a \cdot \frac{1}{b}$
5. $\therefore \frac{1}{a} = \frac{1}{b}$

8. If $a \neq 0$ and $b \neq 0$, then $b\left(\frac{1}{a} \cdot \frac{1}{b}\right) = \frac{1}{a}$.

Proof

1. $a \neq 0$ and $b \neq 0$
2. $\frac{1}{a}$ and $\frac{1}{b}$ are real numbers.
3. $b \cdot \left(\frac{1}{a} \cdot \frac{1}{b}\right) = b \cdot \left(\frac{1}{b} \cdot \frac{1}{a}\right)$
4. $b \cdot \left(\frac{1}{a} \cdot \frac{1}{b}\right) = \left(b \cdot \frac{1}{b}\right) \frac{1}{a}$
5. $b \cdot \left(\frac{1}{a} \cdot \frac{1}{b}\right) = 1 \cdot \frac{1}{a}$
6. $\therefore b \cdot \left(\frac{1}{a} \cdot \frac{1}{b}\right) = \frac{1}{a}$

NAME ______________________________ DATE ______________

2–7 Theorems about Order and Absolute Value

Objective: To prove theorems about inequalities and absolute value.

CAUTION In order to prove the theorems in this lesson, you must be familiar with the properties of order from Lesson 2-1.

Example 1 Name the property of order illustrated.

a. If $7 > 3$ and $3 > -1$, then $7 > -1$.

b. If $2 < 5$, then $2(-6) > 5(-6)$.

Solution

a. Transitive property of order

b. Second multiplication property of order

Name the property of order illustrated in each statement.

1. If $w < 5$, then $2w < 2(5)$.

2. If $-2 < 4$, then $-2 + 3 < 4 + 3$.

3. If $5 > -1$ and $-1 > -6$, then $5 > -6$.

4. If $r > 2$, then $r(-3) < 2(-3)$.

5. If $x < 2$, then $x + (-4) < 2 + (-4)$.

6. Either $x < 0$, $x = 0$, or $x > 0$.

Example 2 Prove for all real numbers a and b:
If $|a| \geq b > 0$, then $a \geq b$ or $a \leq -b$.

Solution

Statements	Reasons
1. $|a| \geq b > 0$	1. Given
2. *Case 1:* If $a \geq 0$, then $|a| = a$.	2. Definition of absolute value
3. $\therefore a \geq b$	3. Substitution principle
4. *Case 2:* If $a < 0$, then $|a| = -a$.	4. Definition of absolute value
5. $-a \geq b$	5. Substitution principle
6. $(-1)a \geq b$	6. Multiplicative property of -1
7. $(-1)(-1)a \leq (-1)b$	7. Second multiplication property of order
8. $1 \cdot a \leq -b$	8. Multiplicative property of -1
9. $\therefore a \leq -b$	9. Identity property of multiplication
10. $a \geq b$ or $a \leq -b$	10. Steps 3 and 9

NAME ____________________ DATE ____________

2–7 Theorems about Order and Absolute Value *(continued)*

Give reasons for the steps shown in the proof of each theorem. The domain of each variable is the set of real numbers unless otherwise stated.

7. If $a > 0$, $b > 0$, and $c > 0$, then $abc > 0$.

Proof

1. $a > 0$, $b > 0$, and $c > 0$
2. $ab > 0 \cdot b$
3. $ab > 0$
4. $(ab)c > 0 \cdot c$
5. $\therefore abc > 0$

8. If $a - b > c$, then $a > c + b$.

Proof

1. $a - b > c$
2. $a - b + b > c + b$
3. $a + (-b) + b > c + b$
4. $a + [(-b) + b] > c + b$
5. $a + 0 > c + b$
6. $\therefore a > c + b$

9. $|a|^2 = a^2$

Proof

1. *Case 1:* If $a \geq 0$, then $|a| = a$.
2. $\therefore |a|^2 = a^2$
3. *Case 2:* If $a < 0$, then $|a| = -a$.
4. $|a^2| = (-a)^2$
5. $|a|^2 = (-a)(-a)$
6. $|a|^2 = a(-1) \cdot (-1)a$
7. $|a|^2 = a \cdot (-1 \cdot -1) \cdot a$
8. $|a|^2 = a \cdot 1 \cdot a$
9. $|a|^2 = a \cdot a$
10. $\therefore |a|^2 = a^2$

10. $ab \leq |ab|$

Proof

1. *Case 1:* If $ab \geq 0$, then $|ab| = ab$.
2. *Case 2:* If $ab < 0$, then $|ab| = -ab$.
3. $-ab > 0$
4. $-ab > 0 > ab$
5. $|ab| > 0 > ab$
6. $|ab| > ab$
7. $|ab| \geq ab$
8. $\therefore ab \leq |ab|$

11. Prove: If $a < b$, then $a - b < 0$.

Mixed Review Exercises

Tell whether each statement is true for all real numbers.

1. If $a < b$, then $b > a$.

2. If $a > b$ and $c < 0$, then $ac > bc$.

3. If $a < b < 0$, then $a^2 > b^2$.

Solve each open sentence and graph each solution set that is not empty.

4. $|x + 2| = 3$

5. $3d + 2 \geq 1$

6. $-2 < 3 - y \leq 4$

7. $5a - 3 = 12$

8. $|m| + 2 < 2$

9. $-3x > 9$ or $x - 2 \geq 1$

NAME ______________________ DATE ______________

3 Linear Equations and Functions

3–1 Open Sentences in Two Variables

Objective: To find solutions of open sentences in two variables and to solve problems involving open sentences in two variables.

Vocabulary

Open sentence in two variables An equation or inequality involving two variables.

Domain of x The set whose members may serve as replacements for x.

Ordered pair A pair of numbers having a definite order. A solution to an open sentence in the two variables x and y is given as the ordered pair (x, y). Example: The ordered pair $(3, 4)$ is a solution to the equation $x + 2y = 11$.

Solution set The set of all solutions (ordered pairs) that satisfy an open sentence.

Example 1 Solve the equation $3x + 2y = 6$ if the domain of x is $\{-1, 0, 2\}$.

Solution Substitute each value in the domain of x in the equation. Then solve for y. You're looking for solutions of the form $(-1, ?)$, $(0, ?)$, and $(2, ?)$.

$x = -1$	$x = 0$	$x = 2$
$3x + 2y = 6$	$3x + 2y = 6$	$3x + 2y = 6$
$3(-1) + 2y = 6$	$3(0) + 2y = 6$	$3(2) + 2y = 6$
$-3 + 2y = 6$	$0 + 2y = 6$	$6 + 2y = 6$
$2y = 9$	$2y = 6$	$2y = 0$
$y = \frac{9}{2}$	$y = 3$	$y = 0$

$\therefore$ the solution set is $\left\{\left(-1, \frac{9}{2}\right), (0, 3), (2, 0)\right\}$.

Solve each equation if the domain of x is (a) $\{-1, 0, 2\}$, and (b) $\{-2, 1, 3\}$.

1. $x - 2y = 1$ **2.** $-x + 2y = 0$ **3.** $3x - 2y = 5$ **4.** $2x - \frac{1}{3}y = 1$

Complete each ordered pair to form a solution of the equation.

5. $x - y = 3$ $(-2, \underline{?})$, $(0, \underline{?})$, $(\underline{?}, -1)$ **6.** $2x + 5y = 20$ $(0, \underline{?})$, $(5, \underline{?})$, $(\underline{?}, 0)$

7. $\frac{1}{2}x + y = 1$ $(-2, \underline{?})$, $(0, \underline{?})$, $(\underline{?}, 5)$ **8.** $x + 3y = 2$ $(-1, \underline{?})$, $(2, \underline{?})$, $\left(\underline{?}, \frac{5}{3}\right)$

Example 2 Find the value of k so that $(2, -1)$ satisfies the equation $kx - 2y = k$.

Solution

$$\begin{aligned} kx - 2y &= k \\ k(2) - 2(-1) &= k \\ 2k + 2 &= k \\ 2k &= k - 2 \\ k &= -2 \end{aligned}$$

Find the value of k so that the ordered pair satisfies the equation.

9. $3x + y = k$; $(-1, 2)$ **10.** $5x - ky = 4$; $(2, -3)$ **11.** $3x + ky = k$; $(-1, -2)$

NAME ____________________ DATE ____________

3–1 Open Sentences in Two Variables *(continued)*

Example 3 A child's bank contains \$1.15 in dimes and quarters. Find all possibilities for the number of each type of coin in the bank.

Solution

Step 1 The problem asks for the number of dimes and quarters whose total value is \$1.15.

Step 2 Let d = number of dimes and q = number of quarters.

Step 3 The total value of d dimes is $10d$ cents and of q quarters is $25q$ cents. Write an equation expressing the total value of the coins in cents:

$$10d + 25q = 115$$

Step 4 Solve the equation for one variable, say d, in terms of the other.

$$\begin{aligned} 10d + 25q &= 115 \\ 2d + 5q &= 23 \\ 2d &= 23 - 5q \\ d &= \frac{23 - 5q}{2} \end{aligned}$$

Remember: The number of each type of coin must be a whole number. If q is even, then $23 - 5q$ is odd and d is not a whole number. Therefore, q must be an odd whole number.

q	$\frac{23-5q}{2}$	d	(q, d)
1	$\frac{23-5(1)}{2}$	9	(1, 9)
3	$\frac{23-5(3)}{2}$	4	(3, 4)

If $q \geq 5$, then $d < 0$.

Step 5 The check is left for you.

∴ the bank can contain 9 dimes and 1 quarter, or 4 dimes and 3 quarters.

In each problem (a) choose two variables to represent the numbers asked for, (b) write an open sentence relating the variables, and (c) solve the open sentence and give the answer. (Include solutions in which one of the variables is equal to zero.)

12. A cashier needs to refund \$90 using \$10 and \$20 bills. Find all the possibilities for the number of each type of bill the cashier could use.

13. Terry has 85 cents in nickels and quarters. Find all the possibilities for the number of each type of coin that Terry could have.

14. Sam needs \$1.75 in change to make a long-distance phone call from a pay phone. He has a roll of quarters and a roll of dimes. Find all the possibilities for the number of each type of coin that Sam could use.

Mixed Review Exercises

Solve each open sentence and graph each solution set that is not empty.

1. $-2 \leq 2a - 1 < 4$ **2.** $|2 - b| < 1$ **3.** $2c + 3 \leq 5c - 6$

NAME ______________________ DATE ______________

3–2 Graphs of Linear Equations in Two Variables

Objective: To graph a linear equation in two variables.

Vocabulary

Plane rectangular coordinate system The intersection of a vertical number line (y-axis) and a horizontal number line (x-axis). A rectangular coordinate system is also called a *Cartesian coordinate system* or an *xy-coordinate plane*.

Quadrant One of the four regions formed by the intersecting axes in a rectangular coordinate system.

Coordinates The ordered pair of real numbers associated with each point in the xy-coordinate plane.
Example: The coordinates of point P are $(-4, 2)$.

x-coordinate (or abscissa) The first coordinate in an ordered pair of real numbers. It indicates the position of a point along the x-axis. Example: The x-coordinate of point P is -4.

y-coordinate (or ordinate) The second coordinate in an ordered pair of real numbers. It indicates the position of a point along the y-axis. Example: The y-coordinate of point P is 2.

Origin The point (0, 0), where the x-axis intersects the y-axis in a coordinate plane.

Linear equation in two variables Any equation that can be written in the form $Ax + By = C$ (A and B not both 0).

Theorem The graph of every equation of the form $Ax + By = C$ (A and B not both 0) is a line. Conversely, every line in the coordinate plane is the graph of an equation of this form.

Symbol $P(a, b)$ (point P with coordinates (a, b))

Example 1 Graph the ordered pairs (2, 3), (3, 0), (−3, 2), (1, −3), and (−1, −2) in the same xy-coordinate plane.

Solution On the x-axis, the positive direction is to the right, and the negative direction is to the left.

On the y-axis, the positive direction is up, and the negative direction is down.

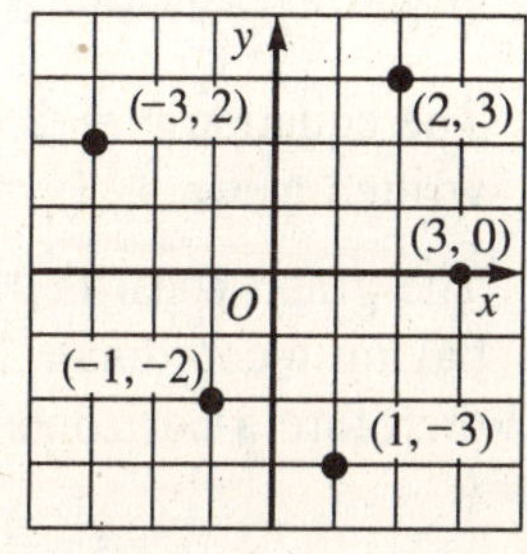

For each exercise, graph the ordered pairs in the same coordinate plane.

1. (2, 6), (3, 0), (−2, 3), (4, −5), (0, −4)

2. (−1, −3), (−4, 0), (−1, 5), (4, 0), (3, 5)

3. (−2, −1), (0, 0), (−1, 2), $\left(\frac{5}{2}, -\frac{3}{2}\right)$, (1, 2)

4. (−1, −1), (−2, 2), (3, 3), $\left(\frac{3}{2}, \frac{3}{2}\right)$, (2, −2)

NAME ______________________________ DATE ______________

3–2 Graphs of Linear Equations in Two Variables *(continued)*

Example 2 Graph $2x - 4y = 12$.

Solution Two points are needed to determine a line. The graph crosses the y-axis at a point whose x-coordinate is 0. The graph crosses the x-axis at a point whose y-coordinate is 0. Find these two points and use them to make your graph.

Let $x = 0$.

$$2(0) - 4y = 12$$
$$-4y = 12$$
$$y = -3$$

Solution $(0, -3)$

Let $y = 0$.

$$2x - 4(0) = 12$$
$$2x = 12$$
$$x = 6$$

Solution $(6, 0)$

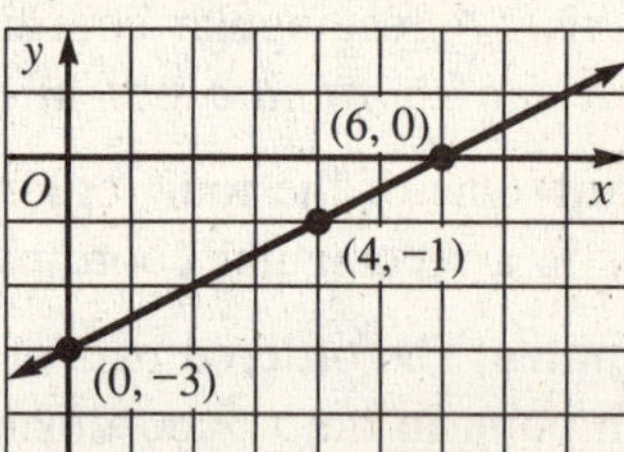

The graph is the line through the points with coordinates $(0, -3)$ and $(6, 0)$.

As a check, note that $(4, -1)$ is a solution of $2x - 4y = 12$ and its graph lies on the line. To avoid errors, it's best to use at least three points to graph a line.

CAUTION When a line passes through the origin, the method used in Example 2 will give you only one solution. In such cases, you will have to let x equal a number besides 0 and solve for y to find another solution.

Graph each equation.

5. $x + y = 6$ **6.** $x - y = 3$ **7.** $2x + y = 2$

8. $x + 3y = 9$ **9.** $3x - y = 6$ **10.** $2x + 7y + 14 = 0$

11. $2x - 3y - 6 = 0$ **12.** $x + y = 0$ **13.** $4x - y = 0$

14. $y = 2 + x$ **15.** $4x - 3y = 0$ **16.** $4x - 2y - 6 = 0$

17. $2x + 3y = 5$ **18.** $x + \frac{1}{2}y = 1$ **19.** $y = \frac{1}{3}x + 2$

Example 3 Graph the equation $y = 2$ in a coordinate plane.

Solution The equation $y = 2$ can be written as $0x + 1y = 2$.

The graph consists of all points having y-coordinate 2 and is therefore a horizontal line.

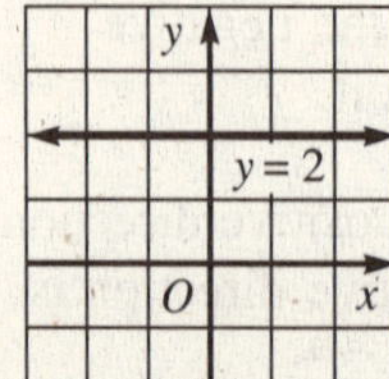

Note: The graph of $Ax + By = C$ is a horizontal line if $A = 0$, and is a vertical line if $B = 0$.

Graph in a coordinate plane.

20. $y = -2$ **21.** $x = 1$ **22.** $2x + 3 = 0$ **23.** $2y - 5 = 0$

NAME ______________________ DATE ______________

3–3 The Slope of a Line

Objective: To find the slope of a line and to graph a line given its slope and a point on it.

Vocabulary

Slope of a line L A measure of the "steepness" of the line.
If points (x_1, y_1) and (x_2, y_2) are on line L, and $(x_1 \neq x_2)$,

$$\text{slope of line } L = \frac{\text{rise}}{\text{run}} = \frac{y_2 - y_1}{x_2 - x_1}.$$

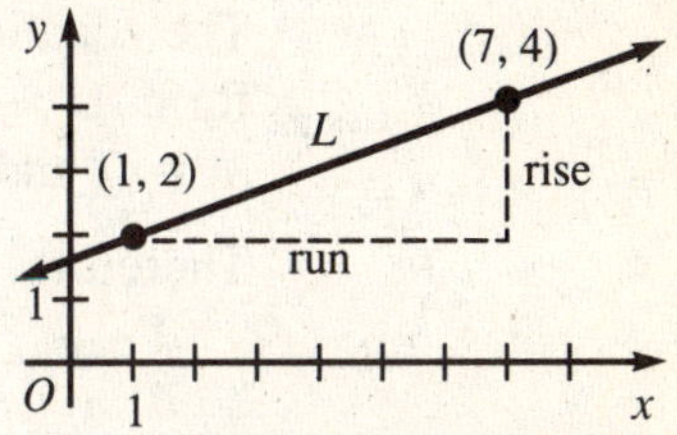

Example: The slope of line L in the diagram is

$$\frac{\text{rise}}{\text{run}} = \frac{4 - 2}{7 - 1} = \frac{2}{6} = \frac{1}{3}.$$

Coefficient The constant (or numerical) factor of a term.
Example: The coefficient of the term $5x^3$ is 5.

Symbol (x_1, y_1) (read "x one, y one" or "x sub one, y sub one")

CAUTION A vertical line has no slope. A horizontal line has slope 0. Having slope 0 is not the same as having no slope.

Example 1 Find the slope of the line containing the given points.

a. $(-3, 2)$ and $(5, -1)$ **b.** $(6, -1)$ and $(-3, -1)$

Solution **a.** slope $= \dfrac{-1 - 2}{5 - (-3)} = -\dfrac{3}{8}$ **b.** slope $= \dfrac{-1 - (-1)}{-3 - 6} = \dfrac{0}{-9} = 0$

$\therefore$ the line is horizontal.

Find the slope of the line containing the given points. If the line has no slope, write "vertical."

1. $(2, 3), (5, -2)$
2. $(3, 1), (0, 1)$
3. $(4, -2), (-2, 1)$
4. $(-5, 2), (-5, 4)$
5. $(4, 2), (-1, -3)$
6. $(-2, 3), (4, -1)$
7. $(-2, -4), (-2, 4)$
8. $\left(\frac{2}{3}, 3\right), \left(\frac{5}{3}, 7\right)$
9. $\left(\frac{1}{3}, -5\right), (0, -4)$
10. $(0.5, 1), (0.75, -2)$
11. $(r, s), (-r, -s)$ $(r \neq 0)$
12. $(-r, -s), (-s, -r)$ $(r \neq s)$

Theorem The slope of the line $Ax + By = C$ $(B \neq 0)$ is $-\dfrac{A}{B}$.
It follows from this theorem that the slope of a line is the coefficient of the x-term when the equation of the line is solved for y: $y = -\dfrac{A}{B}x + \dfrac{C}{B}$.

NAME ______________________ DATE ______________

3–3 The Slope of a Line (continued)

Example 2 Find the slope of the line $7x - 4y = 20$.

Solution

Method 1:

The equation is written in the form $Ax + By = C$, where $A = 7$ and $B = -4$.

Therefore, the slope is

$$-\frac{A}{B} = -\frac{7}{-4} = \frac{7}{4}.$$

Method 2:

Solve the given equation for y.

$$7x - 4y = 20$$
$$-4y = -7x + 20$$
$$y = \frac{7}{4}x - 5$$

The slope is the coefficient of x, $\frac{7}{4}$.

Find the slope of each line.

13. $x + y = 3$ **14.** $x - y - 2 = 0$ **15.** $3x + 2y = 4$ **16.** $3y + 2 = 9x$

17. $-x = y - 6$ **18.** $3(-2x + 1) = 8y$ **19.** $\frac{1}{3}x - \frac{1}{2}y = 1$ **20.** $\frac{x}{4} - \frac{y}{-6} = 1$

Example 3 Graph the line through the point $P(-6, 5)$ having slope $m = -\frac{2}{3}$.

Solution To graph the line, start at a known point and use the slope to find a second point.

$$m = \frac{\text{rise}}{\text{run}} = -\frac{2}{3} = \frac{-2}{3}$$

You will start at $P(-6, 5)$ and move 2 units *down* and 3 units *to the right* to reach a second point $Q(-3, 3)$. Then you can draw the line through P and Q. To find other points on the line, repeat this process. Another point is $R(0, 1)$.

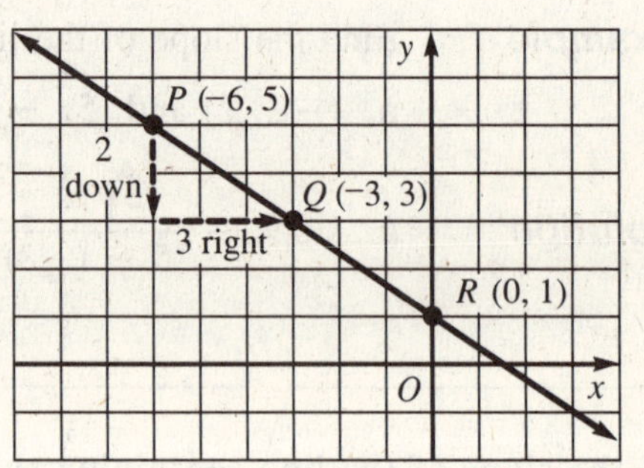

Graph the line through point P having slope m. Find the coordinates of two other points on the line.

21. $P(-1, -2)$, $m = 3$ **22.** $P(0, 0)$, $m = -2$ **23.** $P(1, 2)$, $m = \frac{1}{2}$

24. $P(3, -1)$, $m = -\frac{3}{2}$ **25.** $P(-3, 2)$, $m = 0$ **26.** $P(2, 2)$, no slope

Mixed Review Exercises

Complete the ordered pair to form a solution of the given equation.

1. $x + 2y = 3$; (? , 1) **2.** $x - 2y = 4$; (? , −5) **3.** $4x - 5y = 2$; (3, ?)

Graph each equation.

4. $x - y = 3$ **5.** $4x - 5y = 8$ **6.** $y = x$ **7.** $y - 3 = 0$ **8.** $y = -\frac{1}{3}x - 2$

NAME ____________________ DATE __________

3–4 Finding an Equation of a Line

Objective: To find an equation of a line given its slope and a point on the line, or two points, or its slope and the y-intercept.

Vocabulary

Standard form (of a linear equation in two variables) The equation of a line written as $Ax + By = C$, where A, B, and C are integers, and A and B are not both zero.

Point-slope form The equation of a line written as $y - y_1 = m(x - x_1)$ where m is the slope and (x_1, y_1) is a point on the line.

Example 1 Find an equation in standard form of the line containing the point $(1, -5)$ and having slope $-\frac{2}{3}$.

Solution Use the point-slope form with $(x_1, y_1) = (1, -5)$ and $m = -\frac{2}{3}$.

$$\begin{aligned} y - y_1 &= m(x - x_1) \\ y - (-5) &= -\frac{2}{3}(x - 1) \\ 3(y + 5) &= -2(x - 1) \\ 3y + 15 &= -2x + 2 \\ 2x + 3y &= -13 \end{aligned}$$

{Multiply both sides by 3 to clear the fractions.

Check: $2(1) + 3(-5) = -13$; $m = -\frac{A}{B} = -\frac{2}{3}$ √

Find an equation in standard form of the line containing point P and having slope m.

1. $P(4, 1)$, $m = -1$ **2.** $P(3, 0)$, $m = -2$ **3.** $P(-2, 1)$, $m = 0$ **4.** $P(-1, -3)$, $m = 3$

5. $P(-2, 2)$, $m = \frac{1}{4}$ **6.** $P(2, -1)$, $m = \frac{1}{2}$ **7.** $P(3, 4)$, $m = -\frac{5}{4}$ **8.** $P(-1, 5)$, $m = 0.2$

Example 2 Find an equation in standard form of the line containing $(2, -3)$ and $(-1, 2)$.

Solution Find the slope of the line: $m = \frac{2 - (-3)}{-1 - 2} = \frac{5}{-3} = -\frac{5}{3}$

Next use the point-slope form with either point.

$$\begin{aligned} y - y_1 &= m(x - x_1) \\ y - (-3) &= -\frac{5}{3}(x - 2) \\ 3(y + 3) &= -5(x - 2) \\ 3y + 9 &= -5x + 10 \\ 5x + 3y &= 1 \end{aligned}$$

$$\begin{aligned} y - y_1 &= m(x - x_1) \\ y - 2 &= -\frac{5}{3}(x - (-1)) \\ 3(y - 2) &= -5(x + 1) \\ 3y - 6 &= -5x - 5 \\ 5x + 3y &= 1 \end{aligned}$$

Find an equation in standard form of the line containing the given points.

9. $(4, 6)$, $(0, 0)$ **10.** $(-1, 3)$, $(3, -1)$ **11.** $(1, -2)$, $(-2, -1)$ **12.** $(7, 2)$, $(-1, -2)$

13. $(-3, 1)$, $(-4, 1)$ **14.** $(4, -3)$, $(4, 5)$ **15.** $\left(\frac{1}{4}, \frac{3}{2}\right)$, $\left(\frac{3}{4}, 1\right)$ **16.** $\left(\frac{1}{3}, -\frac{1}{2}\right)$, $\left(\frac{4}{3}, \frac{3}{2}\right)$

NAME ______________________________ DATE ______________

3–4 Finding an Equation of a Line *(continued)*

Vocabulary

y-intercept The y-coordinate of the point where a line (or curve) intersects the y-axis.

x-intercept The x-coordinate of the point where a line (or curve) intersects the x-axis.

Slope-intercept form The equation of a line written as $y = mx + b$ where m is the slope and b is the y-intercept.

Theorem Let L_1 and L_2 be two different lines, with slopes m_1 and m_2.

1. L_1 and L_2 are *parallel* if and only if $m_1 = m_2$.
2. L_1 and L_2 are *perpendicular* if and only if $m_1 m_2 = -1$. (m_1 and m_2 are *negative reciprocals*.)

Example 3 Find an equation in standard form of the line having slope $-\frac{3}{4}$ and y-intercept 2.

Solution Use $y = mx + b$ with $m = -\frac{3}{4}$ and $b = 2$.

$$y = -\frac{3}{4}x + 2$$
$$4y = -3x + 8$$
$$3x + 4y = 8$$

Multiply both sides by 4 to clear the fraction.

Find an equation in standard form of the line having slope m and y-intercept b.

17. $m = -1, b = 3$ **18.** $m = 2, b = -4$ **19.** $m = -\frac{3}{2}, b = \frac{1}{2}$ **20.** $m = 0.8, b = 0.6$

Example 4 Find equations in standard form of the lines through point $P(-1, 3)$ that are (a) parallel to and (b) perpendicular to the line $x - 3y = 6$.

Solution Solve the equation for y to find the slope:

$x - 3y = 6$; $y = \frac{1}{3}x - 2$ ∴ the slope of the line is $\frac{1}{3}$.

a. A line parallel to $x - 3y = 6$ has the same slope, $\frac{1}{3}$. Use the point-slope form:

$$y - 3 = \frac{1}{3}(x + 1)$$
$$3y - 9 = x + 1$$
$$x - 3y = -10$$

b. The slope of a line perpendicular to $x - 3y = 6$ is the negative reciprocal of $\frac{1}{3}$, or -3. Use the point-slope form:

$$y - 3 = -3(x + 1)$$
$$y - 3 = -3x - 3$$
$$3x + y = 0$$

Find equations in standard form of the lines through point P that are (a) parallel to and (b) perpendicular to line L.

21. $P(0, -5)$; L: $y = \frac{3}{4}x + 2$

22. $P(3, 0)$; L: $x - 4y = 4$

23. $P(2, -3)$; L: $2x + 7y = 14$

24. $P(-5, -1)$; L: $x + 4 = 0$

NAME ______________________ DATE ______________

3–5 Systems of Linear Equations in Two Variables

Objective: To solve systems of linear equations in two variables.

Vocabulary

System of linear equations A set of linear equations in the same two variables.

Transformations that produce equivalent systems (systems with the same solution set)

1. Replacing an equation by an equivalent equation.
2. Substituting for one variable in any equation an equivalent expression for that variable obtained from another equation in the system.
3. Replacing any equation by the sum of that equation and another equation in the system.

Linear combination The equation that results when two linear equations are added.

Example 1 Solve these systems: **a.** $3x + 4y = 1$, $5x - 3y = 21$ **b.** $2x + y = 0$, $x - 3y = -7$

Solution

a. Find a linear combination that eliminates y.

1. Multiply the first equation by 3 and the second equation by 4 so that the coefficients of y will be *opposites*.

$$\begin{aligned} 9x + 12y &= 3 \\ 20x - 12y &= 84 \end{aligned}$$

2. Add the equations in Step 1 and solve the resulting equation for x.

$$\begin{aligned} 29x + 0y &= 87 \\ 29x &= 87 \\ x &= 3 \end{aligned}$$

3. Substitute 3 for x in either of the *original* equations to find y.

$$\begin{aligned} 3(3) + 4y &= 1 \\ 9 + 4y &= 1 \\ 4y &= -8 \\ y &= -2 \end{aligned}$$

4. Check that $(3, -2)$ satisfies both of the original equations.

$$\begin{aligned} 3(3) + 4(-2) &= 1 \checkmark \\ 5(3) - 3(-2) &= 21 \checkmark \end{aligned}$$

$\therefore$ the solution is $(3, -2)$.

b. Use the substitution method.

1. Express y in terms of x in the first equation (since the coefficient of y is 1).

$$\begin{aligned} 2x + y &= 0 \\ y &= -2x \end{aligned}$$

2. Substitute $-2x$ for y in the second equation. Solve for x.

$$\begin{aligned} x - 3y &= -7 \\ x - 3(-2x) &= -7 \\ x + 6x &= -7 \\ 7x &= -7 \\ x &= -1 \end{aligned}$$

3. Substitute -1 for x in either of the original equations to find y.

$$\begin{aligned} -1 - 3y &= -7 \\ -3y &= -6 \\ y &= 2 \end{aligned}$$

4. The check is left for you.

$\therefore$ the solution is $(-1, 2)$.

3–5 Systems of Linear Equations in Two Variables *(continued)*

CAUTION The equations of a system are **consistent** and have exactly one solution, as in Example 1, when their graphs are *intersecting lines*. NOT all systems have one solution.

Example 2 Solve these systems:

a. $2x + y = -1 - 2y$
$2x - 9y = 3 + 8x$

b. $2y = x + 3$
$2y - x - 6 = 0$

Solution

a. Transform each equation into standard form. Use a linear combination.

$$\left.\begin{aligned} 2x + y &= -1 - 2y \\ 2x - 9y &= 3 + 8x \end{aligned}\right\} \longrightarrow \begin{aligned} 2x + 3y &= -1 \xrightarrow{\times 3} 6x + 9y = -3 \\ -6x - 9y &= 3 \qquad\quad -6x - 9y = 3 \end{aligned}$$

$$0 = 0$$

Since $0 = 0$ is an identity, all solutions of one equation are solutions of the other. Thus, the solution set is $\{(x, y): 2x + 3y = -1\}$. The equations are consistent and **dependent** and their graphs *coincide*.

b. Solve the first equation for x, and use the substitution method.

$$\begin{aligned} 2y &= x + 3 & 2y - x - 6 &= 0 \\ x &= 2y - 3 \xrightarrow{\text{substitute}} & 2y - (2y - 3) - 6 &= 0 \\ & & 2y - 2y + 3 - 6 &= 0 \\ & & -3 &= 0 \longleftarrow \text{False!} \end{aligned}$$

Since $-3 = 0$ is false, the system has no solution. The equations are **inconsistent** and their graphs are *parallel lines*.

Solve each system. If the system has an infinite solution set, specify it and give three solutions. If the system has no solution, say so.

1. $2x + 5y = 41$
$2x + y = 13$

2. $2p - q = -1$
$3p - 4q = 6$

3. $8x + 3y = 23$
$4x - 5y = 5$

4. $3x + 2y = 22$
$2x - 3y = 6$

5. $5x - 7y = 54$
$2x - 3y = 22$

6. $3d + 7c = 42$
$4d + 5c = 30$

7. $3u + 4v = 9$
$5u - 8v = 4$

8. $2x - 5y = 2$
$7x + \frac{1}{2}y = -1$

9. $8n = 6m - 3$
$9m = 12n + 5$

10. $4x + 3y = x + 6$
$x + 3y - 2 = 2y$

11. $3(5 - x) = y$
$5(3 - x) = -2y + 1$

12. $2b = 2a - b + 4$
$3a = a + 3b + 4$

Mixed Review Exercises

For the line containing the given points, find (a) the slope and (b) an equation in standard form.

1. $(3, 2), (-1, -6)$ **2.** $(0, 3), (-1, 2)$ **3.** $(1, -2), (-1, -2)$ **4.** $(0, -2), (2, -1)$

Find the slope and y-intercept of each line.

5. $2x - 2y = 6$ **6.** $2x + 3y = 9$ **7.** $-x + 2y = 4$ **8.** $y + 7 = 0$

NAME ______________________ DATE ______________

3–6 Problem Solving: Using Systems

Objective: To use systems of equations to solve problems.

Example 1 An algebra test contains 38 problems. Some of the problems are worth 2 points each. The rest are worth 3 points each. A perfect score is 100 points. How many problems are worth 2 points?

Solution

Step 1 The problem asks for the number of 2-point problems on the test.

Step 2 Let a = the number of 2-point problems on the test.
Let b = the number of 3-point problems on the test.

Step 3 Set up a system of two equations.
Total number of problems on the test is 38. ⟶ $a + b = 38$
Total point value of the test is 100. ⟶ $2a + 3b = 100$

Step 4 Solve the system. Multiply the first equation by -2 to eliminate the a's, and solve for b. Then substitute $b = 24$ in either of the original equations and solve for a.

$$\begin{aligned} -2a - 2b &= -76 \\ 2a + 3b &= 100 \\ \hline b &= 24 \end{aligned}$$

$$\begin{aligned} a + 24 &= 38 \\ a &= 14 \end{aligned}$$

There are 14 problems worth 2 points each and 24 problems worth 3 points each.

Step 5 Check your answer against the original problem.
The test contains 38 problems: $14 + 24 = 38$ √
A perfect score is 100 points: $2(14) + 3(24) = 28 + 72 = 100$ √

∴ 14 problems on the test are worth 2 points.

Solve.

1. Tickets for the Senior Prom cost $25 for a single ticket and $40 for a couple. Ticket sales totaled $3800 and 110 tickets were sold. How many tickets of each type were sold?
2. A cashier had to give Sarah $3.45 in change but he had only quarters and dimes in the cash register. If he gave her 15 coins, how many dimes did she receive?

Vocabulary

Air speed The speed of an aircraft in still air.

Wind speed The speed of the wind.

Tail wind A wind blowing in the same direction as the path of the aircraft.

Head wind A wind blowing in the direction opposite to the path of the aircraft.

Ground speed The speed of the aircraft relative to the ground.

With a tail wind: ground speed = air speed + wind speed.
With a head wind: ground speed = air speed − wind speed.

3–6 Problem Solving: Using Systems *(continued)*

Example 2 With a tail wind, a plane traveled 840 miles in 3 hours. With the same wind, as a head wind, the return trip took 30 min longer. Find the plane's air speed and the wind speed.

Solution

Step 1 The problem asks for the plane's air speed and the speed of the wind.

Step 2 Let p = the air speed of the plane in miles per hour.
Let w = the wind speed in miles per hour.

Step 3 Use the formula rate × time = distance to construct a table.

	Rate (ground speed in mi/h) ×	Time (hours) =	Distance (miles)
With tail wind	$p + w$	3	$3(p + w)$
With head wind	$p - w$	$3\frac{1}{2}$ or $\frac{7}{2}$	$\frac{7}{2}(p - w)$

The distance with a tail wind is 840 mi. ⟶ $3(p + w) = 840$

The distance with a head wind is 840 mi. ⟶ $\frac{7}{2}(p - w) = 840$

Step 4 Solve the system.

$3(p + w) = 840$ is equivalent to: $p + w = 280$

$\frac{7}{2}(p - w) = 840$ is equivalent to: $p - w = 240$

$$2p = 520$$
$$p = 260$$
$$3(260 + w) = 840$$
$$260 + w = 280$$
$$w = 20$$

The solution is (260, 20).

Step 5 The check is left for you.

∴ the air speed of the plane is 260 mi/h and the speed of the wind is 20 mi/h.

Solve.

3. With a head wind, a plane traveled 840 miles northward in 2 hours. With the same wind as a tail wind, the return trip southward took 1 hour and 45 minutes. Find the plane's air speed and the wind speed.

4. With a tail wind, a small plane traveled 420 miles in 1 hour and 30 minutes. The return trip against the same wind took 30 minutes longer. Find the wind speed and the air speed of the plane.

NAME ______________________ DATE ____________

3–7 Linear Inequalities in Two Variables

Objective: To graph linear inequalities in two variables and systems of such inequalities.

Vocabulary

Linear inequality in two variables A sentence obtained by replacing the equals sign in a linear equation by an inequality symbol ($>$, $<$, $\geq$, or $\leq$). Examples: $4x - 7y \geq 5$, and $x < 9y$.

Solution of an inequality in two variables An ordered pair of numbers that satisfies the inequality, or makes it true. Example: (5, 9), (0, 0), and (-1, 2) are solutions of the inequality $y \leq 2x + 4$.

Associated equation The linear equation obtained by replacing the inequality symbol in a linear inequality by an equals sign. Example: The associated equation of the inequality $y > 2x + 5$ is $y = 2x + 5$.

Open half-plane The set of all points on one side of a given line (the *boundary*). The boundary appears as a dashed line because it is *not* included in the graph.

Closed half-plane An open half-plane together with its boundary. The boundary appears as a solid line because it *is* included in the graph.

System of linear inequalities A set of linear inequalities in the same two variables. The graph of such a system is the set of points satisfying *all* of its inequalities.

Example 1 Graph $x > -1$ in a coordinate plane.

Solution The graph of the associated equation $x = -1$ is the boundary. Since x is greater than -1, the boundary line is dashed. A point belongs to the graph of the inequality if and only if its x-coordinate is greater than -1. This means that you must shade the points to the *right* of the boundary line.

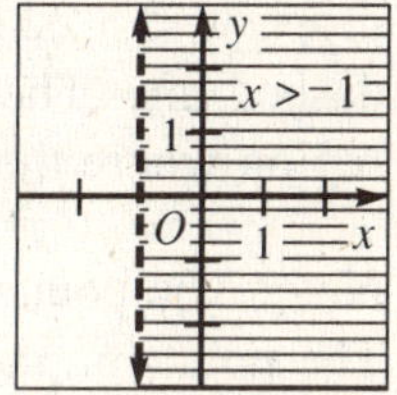

Example 2 Graph $3x - 2y \geq 6$.

Solution Solve the inequality for y.

$$3x - 2y \geq 6$$
$$-2y \geq -3x + 6$$
$$y \leq \frac{3}{2}x - 3$$

Graph the associated equation $y = \frac{3}{2}x - 3$ as a solid line. The graph of the inequality is a closed half-plane that includes this boundary and the points *below* it.

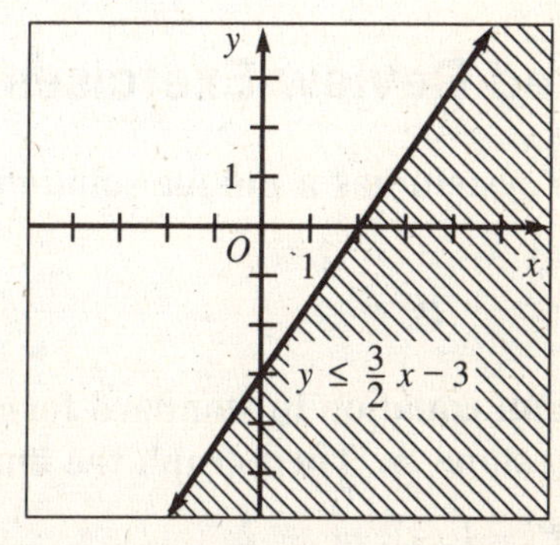

3–7 Linear Inequalities in Two Variables *(continued)*

CAUTION Don't assume that you should shade below for an inequality having the symbol $<$, and above for the symbol $>$. This is only true when an inequality has been solved for y in terms of x. A safer way to determine which side of the boundary to shade is to test a point on either side of it. The origin, (0, 0), is a convenient point to use. If the point you test satisfies the inequality, you shade the region containing that point. If not, shade the opposite region.

Graph each inequality in a coordinate plane.

1. $y \geq 1$ **2.** $x + 2 > 0$ **3.** $x - y \geq 0$ **4.** $2x < y$

5. $x + y < 2$ **6.** $3x - y \leq 1$ **7.** $4x - y \geq 2$ **8.** $3x + 4y > 8$

9. $5x - 2y \leq 6$ **10.** $2x < -3(2 - y)$ **11.** $y \leq -\frac{1}{2}(5x - 4)$ **12.** $4(x + 2) > 3y + 2$

Example 3 Graph this system: $2x + y \leq 2$
$x - y > 0$

Solution As in Example 2, solve each inequality for y.

$2x + y \leq 2$ $\quad y \leq -2x + 2$
$x - y > 0$ $\quad y < x$

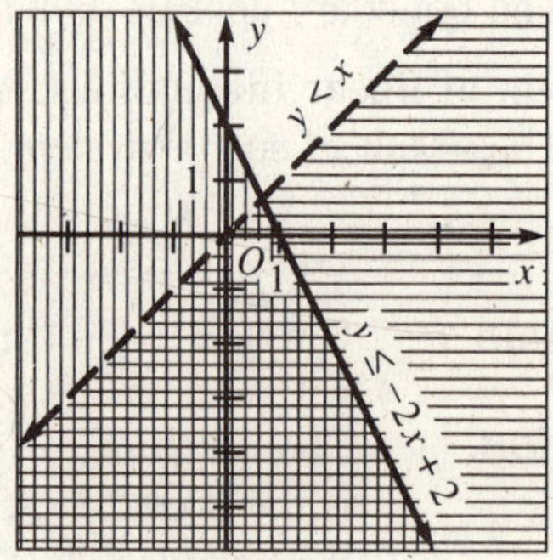

Graph $y \leq -2x + 2$, showing the boundary as a solid line. In the same coordinate plane, graph $y < x$, showing the boundary as a dashed line. The graph of the system is the region where the two graphs overlap.

Note: The points on the solid boundary are part of the solution set. The points on the dashed boundary are not.

Graph each system of inequalities.

13. $y \geq 1$, $y < x + 1$ **14.** $x + y < 0$, $2x - y > 0$ **15.** $y < 2x - 1$, $1 - y \geq 0$ **16.** $2x + y \geq 3$, $2x - y \leq 3$ **17.** $5x - 2y < 10$, $2x - 5y > 10$

Mixed Review Exercises

Each system has a unique solution; find it.

1. $4x - y = 2$, $x - 2y = 4$ **2.** $2x - 3y = -1$, $x + y = 7$ **3.** $-2x + y = 0$, $3x + 2y = -7$

Find an equation in standard form of the line through *P* having the given slope *m*. Then graph the line.

4. $P(-1, 4)$, $m = 2$ **5.** $P(3, 5)$, $m = 0$ **6.** $P(0, 0)$, $m = -2$

7. $P(-5, -1)$, $m = \frac{1}{3}$ **8.** $P(-2, 3)$, $m = -\frac{4}{5}$ **9.** $P(2, 1)$, no slope

NAME ______________________ DATE ____________

3–8 Functions

Objective: To find values of functions and to graph functions.

Vocabulary

Mapping diagram A diagram such as the one on the right that pictures a correspondence between two sets.

Function A correspondence between two sets, D and R, that assigns to each member of D exactly one member of R. Functions are sometimes specified by a rule. The rule for the function in the diagram is "double the number and subtract 1."

Domain The domain of a function is the set D. In the diagram, $D = \{\text{integers}\}$.

Range The range of a function is the set of all members of R assigned to at least one member of D. In the diagram, $R = \{\text{odd integers}\}$.

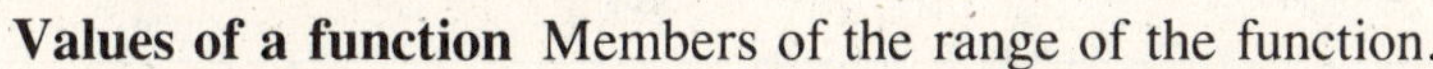

Values of a function Members of the range of the function.

Graph of a function The set of all points (x, y) such that x is in the domain of the function and the rule of the function assigns y to x.

Arrow notation Notation used to describe the rule of a function.
Example: The arrow notation for the function in the diagram is $f\colon x \to 2x - 1$. This is read "f, the function that assigns $2x - 1$ to the number x."

Functional notation Another type of notation used to describe the rule of a function.
Example: The functional notation for the function in the diagram is $f(x) = 2x - 1$. $f(3) = 5$ is read "the value of the function f at 3 is 5" or "f of 3 is 5."

Example 1 Given $f\colon x \to x^2 - 3x$ with domain $D = \{-1, 0, 1, 2, 3\}$.

a. Find the range of f.

b. Graph the function.

Solution

a. Make a table showing the numbers that the rule of f assigns to each member of the domain.

x	$x^2 - 3x$	(x, y)
-1	$(-1)^2 - 3(-1) = 4$	$(-1, 4)$
0	$0^2 - 3(0) = 0$	$(0, 0)$
1	$1^2 - 3(1) = -2$	$(1, -2)$
2	$2^2 - 3(2) = -2$	$(2, -2)$
3	$3^2 - 3(3) = 0$	$(3, 0)$

The range R is $\{4, 0, -2\}$.

Note: You list each member of the range only once.

b. Graph the ordered pairs from the table.

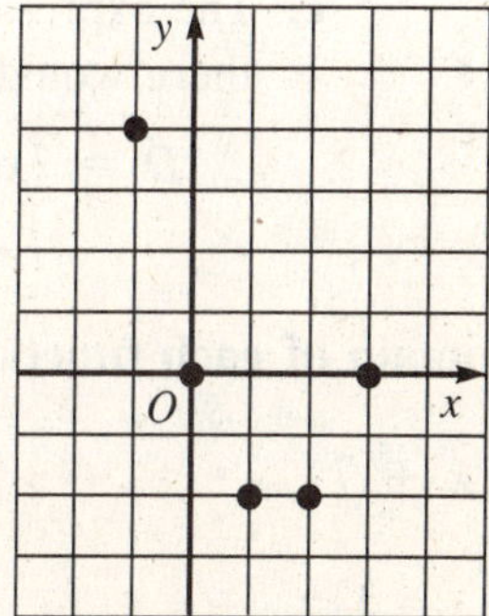

3–8 Functions *(continued)*

Find the range of each function given the domain D and the rule of the function.

1. $F: x \rightarrow 3x - 2;\ D = \{-1, 0, 1, 2\}$

2. $G: x \rightarrow 4x + 2;\ D = \{0, 1, 2, 3\}$

3. $g: x \rightarrow 3 - x;\ D = \{0, 1, 2, 3\}$

4. $f: x \rightarrow x^2 + 1;\ D = \{-2, -1, 0, 1\}$

5. $h: x \rightarrow 4 - 4x + x^2;\ D = \{0, 1, 2, 3, 4\}$

6. $\phi: x \rightarrow 2x^2 - 3x + 1;\ D = \{-1, 0, 1, 2\}$

7. $g: x \rightarrow 2x - x^2;\ D = \{-2, -1, 0, 1, 2\}$

8. $H: x \rightarrow x^3 - x^2;\ D = \{-1, 0, 1, 2\}$

9. $h: x \rightarrow |x - 2|;\ D = \{0, 1, 2, 3, 4\}$

10. $f: x \rightarrow |x| - x;\ D = \{-2, -1, 0, 1, 2\}$

11–20. Graph the functions in Exercises 1–10.

Example 2 Give the domain of each function.

a. $f(x) = \dfrac{3}{x^2(x - 5)}$ **b.** $g(x) = \dfrac{5}{x^2 + 1}$ **c.** $h(x) = \sqrt{x + 1}$

Solution Begin with the set of real numbers. Then eliminate any of these that don't produce a real number when substituted into the rule for the function.

a. The expression $\dfrac{3}{x^2(x - 5)}$ is not defined when the denominator equals zero, in other words, when $x = 0$ or $x = 5$.

$\therefore D = \{\text{all real numbers except 0 and 5}\}$

b. There is no real number that will make the denominator zero, since $x^2 > 0$ for all real numbers.

$\therefore D = \{\text{real numbers}\}$

c. The expression $\sqrt{x + 1}$ is defined only when $x + 1 \geq 0$, otherwise there would be a negative number inside the square root symbol.

$\therefore D = \{x: x \geq -1\}$

Give the domain of each function.

21. $f(x) = x + 7$

22. $g(x) = \dfrac{3}{x + 1}$

23. $h(x) = \dfrac{3}{(x - 2)(x + 2)}$

24. $F(x) = \dfrac{3}{x^2 - 9}$

25. $G(x) = \sqrt{2x - 1}$

26. $H(x) = \sqrt{x^2 + 6}$

NAME ______________________________ DATE ______________

3–9 Linear Functions

Objective: To find equations of linear functions and to apply properties of linear functions.

Vocabulary

Linear function A function f that can be defined by $f(x) = mx + b$ where x, m, and b are any real numbers. The graph of f is the graph of $y = mx + b$, a line with slope m and y-intercept b.

Constant function A linear function where $m = 0$. Its equation is $f(x) = b$.

Rate of change of $f(x)$ A number that tells you how much $f(x)$ changes for any given change in x. In a linear function $f(x) = mx + b$, the constant m is the rate of change. The rate of change can be computed using the formula:

$$\text{rate of change } m = \frac{\text{change in } f(x)}{\text{change in } x}$$

Example: If $f(x)$ decreases by 3 when x increases by 5, then $m = \frac{-3}{5} = -\frac{3}{5}$.

Example 1 Find equations of the linear functions g and h using the given information.

a. $m = -2$ and $g(3) = 5$ **b.** $h(1) = 4$ and $h(-2) = 10$

Solution

a. Since g is linear and $m = -2$, substitute -2 for m in $g(x) = mx + b$. → $g(x) = -2x + b$

$g(3) = 5$ means that $g(x) = 5$ when $x = 3$. Substitute and solve for b. →
$$5 = -2(3) + b$$
$$5 = -6 + b$$
$$11 = b$$

$\therefore\ g(x) = -2x + 11$

b. Since h is linear, $h(x) = mx + b$. First you need to find m.

$$m = \frac{\text{change in } h(x)}{\text{change in } x} = \frac{h(1) - h(-2)}{1 - (-2)} = \frac{4 - 10}{1 + 2} = \frac{-6}{3} = -2$$

Substitute -2 for m in $h(x) = mx + b$. → $h(x) = -2x + b$

To find the value of b, use either $h(1) = 4$ or $h(-2) = 10$. →
$$4 = -2(1) + b$$
$$4 = -2 + b$$
$$6 = b$$

$\therefore\ h(x) = -2x + 6$

Find an equation of the linear function f using the given information.

1. $m = 4$, $b = 1$
2. $m = -2$, $b = 3$
3. $f(0) = 1$, slope of graph $= -3$
4. $f(0) = 3$, slope of graph $= -\frac{1}{2}$
5. $f(0) = -2$; $f(x)$ increases by 12 when x increases by 3
6. $f(0) = -1$; $f(x)$ decreases by 2 when x increases by 3
7. $m = 6$, $f(1) = 5$
8. $m = -\frac{2}{3}$, $f(3) = 2$
9. $f(1) = 3$, $f(-2) = -6$
10. $f(-3) = 1$, $f(2) = 2$
11. $f(4) = -5$, $f(-3) = 3$
12. $f(-3) = 5$, $f(2) = 5$

3–9 Linear Functions *(continued)*

Example 2 The prom committee asked a local restaurant to cater the senior prom. The restaurant charges \$1700 for a buffet for 100 people. It charges \$2450 for a buffet for 150 people.

a. Write a linear function to describe this situation.
b. Find the cost of a buffet for 175 people.

Solution **a.** Let $f(n)$ = the cost of a buffet for n people.
Since f is linear, $f(n) = mn + b$.

To find the values of m and b use the facts that $f(100) = 1700$ and $f(150) = 2450$ to write a system of linear equations:

$$f(100) = 100m + b = 1700$$
$$f(150) = 150m + b = 2450$$

Multiply the first equation by -1 and solve for m.

$$-100m - b = -1700$$
$$150m + b = 2450$$
$$50m = 750$$
$$m = 15$$

Substitute m in one of the original equations and solve for b.

$$100(15) + b = 1700$$
$$1500 + b = 1700$$
$$b = 200$$

$\therefore f(n) = 15n + 200$

b. Substitute $n = 175$ in the equation obtained in part (a).

$f(175) = 15(175) + 200 = 2625 + 200 = 2825$

$\therefore$ the cost of a buffet for 175 people is \$2825.

Solve the following problems. Assume that each situation can be described by a linear function.

13. An electrician charged \$90 for a two-hour job and \$140 for a four-hour job. At this rate, how much would be charged for an eight-hour job?

14. The relationship between hours spent studying and the average grade on an algebra exam was found to be a linear function. The average grade for students who spent 1 hour studying was 76, and it was 88 for students who spent 3 hours studying.

a. What was the average grade for students who spent 4 hours studying?
b. How many hours of studying would have resulted in an average grade of 80?

Mixed Review Exercises

Graph each open sentence in a coordinate plane.

1. $2x - y > 3$ **2.** $y = -x$ **3.** $x - 3 \leq 0$ **4.** $x - 3y \leq -4$

Find the range of each function with domain *D*.

5. $g: x \rightarrow 3 - 2x^2, D = \{-2, -1, 0\}$ **6.** $f(x) = |x| + 3, D = \{-2, -1, 1\}$

NAME ______________________ DATE ______________

3–10 Relations

Objective: To graph relations and to determine when relations are also functions.

Vocabulary

Relation Any set of ordered pairs. The set of first coordinates in the ordered pairs is the *domain*; the set of second coordinates is the *range*.
Example: The set $\{(1, 2), (3, 4), (3, 5)\}$ is a relation, where $D = \{1, 3\}$ and $R = \{2, 4, 5\}$.

Function A relation in which different ordered pairs have different *first* coordinates.
Example: The set $\{(0, -1), (2, 3), (5, 9)\}$ is a function. The set $\{(1, 2), (3, 4), (3, 5)\}$ is not.

Vertical-line test A test to determine whether a given relation is a function. This test says that a relation is a function if and only if no vertical line can intersect its graph more than once.

Symbols $\{(x, y): |x| + |y| = 1\}$ (An example of set notation for ordered pairs. It reads "the set of all ordered pairs (x, y) such that $|x| + |y| = 1$.")

CAUTION A relation can contain different ordered pairs with the same x-value or y-value. A function can contain different ordered pairs with the same y-value, but *not* with the same x-value. (The latter are points on a vertical line.)

Example 1 Give the domain and range of each relation, and tell whether it's a function.

a. 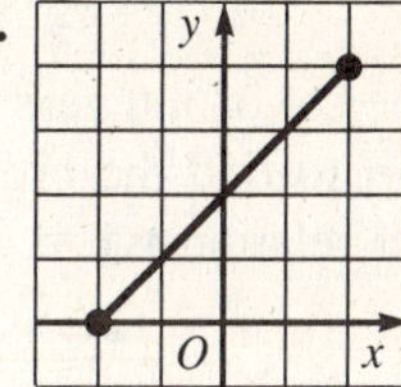

b. 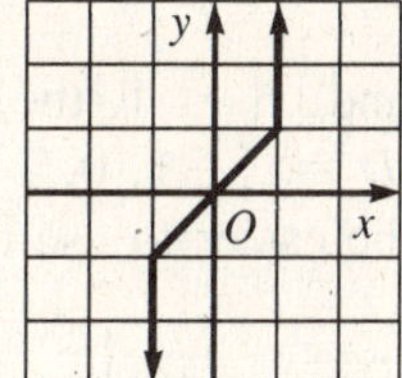

Solution

a. The domain is $\{x: -2 \le x \le 2\}$, since the graph goes from -2 to 2 with respect to the x-axis. The range is $\{y: 0 \le y \le 4\}$, since the graph goes from 0 to 4 with respect to the y-axis. The relation *is* a function, since each value of x in the graph is paired with exactly one value of y.

b. The domain is $\{x: -1 \le x \le 1\}$, since the graph goes from -1 to 1 with respect to the x-axis. The range is {real numbers}, since the graph spans the entire y-axis. The relation is *not* a function, since it includes different ordered pairs with the same first coordinate, like (1, 1) and (1, 2).

Give the domain and range of each relation graphed below. Is the relation a function?

1. 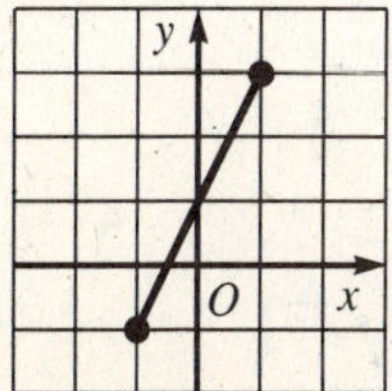

2. 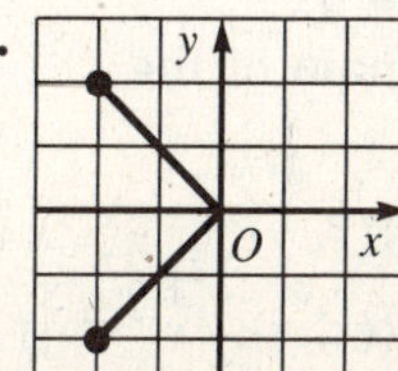

3. 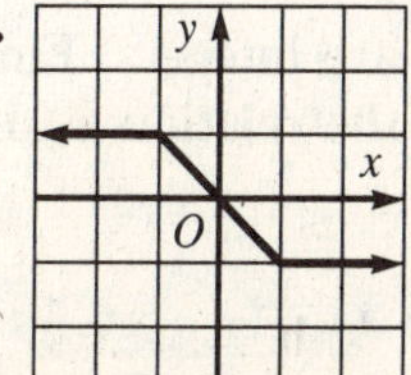

4.

3–10 Relations *(continued)*

Example 2 Graph the relation $\{(2, 1), (3, -1), (2, 3), (-1, 2), (-3, 0)\}$, and tell whether it is a function. If it is not a function, draw a vertical line that intersects the graph more than once.

Solution

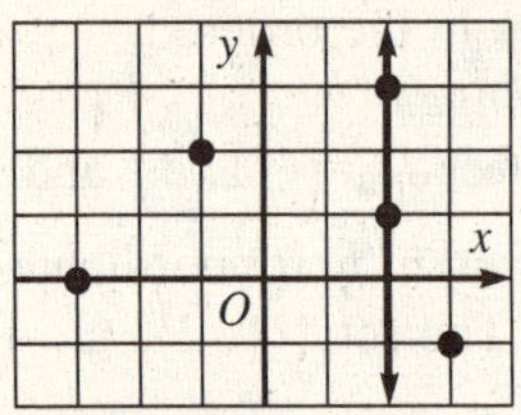

The relation is not a function. The vertical line $x = 2$ intersects the graph at two points, (2, 1), and (2, 3).

Graph each relation. Then tell whether it is a function. If it is not a function, draw a vertical line that intersects the graph more than once.

5. $\{(2, 1), (0, 0), (2, 2)\}$
6. $\{(-2, 1), (0, 1), (2, 1)\}$
7. $\{(3, 1), (2, -1), (0, 0), (-1, 2)\}$
8. $\{(-2, 2), (-1, 2), (0, 1), (-2, 1)\}$
9. $\{(-1, 2), (3, -1), (2, -2), (-1, -2)\}$
10. $\{(1, 3), (3, -1), (1, -1), (-2, -1)\}$

Example 3 In the relation $\{(x, y): |x| + |y| = 3 \text{ and } |x| \le 1\}$, x and y are integers. Find the domain D of the relation and draw its graph. Is the relation a function?

Solution Since x is an integer and $|x| \le 1$, the domain is the set of integers within one unit of the origin, so $D = \{-1, 0, 1\}$. Use a chart to find the ordered pairs for the graph. It will be easier to use the equivalent relation $|y| = 3 - |x|$.

x	$\lvert y\rvert = 3 - \lvert x\rvert$	y
-1	$\lvert y\rvert = 3 - \lvert -1\rvert = 2$	2 or -2
0	$\lvert y\rvert = 3 - \lvert 0\rvert = 3$	3 or -3
1	$\lvert y\rvert = 3 - \lvert 1\rvert = 2$	2 or -2

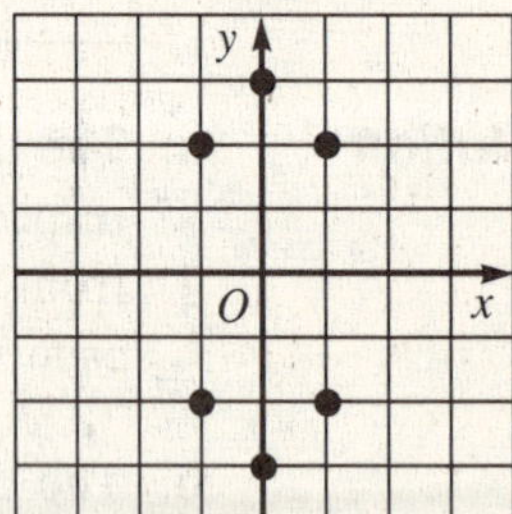

Since a vertical line, such as $x = -1$, intersects the graph at more than one point, the relation is not a function.

CAUTION To find the domain of a relation that is defined by a conjunction, as in Example 3, be sure to consider *both* open sentences. For example, if x is an integer, the domain of $\{x: |x| \le 1 \text{ and } |y| = x\}$ cannot include -1 since $|y|$ is always nonnegative.

In each relation below, x and y are integers. Find the domain of the relation and draw its graph. Is the relation a function?

11. $\{(x, y): |x| + |y| = 2 \text{ and } |x| \le 1\}$
12. $\{(x, y): |y| = x \text{ and } x \le 2\}$
13. $\{(x, y): |x| = y \text{ and } y \le 2\}$
14. $\{(x, y): |x| = |y| \text{ and } |x| \le 3\}$
15. $\{(x, y): |x + y| = 0 \text{ and } |x| \le 3\}$
16. $\{(x, y): |y| + |x| \le 2 \text{ and } |y| \le 1\}$

NAME ______________________ DATE ______________

4 Products and Factors of Polynomials

4–1 *Polynomials*

Objective: To simplify, add, and subtract polynomials.

Vocabulary

Monomial A constant (a number), a variable, or a product of a constant and one or more variables. Examples: $9, t, -b^2, \frac{1}{4}x^2y$

Coefficient (or **numerical coefficient**) The numerical factor in a monomial. Example: The coefficient of $-2x^3$ is -2.

Degree of a variable in a monomial The number of times the variable occurs as a factor in the monomial.
Example: In the monomial $3x^2y$, the degree of x is 2, and the degree of y is 1.

Degree of a monomial The sum of the degrees of the variables in the monomial. A nonzero constant has degree 0. The constant 0 has *no degree*.
Example: $5ab^2$ has degree $1 + 2$, or 3.

Similar (or **like**) **monomials** Monomials that are identical or that differ only in their coefficients. Example: $2mn^2$ and $-mn^2$

Polynomial A monomial or a sum of monomials. The monomials are called *terms* of the polynomials.
Example: The terms of the polynomial $y^2 - 3y - 7$ are y^2, $-3y$, and -7.

Simplified polynomial A polynomial in which no two terms are similar.

Degree of a polynomial The greatest of the degrees of its terms after the polynomial has been simplified.

Example 1 Simplify, arranging terms in order of decreasing degree of x. Then write the degree of the polynomial $3 - x + x^2 + xy + 2x^2 + 3x - 6 - 8xy$.

Solution

$$3 - x + x^2 + xy + 2x^2 + 3x - 6 - 8xy =$$
$$(x^2 + 2x^2) + (-8xy + xy) + (-x + 3x) + (3 - 6) =$$
$$\underbrace{3x^2}_{2}\ \underbrace{-\ 7xy}_{2}\ \underbrace{+\ 2x}_{1}\ \underbrace{-\ 3}_{0}$$

$3x^2 - 7xy + 2x - 3$ ←simplified polynomial
2 2 1 0 ←the degree of each term

The greatest degree of any term is 2, so the degree of the polynomial is 2.

Simplify, arranging terms in order of decreasing degree of x. Then write the degree of the polynomial.

1. $1 - 5x^3 + 3x^2 - 2 + 5x^2$

2. $x^4 - x^3 + 3x^4 - 2x^3 + 3x^2$

3. $3 - x^2 + 3x^2 - 5 + 2x^2 + 3x^3$

4. $7xy^3 + 3x^2y - 2x^3 + 6xy^3 - x^2y - 3x^3$

5. $x^2y^2 + x^3y^2 - 2x^2y^2 + 3x^3y + 4x^3y^2$

6. $2x^2y + 3xy^2 - 6xy^2 + 2x^2y - 4xy$

NAME ______________________ DATE ______________

4–1 Polynomials *(continued)*

Example 2 Add $4x^2 - 2x - 3$ and $3x^3 - 2x^2 + 3x - 4$.

Solution 1 $(4x^2 - 2x - 3) + (3x^3 - 2x^2 + 3x - 4) =$
$3x^3 + [4x^2 + (-2x^2)] + (-2x + 3x) + [-3 + (-4)] = 3x^3 + 2x^2 + x - 7$

Solution 2 You can add vertically:

$$\begin{array}{r} 4x^2 - 2x - 3 \\ 3x^3 - 2x^2 + 3x - 4 \\ \hline 3x^3 + 2x^2 + x - 7 \end{array}$$

Example 3 Subtract $3x^2 + 5x - 3$ from $2x^4 - 3x^3 - 2x^2 - x + 2$.

Solution $(2x^4 - 3x^3 - 2x^2 - x + 2) - (3x^2 + 5x - 3) =$
$2x^4 - 3x^3 - 2x^2 - x + 2 + (-3x^2 - 5x + 3) =$
$2x^4 - 3x^3 + [-2x^2 + (-3x^2)] + [-x + (-5x)] + (2 + 3) =$
$2x^4 - 3x^3 - 5x^2 - 6x + 5$

In Exercises 7–14, (a) add the polynomials and (b) subtract the second polynomial from the first.

7. $2x + 3,\ 5x - 6$

8. $5a - 2,\ 6a - 1$

9. $x^2 - 2x + 1,\ 3x^2 + 2x + 4$

10. $5x^2 + 2x + 3,\ x^2 - 7x$

11. $x^3 - 2x^2,\ 3x^2 + 5x + 1$

12. $4x^2 + 3x - 6,\ 5x + 8$

13. $3x^2 + 2,\ 5x^2 + 4x + 1$

14. $x^3 + 2x^2 - 3x + 2,\ 3x^3 - 2x^2 + 4x - 5$

Example 4 Simplify $4x(3x + 2) - 3x(2x - 1) - 4$.

Solution $4x(3x + 2) - 3x(2x - 1) - 4 =$ First use the distributive property.
$4x(3x) + 4x(2) - 3x(2x) - 3x(-1) - 4 =$
$12x^2 + 8x - 6x^2 + 3x - 4 =$ Then combine similar terms.
$(12x^2 + (-6x^2)) + (8x + 3x) - 4 =$
$6x^2 + 11x - 4$

Simplify.

15. $4(2a^2 + 3a - 2) + 2(a^2 - 6)$

16. $3(x^2 - 3x + 4) - 2(3x^2 - 4x - 1)$

17. $2(x^3 - 2x^2) + 4(x^4 - 3x^3 + 2x^2)$

18. $5(x^2 - x - 2) - 3(3x^2 - 2x - 6)$

Mixed Review Exercises

For the line containing the given points, find (a) the slope and (b) an equation in standard form.

1. $(1, 3), (-2, -3)$

2. $(4, -2), (0, 6)$

3. $(3, -1), (-2, -3)$

4. $(-1, -2), (3, -3)$

5. $(3, 8), (6, -2)$

6. $(3, 2), (-1, 2)$

NAME ______________________ DATE ______________

4–2 Using Laws of Exponents

Objective: To use laws of exponents to multiply a polynomial by a monomial.

Vocabulary

Laws of Exponents Let a and b be real numbers and m and n be positive integers. Then:

1. $a^m \cdot a^n = a^{m+n}$ Example: $a^3 \cdot a^4 = a^{3+4} = a^7$
2. $(ab)^m = a^m b^m$ Example: $(ab)^2 = a^2b^2$
3. $(a^m)^n = a^{mn}$ Example: $(a^4)^3 = a^{4 \cdot 3} = a^{12}$

Symbol x^4 means $x \cdot x \cdot x \cdot x$

Example 1 Simplify: **a.** $(-2x^3y^2)(-3x^2y)$ **b.** $(p^5q)^4$ **c.** $(-a^2)^3$

Solution **a.** $(-2x^3y^2)(-3x^2y) = (-2)(-3)(x^3 \cdot x^2)(y^2 \cdot y^1)$ $(y = y^1)$

$= 6 \cdot x^{3+2} \cdot y^{2+1}$ (Law 1)

$= 6x^5y^3$

b. $(p^5q)^4 = (p^5)^4 \cdot q^4$ (Law 2)

$= p^{5 \cdot 4} \cdot q^4$ (Law 3)

$= p^{20} \cdot q^4$

c. $(-a^2)^3 = [(-1)a^2]^3$ $[-a^2 = (-1)a^2]$

$= (-1)^3(a^2)^3$ (Law 2)

$= -1 \cdot a^{2 \cdot 3}$ (Law 3)

$= -a^6$

Simplify.

1. $a^3 \cdot a^3$ **2.** $c^4 \cdot c^2$ **3.** $(a^3)^3$ **4.** $(c^4)^2$ **5.** $(mn^2)^4$

6. $(x^2y^3)^5$ **7.** $5x^3 \cdot 2x^2$ **8.** $-2r^5 \cdot 6r^8$ **9.** $(-x^5)^2$ **10.** $(-x^2)^5$

11. $(4bd^2)(b^3d^2)$ **12.** $(-6m^4n^3)(2mn)$ **13.** $(-p^3)(pq^3)(-p^2q)$ **14.** $(4r^4s^2)(-3rs^3)(rs)$

Example 2 Simplify: **a.** $r(r^3)^2r^8$ **b.** $(2x^2yz^4)^3$ **c.** $(3t^2)^3(2t^4)^2$

Solution **a.** $r(r^3)^2r^8 = r^1 \cdot r^6 \cdot r^8$ (Law 3)

$= r^{1+6+8}$ (Law 1)

$= r^{15}$

b. $(2x^2yz^4)^3 = 2^3(x^2)^3(y^1)^3(z^4)^3$ (Law 2)

$= 8x^6y^3z^{12}$ (Law 3)

c. $(3t^2)^3(2t^4)^2 = 3^3(t^2)^3 \cdot 2^2(t^4)^2$ (Law 2)

$= 27t^6 \cdot 4t^8$ (Law 3)

$= 108t^{14}$ (Law 1)

NAME ______________________ DATE ____________

4–2 Using Laws of Exponents *(continued)*

Simplify.

15. $(2t^5)^3$ **16.** $(3s^6)^2$ **17.** $(3xy^2z)^2$

18. $(-abc^3)^4$ **19.** $(-2r^3st^2)^3$ **20.** $(6c^2d^4e^5)^2$

21. $(-m^2)(-m)^3$ **22.** $(-z^4)(-z^2)$ **23.** $(-t^3)(-t^4)(-t)^2$

24. $(-a)^3(-a)^5(-a^4)$ **25.** $(2r^2)^3(3r)^2$ **26.** $(5k^4)^2(2k)^4$

27. $(xy^2)^2(xy)^3$ **28.** $(g^3h)^3(-g^2h^2)^3$ **29.** $(4xy)^2(-xy^2)^3$

30. $(r^3s^5)(2r^2s)^6$ **31.** $(-2a^2b)^2(-4ab^3)^2$ **32.** $(-3p^2q^4)^3(2p^3q)^4$

Example 3 Simplify.

a. $5x^3(5x^3 - 6x^2 + 3x - 1)$

b. $-2ab^2(3a^2 - 2ab + 5b^2)$

Solution First use the distributive law. Then simplify.

a. $5x^3(5x^3 - 6x^2 + 3x - 1) = 5x^3(5x^3) - 5x^3(6x^2) + 5x^3(3x) - 5x^3(1)$
$= 25x^6 - 30x^5 + 15x^4 - 5x^3$

b. $-2ab^2(3a^2 - 2ab + 5b^2) = (-2ab^2)(3a^2) - (-2ab^2)(2ab) + (-2ab^2)(5b^2)$
$= -6a^3b^2 + 4a^2b^3 - 10ab^4$

Simplify.

33. $x^2(7x^3 - x^2 + 10x + 8)$ **34.** $-a^3(-4a^4 + 5a^3 + 9a^2 - a)$

35. $3x^2(5x^2 + 3x - 2)$ **36.** $-8y(2y^5 - 6y^3 - 3y^2)$

37. $xy^2(x^2 - 5xy - 2y^2)$ **38.** $a^2b^2(3a^3 - 4a^2 - 2a + 3)$

Example 4 Simplify. Assume that variable exponents represent positive integers.

a. $x^k \cdot x^{k-1} \cdot x^5$ **b.** $(y^3)^m(y^m)^4$

Solution **a.** $x^k \cdot x^{k-1} \cdot x^5 = x^{k+k-1+5} = x^{2k+4}$

b. $(y^3)^m(y^m)^4 = y^{3m} \cdot y^{4m} = y^{3m+4m} = y^{7m}$

Simplify. Assume that variable exponents represent positive integers.

39. $t^3 \cdot t^{k-1}$ **40.** $x^{p+5} \cdot x^p$

41. $(a^t)^2a^{3t}$ **42.** $y^4(y^{2m})^3$

43. $x^{m-2} \cdot x^{m+2} \cdot x$ **44.** $z^{n+3} \cdot z^{n+2} \cdot z^{4-2n}$

NAME ______________________ DATE ______________

4–3 Multiplying Polynomials

Objective: To multiply polynomials.

Vocabulary

Binomial A polynomial that has two terms. **Trinomial** A polynomial that has three terms.

Example 1 Multiply $(2x + 3y)(5x - 4y)$.

Solution $(2x + 3y)(5x - 4y) = 10x^2 - 8xy + 15xy - 12y^2$

1. Multiply the *first* terms. → $10x^2$
2. Multiply the *outer* terms. → $-8xy$
3. Multiply the *inner* terms. → $15xy$
4. Multiply the *last* terms. → $-12y^2$
5. Combine any *similar* terms.

$= 10x^2 + 7xy - 12y^2$

O I

$(2x + 3y)(5x - 4y) = 10x^2 + 7xy - 12y^2$

F L

This is sometimes called the FOIL method, as shown by the diagram. The word FOIL reminds you to multiply the **F**irst, **O**uter, **I**nner, and **L**ast terms. This method can be used only when multiplying two binomials.

Multiply.

1. $(x + 2)(x + 4)$ **2.** $(x - 5)(x + 6)$ **3.** $(y + 6)(y - 10)$

4. $(x - 8)(x - 7)$ **5.** $(a + 2)(3a - 5)$ **6.** $(z - 4)(6z + 1)$

7. $(2b + 1)(3b + 2)$ **8.** $(5m + 6)(2m - 3)$ **9.** $(r + 2s)(2r - s)$

10. $(4x - 3y)(x - 2y)$ **11.** $(7y + z)(2y + 5z)$ **12.** $(3m - 2n)(2m + 3n)$

Example 2 Multiply $(3x - 4)(x^2 + 2x - 5)$.

Solution 1

$(3x - 4)(x^2 + 2x - 5) =$

$3x(x^2 + 2x - 5) - 4(x^2 + 2x - 5) =$ Use the distributive property.

$3x^3 + 6x^2 - 15x - 4x^2 - 8x + 20 =$ Use the distributive property again.

$3x^3 + 2x^2 - 23x + 20$ Combine similar terms.

Solution 2 You can multiply vertically:

$$\begin{array}{rrrr} & x^2 & +\,2x & -\,5 \\ & & 3x & -\,4 \\ \hline 3x^3 & +\,6x^2 & -\,15x & \\ & -\,4x^2 & -\,8x & +\,20 \\ \hline 3x^3 & +\,2x^2 & -\,23x & +\,20 \end{array}$$

First multiply $x^2 + 2x - 5$ by $3x$. → $3x^3 + 6x^2 - 15x$

Then multiply $x^2 + 2x - 5$ by -4. → $-4x^2 - 8x + 20$

Add. → $3x^3 + 2x^2 - 23x + 20$

NAME ______________________ DATE ____________

4–3 Multiplying Polynomials *(continued)*

Multiply.

13. $(a + 3)(a^2 + 4a - 1)$

14. $(t - 2)(t^2 + 3t + 4)$

15. $(4x - 2)(3x^2 - 2x - 5)$

16. $(5y + 1)(2y^2 - y + 3)$

17. $(x^2 - 3x + 5)(2x - 1)$

18. $(m^2 + 4m + 6)(3m + 2)$

19. $(z^4 + 2z^2 - 1)(z^2 + 3)$

20. $(2 - y^2)(3 - y^2 - y^4)$

Example 3 Multiply: **a.** $(x + 2)^2$ **b.** $(5x - 3)^2$ **c.** $(3x + 1)(3x - 1)$ **d.** $xy(x + y)(x - y)$

Solution You can use the FOIL method or the special product formulas given below.

$$(a + b)^2 = a^2 + 2ab + b^2$$
$$(a - b)^2 = a^2 - 2ab + b^2$$
$$(a + b)(a - b) = a^2 - b^2$$

a. $(x + 2)^2 = (x)^2 + 2(x)(2) + (2)^2 = x^2 + 4x + 4$

b. $(5x - 3)^2 = (5x)^2 - 2(5x)(3) + (3)^2 = 25x^2 - 30x + 9$

c. $(3x + 1)(3x - 1) = (3x)^2 - (1)^2 = 9x^2 - 1$

d. $xy(x + y)(x - y) = xy(x^2 - y^2) = x^3y - xy^3$

Find each special product.

21. $(y + 6)^2$

22. $(t - 1)^2$

23. $(2a - 3)^2$

24. $(4z - 5)^2$

25. $(x + 2y)^2$

26. $(m + 9)(m - 9)$

27. $(4k + 5)(4k - 5)$

28. $(8 - p)(8 + p)$

29. $(2x + y)(2x - y)$

Multiply.

30. $k(k - 3)(k + 2)$

31. $g(g + 6)(g - 6)$

32. $(x^3 + 7)(x^3 - 7)$

33. $(x^2 + xy + y^2)(x - y)$

34. $(5r^2 + 1)(2r^2 + 3)$

35. $m^2(m - 4)^2$

36. $y^2(2y + 9)(2y - 9)$

37. $st(s - t)(2s - t)$

38. $cd(c + 2d)^2$

Mixed Review Exercises

Simplify.

1. $(5x^2 - 3x + 2) - (3x^2 - 2x - 1)$

2. $(a^3b^2)^4$

3. $(5r^2s^2t^3)(-3r^3st^2)$

4. $3(3a + 6) - 2(a - 4)$

5. $3r(s - 4) + 5s(2r + 4)$

6. $(-x^2)^3(-x)^4$

7. $(3x^2 - 4x - 13) - (12 - 2x + x^2)$

8. $(-36x^4)\left(-\frac{1}{4}x^3\right)$

NAME ______________________ DATE ______________

4–4 Using Prime Factorization

Objective: To find the GCF and LCM of integers and monomials.

Vocabulary

Factor a number To express a number as a product of other numbers.

Factor set The set of numbers from which factors of a number are chosen.

Prime number (or **prime**) An integer greater than 1 whose only positive integral factors are itself and 1.

Examples: 2, 3, 5, 7, 11, 13, 17, 19

Prime factorization Expressing a positive integer as a product of primes. There is a unique prime factorization for every positive integer.

Greatest Common Factor (GCF) The GCF of two or more integers is the greatest integer that is a factor of each. The GCF of two or more monomials is the common factor that has the greatest degree and the greatest numerical coefficient.

Examples: The GCF of 8 and -12 is 4. ← The GCF is always positive.
The GCF of $8x$ and $12x^2$ is $4x$.

Least Common Multiple (LCM) The LCM of two or more integers is the least positive integer having each as a factor. The LCM of two or more monomials is the common multiple having the least degree and the least positive numerical coefficient.

Examples: The LCM of 4 and 6 is 12.
The LCM of $-4x$ and $6x^2$ is $12x^2$. ← The LCM is always positive.

Example 1 Find the prime factorization of 720.

Solution

$780 = 2 \cdot 390$ — Try the first prime, 2, as a factor. Keep trying 2 until it is no longer a factor.

$= 2 \cdot 2 \cdot 195$ — 2 is not a factor of 195. Try the next prime, 3, as a factor.

$= 2 \cdot 2 \cdot 3 \cdot 65$ — 3 is not a factor of 65. Try the next prime, 5, as a factor.

$= 2 \cdot 2 \cdot 3 \cdot 5 \cdot 13$ — Since 13 is prime, you cannot factor again.

$\therefore 780 = 2^2 \cdot 3 \cdot 5 \cdot 13$

Find the prime factorization of each integer.

1. 84 **2.** 126 **3.** 360 **4.** 29

5. 616 **6.** 133 **7.** 1456 **8.** 2925

NAME ______________________ DATE ______________

4–4 Using Prime Factorization *(continued)*

Example 2 Find (a) the GCF and (b) the LCM of 60, 72, and 84.

Solution First find the prime factorization of each number.

$60 = 2^2 \cdot 3 \cdot 5$ $\quad 72 = 2^3 \cdot 3^2$ $\quad 84 = 2^2 \cdot 3 \cdot 7$

a. To find the GCF, take the *least* power of each *common* prime factor.

$\text{GCF} = 2^2 \cdot 3 \leftarrow$ 2 and 3 are the only common factors.
$= 12$

b. To find the LCM, take the *greatest* power of *each* prime factor.

$\text{LCM} = 2^3 \cdot 3^2 \cdot 5 \cdot 7 \leftarrow$ 2, 3, 5, and 7 are all the prime factors.
$= 2520$

Find (a) the GCF and (b) the LCM of the following integers.

9. 30, 55 **10.** 54, 90 **11.** -66, 110 **12.** 245, -385

13. 2, 5, 13 **14.** 5, 6, 8 **15.** 6, 10, 14 **16.** 36, 40, 90

Example 3 Find (a) the GCF and (b) the LCM of the following monomials:
$54r^2s^3$ and $72rs^2t$

Solution First find the prime factorization of the coefficients.

$54 = 2 \cdot 3^3 \qquad 72 = 2^3 \cdot 3^2$

a. The GCF of 54 and 72 is $2 \cdot 3^2 = 18$.

Compare the powers of each variable occurring in *both* monomials.
Use the power with the *least* exponent.

Compare r^2 and r. Use r.
Compare s^3 and s^2. Use s^2.
Since t does not occur in both monomials, ignore it.
$\therefore \text{GCF} = 18rs^2$

b. The LCM of 54 and 72 is $2^3 \cdot 3^3 = 216$.

Compare the powers of each variable occurring in *either* monomial.
Use the power with the *greatest* exponent.

Compare r^2 with r. Use r^2.
Compare s^3 with s^2. Use s^3.
Use t.
$\therefore \text{LCM} = 216r^2s^3t$

Find (a) the GCF and (b) the LCM of the following monomials.

17. $12xy$, $15xy^2$ **18.** $49pq^2$, $56q^3$

19. $-55mn^3$, $85m^2n^2$ **20.** $42xy$, $-70yz$

21. $60a^2bc$, $280a^3b^2$ **22.** $108pqr^2$, $144p^2qr^2$

23. $10rs$, $12rst$, $14r^2st^2$ **24.** $26xy$, $39xz$, $65yz$

NAME ______________________ DATE ____________

4–5 Factoring Polynomials

Objective: To factor polynomials by using the GCF, by recognizing special products, and by grouping terms.

Vocabulary

Factor a polynomial To express a polynomial as a product of other polynomials.

Greatest monomial factor The GCF of the terms of a polynomial.

Special factoring patterns

Perfect square trinomials	$a^2 + 2ab + b^2 = (a + b)^2$	$a^2 - 2ab + b^2 = (a - b)^2$
Difference of squares	$a^2 - b^2 = (a + b)(a - b)$	
Sum of cubes	$a^3 + b^3 = (a + b)(a^2 - ab + b^2)$	
Difference of cubes	$a^3 - b^3 = (a - b)(a^2 + ab + b^2)$	

Example 1 Factor: **a.** $3x^4 - 6x^3 + 12x^2$ **b.** $8a^3b - 12a^2b^2$

Solution **a.** $3x^4 - 6x^3 + 12x^2 = 3x^2(x^2 - 2x + 4)$ ← The GCF of the terms is $3x^2$.

b. $8a^3b - 12a^2b^2 = 4a^2b(2a - 3b)$ ← The GCF of the terms is $4a^2b$.

Factor each polynomial.

1. $y^2 + y$ **2.** $4x - 28$ **3.** $8a^4 - 14a^2b$

4. $6x^2 - 8x^3 - 10x^4$ **5.** $10x^8 + 15x^7 - 35x^5$ **6.** $11a^3b - 22a^2b^2 + 55ab^3$

Example 2 Factor: **a.** $z^2 + 8z + 16$ **b.** $9x^2 - 6xy + y^2$ **c.** $36m^2 - 49n^2$

Solution **a.** $z^2 + 8z + 16 = z^2 + 2(z)(4) + (4)^2$ ← perfect square trinomial
$= (z + 4)^2$

b. $9x^2 - 6xy + y^2 = (3x)^2 - 2(3x)(y) + y^2$ ← perfect square trinomial
$= (3x - y)^2$

c. $36m^2 - 49n^2 = (6m)^2 - (7n)^2$ ← difference of squares
$= (6m + 7n)(6m - 7n)$

Example 3 Factor $2a^5 - 162a$.

Solution Always begin by looking for the GCF of the terms. If the GCF is not 1, then factor the GCF out. The GCF of $2a^5$ and $-162a$ is $2a$.

$2a^5 - 162a = 2a(a^4 - 81)$
$= 2a[(a^2)^2 - (9)^2]$ ← difference of squares
$= 2a(a^2 + 9)(a^2 - 9)$
$= 2a(a^2 + 9)(a^2 - 3^2)$ ← difference of squares
$= 2a(a^2 + 9)(a + 3)(a - 3)$

NAME ______________________ DATE ____________

4–5 Factoring Polynomials *(continued)*

Factor each polynomial.

7. $x^2 + 10x + 25$ **8.** $a^2 - 16a + 64$ **9.** $4y^2 - 12y + 9$

10. $4b^2 + 28b + 49$ **11.** $x^2 - 16$ **12.** $y^2 - 100$

13. $4k^2 - 25$ **14.** $9m^2 - 64$ **15.** $25x^2 + 20xy + 4y^2$

16. $81p^2 - 49q^2$ **17.** $3x^2 + 12x + 12$ **18.** $5c^3 + 30c^2d + 45cd^2$

19. $rt^2 - r$ **20.** $4x^2y - 36y$ **21.** $16n^4 - 1$

Example 4 Factor: **a.** $a^3 - 8$ **b.** $27x^3 + 1$

Solution **a.** $a^3 - 8 = a^3 - 2^3 = (a - 2)(a^2 + 2a + 4)$ ← difference of cubes

b. $27x^3 + 1 = (3x)^3 + 1^3 = (3x + 1)(9x^2 - 3x + 1)$ ← sum of cubes

Factor each polynomial.

22. $x^3 + 1$ **23.** $64 - a^3$ **24.** $t^3 + 125$ **25.** $1000c^3 - 27$

Example 5 Factor: **a.** $2a^3 - 3a^2 - 4a + 6$ **b.** $12x^3 + 4x^2y - 3x - y$

Solution **a.** The first and second terms have a common factor of a^2, and the third and fourth terms have a common factor of -2. Factor by grouping terms.

$$2a^3 - 3a^2 - 4a + 6 = (2a^3 - 3a^2) + (-4a + 6)$$
$$= a^2(2a - 3) - 2(2a - 3) \quad \text{Common factor is } 2a - 3.$$
$$= (a^2 - 2)(2a - 3) \quad \text{Factor out } (2a - 3).$$

b. The first and third terms have a common factor of $3x$, and the second and fourth terms have a common factor of y. Factor by grouping terms.

$$12x^3 + 4x^2y - 3x - y = (12x^3 - 3x) + (4x^2y - y)$$
$$= 3x(4x^2 - 1) + y(4x^2 - 1)$$
$$= (3x + y)(4x^2 - 1) \leftarrow \text{difference of squares}$$
$$= (3x + y)(2x + 1)(2x - 1)$$

Factor each polynomial.

26. $a(b + 2) - 3(b + 2)$ **27.** $m(n - 2) - (2 - n)$ **28.** $20a^3 - 5a^2 + 8a - 2$

29. $10y^3 + 10y^2 + 3y + 3$ **30.** $9a^2b - 8a^2 - 9b + 8$ **31.** $5x^2y - 7x^2 - 7 + 5y$

Mixed Review Exercises

Write as a simplified polynomial.

1. $(a + 2)^2$ **2.** $(3a - 2)(4a + 3)$ **3.** $r^2s^2(4r - 5s)$

4. $(5a - 2) - (3 - a)$ **5.** $(x^2 - 2)(x + 5)$ **6.** $(-a)^2(2a^2)^3$

NAME ______________________ DATE ______________

4–6 Factoring Quadratic Polynomials

Objective: To factor quadratic polynomials.

Vocabulary

Quadratic polynomial (or **second-degree polynomial**) A polynomial of the form $ax^2 + bx + c$ $(a \neq 0)$. The term ax^2 is the *quadratic term, bx* is the *linear term,* and c is the *constant term*. Examples: $x^2 + 6$ $\quad 2x^2 + 3x$ $\quad x^2 - 4x + 3$

Quadratic trinomial A quadratic polynomial for which a, b, and c are all nonzero integers. Examples: $x^2 + 5x + 6$ $\quad 2x^2 - x + 3$

Irreducible polynomial A polynomial that has more than one term and cannot be expressed as a product of polynomials of lower degree taken from a given factor set. Examples: $3x^2 + 3x - 9$ $\quad x^2 + 2x + 4$

Prime polynomial An irreducible polynomial with integral coefficients is prime if the GCF of its coefficients is 1. Examples: $x^2 + 2x + 4$ $\quad 2x^2 - 3x + 5$

Factor completely a polynomial Express the polynomial as a product of factors where each factor is either a monomial, a prime polynomial, or a power of a prime polynomial.

Example: $3x^3 + 12x^2 + 12x = 3x(x^2 + 4x + 4)$ ← not factored completely

$3x^3 + 12x^2 + 12x = 3x(x + 2)^2$ ← factored completely

Example 1 Factor $x^2 + 5x - 14$.

Solution Recall that multiplying two binomials often gives a trinomial. The FOIL method can be used to multiply two binomials:

$$(2x + 3)(x - 5) = \overset{F}{(2x)(x)} + \overset{O \quad I}{(-10x + 3x)} + \overset{L}{(3)(-5)}$$
$$= 2x^2 - 7x - 15$$

So one way to factor a trinomial is to try using the FOIL method in reverse.

1. $\overset{F}{x^2} + 5x - 14 = (x \quad)(x \quad)$ $\quad$ $x^2 = (x)(x)$ $\quad$ There is only one possible product that equals the quadratic term.

2. $x^2 + 5x - \overset{L}{14} = (\quad ?)(\quad ?)$ $\quad$ Four possible products equal the constant term.

$(-1)(14) \quad (1)(-14)$

$(-2)(7) \quad (2)(-7)$

The possible factorizations are:

$(x - 1)(x + 14)$	$(x + 1)(x - 14)$	$(x - 2)(x + 7)$	$(x + 2)(x - 7)$
$O + I = 13x$	$O + I = -13x$	$O + I = 5x$	$O + I = -5x$

3. The linear term of the given trinomial is $5x$. Of the four possible factorizations above, only $(x - 2)(x + 7)$ gives the required linear term.

$$\therefore x^2 + 5x - 14 = (x - 2)(x + 7)$$

NAME ______________________ DATE ______________

4-6 Factoring Quadratic Polynomials *(continued)*

Example 2 Factor $9x^2 - 15x + 4$.

Solution

1. $9x^2 - 15x + 4 = (x \quad)(9x \quad)$ or $(3x \quad)(3x \quad)$

   $(x)(9x) \quad (3x)(3x)$ ⟵ Two possible products equal the quadratic term.

2. $9x^2 - 15x + 4 = (\quad ?)(\quad ?)$

   $(1)(4)$ $\boxed{(-1)(-4)}$
   $(2)(2)$ $\boxed{(-2)(-2)}$ Only two of these products are possible since the coefficient of the linear term is negative.

The possible factorizations are:

$(x - 1)(9x - 4)$ $O + I = -13x$; $(x - 4)(9x - 1)$ $O + I = -37x$; $(x - 2)(9x - 2)$ $O + I = -20x$; $(3x - 1)(3x - 4)$ $O + I = -15x$; $(3x - 2)(3x - 2)$ $O + I = -12x$

3. The linear term of the given trinomial is $-15x$. Of the possible factorizations above, only $(3x - 1)(3x - 4)$ gives the required linear term.

$$\therefore 9x^2 - 15x + 4 = (3x - 1)(3x - 4)$$

Example 3 Factor completely: **a.** $x^2 + 2x + 3$ **b.** $4x^4 + 10x^3 - 6x^2$

Solution

a. $x^2 + 2x + 3 = (x \quad ?)(x \quad ?)$

$\boxed{(1)(3)}$ $(-1)(-3)$ Only one of these products is possible since the coefficient of the linear term is positive.

The only possible factorization $(x + 1)(x + 3)$ does not check since $O + I = 3x + x = 4x$ and the required linear term is $2x$.

$$\therefore x^2 + 2x + 3 \text{ is } \textit{prime}.$$

b. $4x^4 + 10x^3 - 6x^2 = 2x^2(2x^2 + 5x - 3)$ Factor out $2x^2$.
$= 2x^2(2x - 1)(x + 3)$ Reverse the FOIL method.

Factor completely. If the polynomial is prime, say so.

1. $x^2 - 7x + 6$
2. $k^2 + 7k + 12$
3. $y^2 - 11y + 24$
4. $p^2 + 2p - 3$
5. $z^2 + 5z + 6$
6. $a^2 - 5a - 6$
7. $t^2 + 9t - 22$
8. $v^2 - 11v - 60$
9. $x^2 + 4x + 5$
10. $s^2 + 5s - 36$
11. $h^2 - 15h + 36$
12. $x^2 + 3x - 8$
13. $8u^2 + 6u + 1$
14. $12y^2 - y - 1$
15. $2p^2 + p - 3$
16. $3r^2 + r - 10$
17. $6n^2 + n - 2$
18. $4w^2 - 11w + 6$
19. $2t^2 - 4t - 70$
20. $3x^2 + 9x - 30$
21. $2k^3 - 2k^2 - 112k$
22. $5z^3 - 25z^2 + 30z$
23. $27x^4 + 48x^3 - 12x^2$
24. $36x^4 + 44x^3 + 24x^2$

NAME ______________________________ DATE ______________

4–7 Solving Polynomial Equations

Objective: To solve polynomial equations.

Vocabulary

Polynomial equation An equation that is equivalent to one with a polynomial as one side and 0 as the other side.

Examples: $x^2 + 5x - 6 = 0 \qquad x^2 = 6 - 5x$

Root (solution) of a polynomial equation A value of the variable that satisfies the equation.

Examples: -6 and 1 are the roots of $x^2 + 5x - 6 = 0$ because

$$(-6)^2 + 5(-6) - 6 = 0 \quad \text{and} \quad (1)^2 + 5(1) - 6 = 0.$$

Double root A root of an equation that arises from a factor that occurs twice.

Example: 2 is a double root of the equation $(x - 2)(x - 2) = 0$.

Zero-product property $ab = 0$ if and only if $a = 0$ or $b = 0$.

CAUTION You cannot use the zero-product property to solve a polynomial equation unless one side of the equation is zero.

Example 1 Solve $(5x + 4)(x - 3) = 0$.

Solution One side of the equation is a product of factors, and the other side is 0. You can use the zero-product property.

$$(5x + 4)(x - 3) = 0 \qquad \text{Set each factor equal to 0.}$$

$$5x + 4 = 0 \quad \text{or} \quad x - 3 = 0 \qquad \text{Solve.}$$

$$x = -\frac{4}{5} \quad \text{or} \quad x = 3$$

Check $x = -\frac{4}{5}$: $\left[5\left(-\frac{4}{5}\right) + 4\right]\left(-\frac{4}{5} - 3\right) \stackrel{?}{=} 0$

$$(-4 + 4)\left(-\frac{4}{5} - \frac{15}{5}\right) \stackrel{?}{=} 0$$

$$0 \cdot \left(-\frac{19}{5}\right) \stackrel{?}{=} 0$$

$$0 = 0 \ \checkmark$$

Check $x = 3$: $(3 + 4)(3 - 3) \stackrel{?}{=} 0$

$$7 \cdot 0 \stackrel{?}{=} 0$$

$$0 = 0 \ \checkmark$$

$\therefore$ the solution set is $\left\{-\frac{4}{5}, 3\right\}$.

Note: If the equation were $(x + 4)(x - 3) = 6$, you *could not* write

$$(x + 4) = 6 \quad \text{or} \quad (x - 3) = 6.$$

You could not use the zero-product property immediately because the product of the two factors, $x + 4$ and $x - 3$, is not equal to 0.

Solve.

1. $(x - 2)(x + 4) = 0$
2. $(t + 3)(t + 4) = 0$
3. $(2y + 3)(y - 3) = 0$
4. $(3k - 1)(2k - 1) = 0$
5. $(x + 5)(x + 1)(x - 8) = 0$
6. $r(r + 2)(r - 1) = 0$

NAME ______________________ DATE ______________

4–7 Solving Polynomial Equations *(continued)*

Example 2 Solve $x^2 = 2x + 8$.

Solution

1. $x^2 - 2x - 8 = 0$ Make one side 0.
2. $(x - 4)(x + 2) = 0$ Factor the polynomial.
3. $x - 4 = 0$ or $x + 2 = 0$ Use the zero-product property.
   $x = 4$ or $x = -2$ Solve.

Check: $(4)^2 \stackrel{?}{=} 2(4) + 8$ $(-2)^2 \stackrel{?}{=} 2(-2) + 8$

$16 \stackrel{?}{=} 8 + 8$ $4 \stackrel{?}{=} -4 + 8$

$16 = 16 \checkmark$ $4 = 4 \checkmark$

$\therefore$ the solution set is $\{4, -2\}$.

Example 3 Solve $x^3 - 16x^2 = -64x$.

Solution

1. $x^3 - 16x^2 + 64x = 0$ Make one side 0.
2. $x(x^2 - 16x + 64) = 0$ Factor the polynomial.
   $x(x - 8)^2 = 0$
   $x(x - 8)(x - 8) = 0$
3. $x = 0$ or $x - 8 = 0$ or $x - 8 = 0$ Use the zero-product property.
   $x = 0$ or $x = 8$ or $x = 8$ Solve.

Note that 8 is a *double root* since $(x - 8)$ occurs twice as a factor.

$\therefore$ the solution set is $\{0, 8\}$. The check is left for you.

Solve. Identify all double roots.

7. $x^2 + 6x - 27 = 0$ **8.** $y^2 - 9 = 0$ **9.** $t^2 = -4t$

10. $5z^2 = 35z$ **11.** $x^2 + 25 = 10x$ **12.** $n^2 - 5n = 14$

13. $p + 1 = 6p^2$ **14.** $3s^2 + 20s = 7$ **15.** $4x^2 = 13x - 10$

16. $4x - 1 = 4x^2$ **17.** $5m^2 + 13m - 6 = 0$ **18.** $7y - 12y^2 = 1$

19. $t^3 + 4t^2 = 0$ **20.** $z^3 = 18z - 7z^2$ **21.** $x^3 + 36x = -12x^2$

22. $15u^2 = 8u - 2u^3$ **23.** $(x - 2)(x - 3) = 2$ **24.** $(x + 1)(x - 5) = 7$

25. $-4(3a + 1) = 9a^2$ **26.** $5x(x + 4) = 3(6x + 1)$ **27.** $(y + 2)(y^2 - 3y - 10) = 0$

Mixed Review Exercises

Express as a simplified polynomial.

1. $b^3(b - 1)$ **2.** $(5x - y)^2$ **3.** $(2r - 7) + 2(r + 3)$

4. $(z^2 - 5)(z^2 + 6)$ **5.** $t - (2t + 3)$ **6.** $(-2c^2d^3)^3$

NAME ______________________ DATE ______________

4–8 Problem Solving Using Polynomial Equations

Objective: To solve problems using polynomial equations.

Vocabulary

Mathematical model An equation that represents a real-life problem.

CAUTION After solving the mathematical model (equation), you should always check your answer against the conditions in the problem because an answer that satisfies the equation may not be reasonable in real life.

Example 1 Find two consecutive even integers such that the sum of their squares is 100.

Solution

Step 1 You are asked to find two consecutive even integers whose squares add up to 100.

Step 2 Let n = first consecutive even integer.
Then $n + 2$ = second consecutive even integer.

Step 3 Write an equation that shows the sum of squares of the integers is 100.

$$n^2 + (n + 2)^2 = 100$$

Step 4 Solve the equation.

$n^2 + n^2 + 4n + 4 = 100$	Square $n + 2$.
$2n^2 + 4n - 96 = 0$	Make one side 0.
$n^2 + 2n - 48 = 0$	Divide both sides by 2.
$(n + 8)(n - 6) = 0$	Factor the polynomial.
$n + 8 = 0$ or $n - 6 = 0$	Use the zero-product property.
$n = -8$ or $n = 6$	

Step 5 Check each possible solution.

If $n = -8$, then $n + 2 = -6$.
sum of squares $= (-8)^2 + (-6)^2 = 64 + 36 = 100$

If $n = 6$, then $n + 2 = 8$.
sum of squares $= (6)^2 + (8)^2 = 36 + 64 = 100$

$\therefore$ there are two correct answers: -8 and -6 or 6 and 8.

Solve each problem. If there are two correct answers, give both of them.

1. Find two consecutive integers such that the sum of their squares is 113.
2. Find two consecutive even integers such that the sum of their squares is 340.
3. Find two consecutive odd integers whose product is 195.
4. The sum of a number and its square is 42. Find the number.

NAME ______________________________ DATE ______________

4–8 Problem Solving Using Polynomial Equations *(continued)*

Example 2 A rectangular residential lot with area 7475 m^2 is 50 m longer than it is wide. Find the dimensions of the lot.

Solution

Step 1 You are asked to find the width and length of the lot. Draw a diagram.

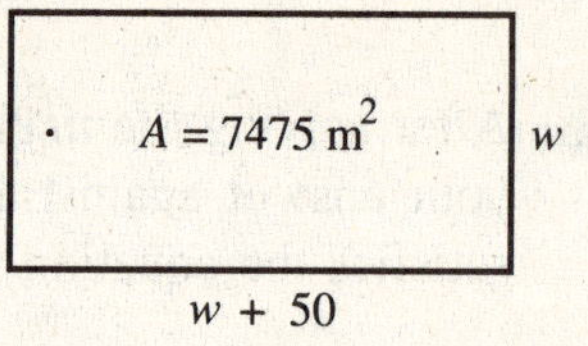

Step 2 Let w = width of the lot in meters.
Then $w + 50$ = length of the lot.
Label your diagram.

Step 3 Write an equation that shows the area of the lot is 7475 m^2.

$$\text{Length} \times \text{Width} = \text{Area}$$
$$(w + 50)w = 7475$$

Step 4 Solve the equation.

$$w^2 + 50w = 7475$$
$$w^2 + 50w - 7475 = 0 \qquad \text{Make one side 0.}$$
$$(w + 115)(w - 65) = 0 \qquad \text{Factor the polynomial.}$$
$$w + 115 = 0 \quad \text{or} \quad w - 65 = 0 \qquad \text{Use the zero-product property.}$$
$$w = -115 \quad \text{or} \quad w = 65$$

Step 5 Although $w = -115$ is a solution of $(w + 50)w = 7475$, it must be rejected because width cannot be negative. The other solution, $w = 65$, gives a lot that is 65 m wide and 115 m long. Check these dimensions:

$$\text{Lot's area} = (115)(65) = 7475 \checkmark$$

$\therefore$ the dimensions are 115 m by 65 m.

Solve each problem. If there are two correct answers, give both of them.

5. A rectangle is 5 cm longer than it is wide, and its area is 176 m^2. Find its dimensions.
6. An entry hall with an area of 72 ft^2 is 14 ft longer than it is wide. Find the dimensions of the hall.
7. The height of a triangle is 3 cm less than the length of its base, and its area is 27 cm^2. Find the height. (Area of a triangle = $\frac{1}{2}$ × Base × Height.)
8. The area of a right triangle is 96 m^2. The length of one leg is 8 m less than twice the length of the other. Find the length of each leg.
9. A rectangular lot has perimeter 78 ft and area 350 ft^2. Find the dimensions of the lot.
10. The hypotenuse of a right triangle is 13 in. long. One leg is 7 in. longer than the other leg. Find the length of each leg.

NAME ______________________ DATE ____________

4–9 Solving Polynomial Inequalities

Objective: To solve polynomial inequalities.

Vocabulary

Polynomial inequality An inequality equivalent to an inequality with a polynomial as one side and 0 as the other side. Examples: $x^2 - 2x - 3 < 0$ $\quad x^2 < 2x + 3$

Sign graph A graph used to help find the solution set of a polynomial inequality.

Example 1 Find and graph the solution set of $x^2 - 6x > x - 10$.

Solution 1

$$x^2 - 6x > x - 10$$

$x^2 - 7x + 10 > 0$ Make one side 0.

$\underbrace{(x - 5)(x - 2) > 0}$ Factor the polynomial.

The product is positive.

The product is positive if and only if both factors have the *same sign*.

Both factors positive	or	**Both factors negative**
$x - 2 > 0$ and $x - 5 > 0$		$x - 2 < 0$ and $x - 5 < 0$
$x > 2$ and $x > 5$		$x < 2$ and $x < 5$
The solution set of this conjunction is $\{x: x > 5\}$.		The solution set of this conjunction is $\{x: x < 2\}$.

$\therefore$ the solution set of the given inequality is $\{x: x < 2 \text{ or } x > 5\}$.

The graph of the solution set is:

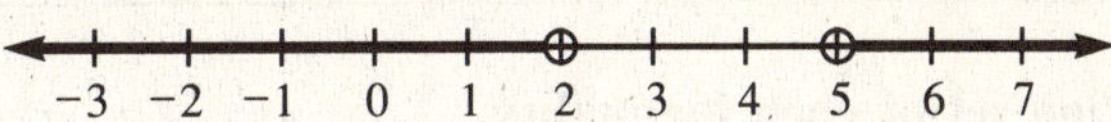

Solution 2 You can use a *sign graph* to plot the signs of the factors $x - 2$ and $x - 5$ and the sign of their product. The sign of each factor depends on the value of x.

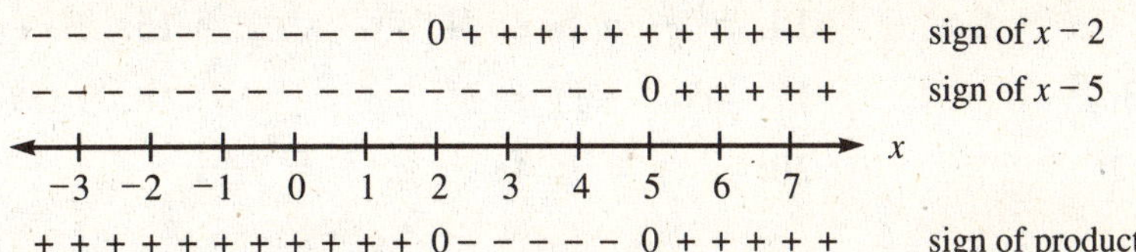

The product is positive when $x < 2$ or $x > 5$.

$\therefore$ the solution set of the given inequality is $\{x: x < 2 \text{ or } x > 5\}$.

The graph of the solution set is as in Solution 1.

Find and graph the solution set of each inequality.

1. $(x + 2)(x - 4) > 0$ **2.** $(x - 3)(x - 1) \geq 0$ **3.** $a^2 - 9 \geq 0$

4. $t^2 - 4t > 0$ **5.** $y^2 - 3y + 2 \geq 0$ **6.** $x^2 - x - 2 > 0$

4-9 Solving Polynomial Inequalities *(continued)*

Example 2 Find and graph the solution set of $x^2 + 3x \le 4x + 12$.

Solution

$$x^2 + 3x \le 4x + 12$$

$x^2 - x - 12 \le 0$ Make one side 0.

$(x - 4)(x + 3) \le 0$ Factor the polynomial.

The product is negative or is equal to zero.

The product is negative if and only if the factors have *opposite signs*.

Use a sign graph to plot the signs of $x - 4$ and $x + 3$.

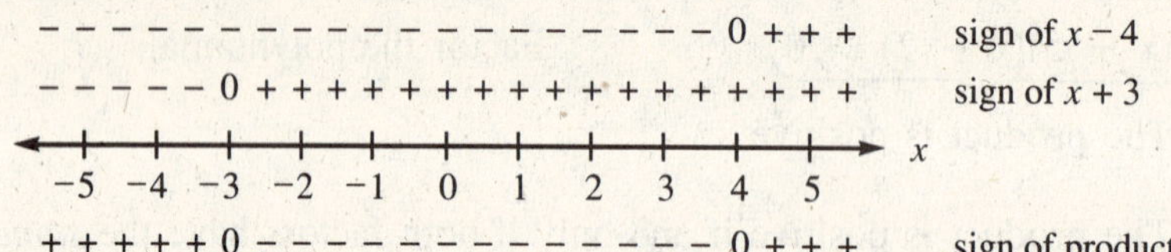

The product is negative when x is between -3 and 4. The product is zero when $x = -3$ or $x = 4$.

$\therefore$ the solution set of the inequality is $\{x: -3 \le x \le 4\}$.

The graph of the solution set is:

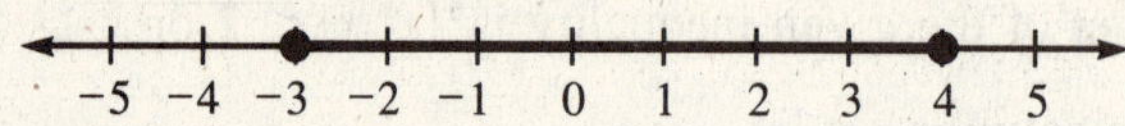

Find and graph the solution set of each inequality.

7. $(x - 2)(x + 3) \le 0$ **8.** $(x + 4)(x + 1) \le 0$ **9.** $y^2 - 1 < 0$

10. $s^2 - 3s < 0$ **11.** $m^2 + 3m + 2 < 0$ **12.** $z^2 + z - 2 \le 0$

13. $t^2 < 9$ **14.** $p^2 - p \ge 2p + 4$ **15.** $x^2 + 15x < 4x - 24$

16. $a^2 - 2a > 3a - 6$ **17.** $4c(c - 1) \le 3$ **18.** $9r(r - 1) > -2$

19. $t^2 + 1 \ge -2t$ **20.** $x^2 + 4 \le 4x$

Mixed Review Exercises

Solve.

1. $(2x + 1)(x - 3) = 0$ **2.** $3z - 8 = -7$ **3.** $|t| - 2 = 3$

4. $2 - |k| = 3$ **5.** $b(b - 2) = b(b + 2)$ **6.** $|3r - 1| = 5$

7. $3y^2 = -2y + 8$ **8.** $|5 + u| = 0$ **9.** $(2x - 3)^2 = 0$

10. $n^3 = n^2$ **11.** $d^2 + 6d + 9 = 0$ **12.** $5(2x + 3) = 3(3x - 4)$

NAME ______________________ DATE ______________

5 Rational Expressions

5–1 Quotients of Monomials

Objective: To simplify quotients using the law of exponents.

Vocabulary

Multiplication rule for fractions Let p, q, r, and s be real numbers with $q \neq 0$ and $s \neq 0$. Then
$$\frac{p}{q} \cdot \frac{r}{s} = \frac{pr}{qs}.$$

Rule for simplifying fractions Let p, q, and r be real numbers with $q \neq 0$ and $r \neq 0$. Then
$$\frac{pr}{qr} = \frac{p}{q}.$$

Law of exponents Let m and n be positive integers and a and b be real numbers, with $a \neq 0$ and $b \neq 0$ when they are divisors. Then:

1. $a^m \cdot a^n = a^{m+n}$
2. $(ab)^m = a^m b^m$
3. $(a^m)^n = a^{mn}$

4a. If $m > n$, $\frac{a^m}{a^n} = a^{m-n}$
4b. If $n > m$, $\frac{a^m}{a^n} = \frac{1}{a^{n-m}}$
5. $\left(\frac{a}{b}\right)^m = \frac{a^m}{b^m}$

Simplifying a quotient of monomials A quotient of monomials having integral coefficients is simplified when:

1. the integral coefficients have no common factors except 1 and -1;
2. each base appears only once; and
3. there are no powers of powers [such as $(a^2)^3$].

Example 1 Simplify: **a.** $\frac{45}{60}$ **b.** $\frac{12xy^3}{21x^2y}$

Solution

a. The GCF of 45 and 60 is 15.
$$\frac{45}{60} = \frac{3 \cdot \cancel{15}}{4 \cdot \cancel{15}} = \frac{3}{4}$$

b. The GCF of $12xy^3$ and $21x^2y$ is $3xy$.
$$\frac{12xy^3}{21x^2y} = \frac{4y^2 \cdot \cancel{3xy}}{7x \cdot \cancel{3xy}} = \frac{4y^2}{7x}$$

Use the method shown in Example 1 to simplify. Assume that no denominator equals 0.

1. $\frac{18}{48}$
2. $\frac{36}{63}$
3. $\frac{100}{24}$
4. $\frac{54}{90}$
5. $\frac{8t}{2t^2}$
6. $\frac{16x^4}{24x}$
7. $\frac{18y}{4y^3}$
8. $\frac{3a^2b}{9ab^2}$
9. $\frac{50st^2}{25st^3}$
10. $\frac{15p^3q^2}{18p^2q}$

Example 2 Simplify: **a.** $\frac{3^6}{3^4}$ **b.** $\frac{7x^2}{x^5}$ **c.** $\left(\frac{y^4}{5}\right)^2$

Solution

a. $\frac{3^6}{3^4} = 3^{6-4} = 3^2 = 9$

b. $\frac{7x^2}{x^5} = \frac{7}{x^{5-2}} = \frac{7}{x^3}$

c. $\left(\frac{y^4}{5}\right)^2 = \frac{(y^4)^2}{5^2} = \frac{y^8}{25}$

NAME ______________________ DATE ____________

5-1 Quotients of Monomials *(continued)*

Use the method shown in Example 2 to simplify. Assume that no denominator equals 0.

11. $\frac{8^7}{8^5}$ **12.** $\frac{4^2}{4^5}$ **13.** $\frac{x^3}{x^4}$ **14.** $\frac{t^9}{t^4}$

15. $\frac{5a^3}{a^5}$ **16.** $\frac{y^6}{8y^3}$ **17.** $\left(\frac{x^2}{9}\right)^2$ **18.** $\left(\frac{4}{z^3}\right)^2$

Example 3 Simplify: **a.** $\frac{24x^5y}{56x^3y^7}$ **b.** $\frac{4r}{t} \cdot \frac{r^3}{t^2}$ **c.** $\frac{3a}{4c}\left(\frac{2c}{a^3}\right)^2$ **d.** $\frac{(uw^2)^3}{(u^2w)^2}$.

Solution **a.** $\frac{24x^5y}{56x^3y^7} = \frac{24}{56} \cdot \frac{x^5}{x^3} \cdot \frac{y^1}{y^7}$

$= \frac{3}{7} \cdot \frac{x^2}{1} \cdot \frac{1}{y^6}$

$= \frac{3x^2}{7y^6}$

b. $\frac{4r}{t} \cdot \frac{r^3}{t^2} = \frac{4r \cdot r^3}{t \cdot t^2}$

$= \frac{4r^4}{t^3}$

c. $\frac{3a}{4c}\left(\frac{2c}{a^3}\right)^2 = \frac{3a}{4c} \cdot \frac{4c^2}{a^6}$

$= \frac{3 \cdot 4}{4} \cdot \frac{a^1}{a^6} \cdot \frac{c^2}{c^1}$

$= \frac{3c}{a^5}$

d. $\frac{(uw^2)^3}{(u^2w)^2} = \frac{u^3w^6}{u^4w^2}$

$= \frac{u^3}{u^4} \cdot \frac{w^6}{w^2}$

$= \frac{1}{u} \cdot \frac{w^4}{1}$

$= \frac{w^4}{u}$

Simplify. Assume that no denominator equals 0.

19. $\frac{27x^3y^5}{18x^4y^2}$ **20.** $\frac{10c^4d^2}{-12cd}$ **21.** $\frac{-8s^4t}{16st^4}$ **22.** $\frac{5u^2v^6}{15u^3v^8}$

23. $\frac{6x^2}{y^2} \cdot \frac{x}{y^2}$ **24.** $\frac{m^3}{2n} \cdot \frac{m^4}{n^3}$ **25.** $\frac{2a}{b} \cdot \frac{3b^3}{4a^2}$ **26.** $\frac{5xy}{2} \cdot \frac{8x^2}{15y^3}$

27. $\frac{p^3}{q}\left(\frac{3q}{p}\right)^2$ **28.** $\frac{2r^3}{t^3}\left(\frac{-t}{2r^2}\right)^2$ **29.** $\frac{(x^2y^3)^2}{(x^3y)^2}$ **30.** $\frac{(ab^2c)^2}{(a^3bc^2)^3}$

Mixed Review Exercises

Find the solution set of each inequality.

1. $x^2 + 7x + 6 > 0$ **2.** $3x + 5 \le 2$ **3.** $r^2 > 4$

4. $3 - 2x < -9$ **5.** $|3x - 2| \le 4$ **6.** $-2 < 3a + 1 < 4$

7. $x^2 - x \le 2$ **8.** $|x + 3| > 2$ **9.** $-3a < -9$ or $a - 3 \le -6$

NAME ______________________________ DATE ______________

5–2 Zero and Negative Exponents

Objective: To simplify expressions involving the exponent zero and negative integral exponents.

Vocabulary

Zero exponent Let a be any *nonzero* real number. Then: $a^0 = 1$

Examples: $5^0 = 1$ $\quad \left(\frac{1}{3}\right)^0 = 1$ $\quad (-200)^0 = 1$

Note that 0^0 is undefined.

Negative exponent Let a be any *nonzero* real number and n be a positive integer. Then:

$$a^{-n} = \frac{1}{a^n}$$

Examples: $2^{-1} = \frac{1}{2^1}$ $\quad 4^{-2} = \frac{1}{4^2}$ $\quad (-5)^{-3} = \frac{1}{(-5)^3}$

Example 1 Write in simplest form without negative or zero exponents.

a. 10^{-3} **b.** $-2x^{-2}$ **c.** $(-2x)^{-2}$ **d.** $3^{-1}x^0y^{-4}$

Solution We will assume that the variables are restricted so that no denominator will be zero.

a. $10^{-3} = \frac{1}{10^3} = \frac{1}{1000}$ **b.** $-2x^{-2} = -2 \cdot \frac{1}{x^2} = -\frac{2}{x^2}$

c. $(-2x)^{-2} = \frac{1}{(-2x)^2} = \frac{1}{4x^2}$ **d.** $3^{-1}x^0y^{-4} = \frac{1}{3^1} \cdot 1 \cdot \frac{1}{y^4} = \frac{1}{3y^4}$

Write in simplest form without negative or zero exponents.

1. 10^{-2} **2.** $(-4)^{-1}$ **3.** 17^0 **4.** $5t^{-3}$

5. $-8a^{-2}$ **6.** $(-8a)^{-2}$ **7.** $2^{-1}h^{-5}k^0$ **8.** $3^2u^{-1}v^{-2}$

Example 2 Write in simplest form without negative or zero exponents.

a. $5^{-3} \cdot 5^{-1}$ **b.** $(-4 \cdot 3^{-2})^{-2}$ **c.** $\frac{2a^{-2}b^3}{5^0ab^{-2}}$

Solution The laws of exponents hold for negative and zero exponents as well as for positive exponents.

a. $5^{-3} \cdot 5^{-1} = 5^{-3+(-1)} = 5^{-4} = \frac{1}{5^4} = \frac{1}{625}$

b. $(-4 \cdot 3^{-2})^{-2} = (-4)^{-2}(3^{-2})^{-2}$

$= \frac{1}{(-4)^2} \cdot 3^4$

$= \frac{81}{16}$

c. $\frac{2a^{-2}b^3}{5^0ab^{-2}} = \frac{2 \cdot a^{-2-1} \cdot b^{3-(-2)}}{1}$

$= 2a^{-3}b^5$

$= \frac{2b^5}{a^3}$

NAME ______________________________ DATE ______________

5–2 Zero and Negative Exponents *(continued)*

Write in simplest form without negative or zero exponents.

9. $3^{-2} \cdot 3^{-1}$ **10.** $(-2)^2 \cdot (-2)^{-5}$ **11.** $(5^{-2})^2$ **12.** $(-2^{-1})^{-2}$

13. $(3 \cdot 8^{-2})^{-1}$ **14.** $(4^{-1} \cdot 6^{-1} \cdot 7^0)^{-1}$ **15.** $\frac{4^2}{4^{-1}}$ **16.** $\frac{x^{-1}}{x^{-2}}$

17. $\frac{z^{-6}}{z^2}$ **18.** $\frac{s^{-3}t^{-4}}{s^{-2}t^0}$ **19.** $\frac{9ab^{-2}}{-3a^{-3}b^{-1}}$ **20.** $\frac{18u^4v^{-5}}{u^{-1}v^7}$

Example 3 Write in simplest form without negative or zero exponents.

a. $\left(\frac{4}{5}\right)^{-3}$ b. $\left(\frac{a^2}{b}\right)^{-3}\left(\frac{b^{-1}}{a}\right)^2$

Solution

a. $\left(\frac{4}{5}\right)^{-3} = \frac{4^{-3}}{5^{-3}} = 4^{-3} \cdot \frac{1}{5^{-3}} = \frac{1}{4^3} \cdot 5^3 = \frac{5^3}{4^3} = \frac{125}{64}$

b. $\left(\frac{a^2}{b}\right)^{-3}\left(\frac{b^{-1}}{a}\right)^2 = \frac{(a^2)^{-3}}{b^{-3}} \cdot \frac{(b^{-1})^2}{a^2} = \frac{a^{-6}}{b^{-3}} \cdot \frac{b^{-2}}{a^2} = \frac{a^{-6}b^{-2}}{a^2b^{-3}} = a^{-8}b^1 = \frac{b}{a^8}$

Note: $\frac{1}{a^{-n}} = a^n$ since a^{-n} and a^n are reciprocals.

Write in simplest form without negative or zero exponents.

21. $\left(\frac{5}{2}\right)^{-2}$ **22.** $\left(\frac{2}{3}\right)^{-3}$ **23.** $\left(\frac{2}{x^2y^{-3}}\right)^{-2}$ **24.** $\left(\frac{3pq^{-1}}{p^{-2}q}\right)^{-1}$

25. $4z(2y^2z)^{-2}$ **26.** $\frac{(2a^{-1})^{-3}}{2(a^{-1})^{-2}}$ **27.** $\left(\frac{x^3}{y^{-1}}\right)^{-2}\left(\frac{y^3}{x^{-2}}\right)^2$ **28.** $\left(\frac{c^2}{d^{-6}}\right)^0\left(\frac{c^{-5}}{d^{-1}}\right)^2$

Example 4 Write without using fractions: a. $\frac{7}{1000}$ b. $\frac{3a}{b^2c^3}$

Solution a. $\frac{7}{1000} = \frac{7}{10^3} = 7 \times 10^{-3}$ b. $\frac{3a}{b^2c^3} = 3ab^{-2}c^{-3}$

Write without using fractions.

29. $\frac{9}{10{,}000}$ **30.** $\frac{3}{100}$ **31.** $\frac{11}{100{,}000}$

32. $\frac{a^3}{b^2}$ **33.** $\frac{6x^3}{yz^2}$ **34.** $\frac{2hk^3}{j^4}$

NAME ______________________ DATE ____________

5–3 Scientific Notation and Significant Digits

Objective: To use scientific notation and significant digits.

Vocabulary

Scientific notation A number represented in the form $m \times 10^n$, where $1 \le m < 10$ and n is an integer.

Examples: $8{,}320{,}000 = 8.32 \times 10^6$ $\quad 0.00079 = 7.9 \times 10^{-4}$

Significant digit A significant digit of a number written in decimal form is any nonzero digit or any zero that has a purpose other than placing the decimal point.

Examples: 0.602 3054 81.0 0.00090 (brackets indicate significant digits)

Symbol $\approx$ (is approximately equal to)

Example 1 Write each number in scientific notation. If the number is an *integer* and ends in zeros, assume that the zeros are not significant.

a. 34.070 **b.** 0.000242 **c.** 5,070,000 **d.** 0.068×10^3

Solution

a. $34.070 = 3.4070 \times 10^1$ decimal point moved 1 place to the *left*

b. $0.000242 = 2.42 \times 10^{-4}$ decimal point moved 4 places to the *right*

c. $5{,}070{,}000 = 5.07 \times 10^6$ decimal point moved 6 places to the *left*

d. $0.068 \times 10^3 = (6.8 \times 10^{-2}) \times 10^3$ decimal point moved 2 places to the *right*

$= 6.8 \times (10^{-2} \times 10^3)$ associative property used

$= 6.8 \times 10^1$ $\quad a^m \cdot a^n = a^{m+n}$

Write each number in scientific notation. If the number is an integer and ends in zeros, assume that the zeros are not significant.

1. 750 **2.** 347,000 **3.** 89.2 **4.** 0.037

5. 2100 **6.** 34 **7.** 0.00086 **8.** 51.080

9. 9,006,000 **10.** 0.07 **11.** 0.00401 **12.** 958.05

13. 0.8490 **14.** 0.0000265 **15.** 70,030 **16.** 2570.20

17. 302×10^2 **18.** 0.51×10^{-2} **19.** 0.840×10^3 **20.** 6376×10^{-1}

Example 2 Write each number in decimal form: **a.** 5×10^{-2} **b.** 8.46×10^3

Solution **a.** $5 \times 10^{-2} = 0.05$ move decimal point 2 places to the *left*

b. $8.46 \times 10^3 = 8460$ move decimal point 3 places to the *right*

Write each number in decimal form.

21. 3×10^4 **22.** 10^{-5} **23.** 6.80×10^{-3} **24.** 2.7×10^6

25. 5.02×10^{-4} **26.** 7×10^{-4} **27.** 9.000×10^2 **28.** 1.40×10^5

NAME ______________________ DATE ______________

5–3 Scientific Notation and Significant Digits *(continued)*

Example 3

a. $8.97 \times 10^5 < 1.36 \times 10^8$ — The number with the larger exponent is greater.

b. $5.72 \times 10^{-1} > 4.88 \times 10^{-1}$ — The exponents are equal, so compare decimals: $5.72 > 4.88$.

c. $6.13 \times 10^6 < (3 \times 10^3)^2$ — Since $(3 \times 10^3)^2 = 9 \times 10^6$, and $6.13 < 9$.

Replace the ? with > or < to make a true statement.

29. 5.3×10^6 ? 2.56×10^6

30. 6.17×10^4 ? 3.27×10^5

31. 4.3×10^{-2} ? 1.2×10^{-1}

32. $(2 \times 10^3)^2$? 9.24×10^3

Example 4 Find a one-significant-digit estimate of x, where $x = \dfrac{9300 \times 78.4}{0.0018 \times 226}$.

Solution

$x = \dfrac{9.3 \times 10^3 \times 7.84 \times 10^1}{1.8 \times 10^{-3} \times 2.26 \times 10^2}$ — Write each number in scientific notation.

$= \dfrac{9 \times 10^3 \times 8 \times 10^1}{2 \times 10^{-3} \times 2 \times 10^2}$ — Round each decimal to a whole number.

$= \dfrac{9 \times 8}{2 \times 2} \times 10^{3+1-(-3)-2}$ — Compute, and give the result to one significant digit.

$= 18 \times 10^5$

$\approx 20 \times 10^5 = 2 \times 10^6$ — Note that $20 = 2 \times 10^1$.

So, $x \approx 2 \times 10^6$, or 2,000,000.

Find a one-significant-digit estimate of the following.

33. $\dfrac{26.1 \times 0.73}{0.00012 \times 3800}$

34. $\dfrac{0.642 \times 3890}{12.6 \times 0.00024}$

35. $\dfrac{0.0373 \times 0.561}{0.0017 \times 41.5}$

36. $\dfrac{88.3 \times 0.057}{46{,}000 \times 0.0019}$

Mixed Review Exercises

Simplify.

1. $\dfrac{24a^4b}{30ab^2}$

2. $\left(\dfrac{5r}{3s}\right)^2$

3. $\dfrac{4x}{y} \cdot \dfrac{x^2}{3y} \cdot \dfrac{9x}{y}$

4. $\dfrac{(a^2b)^2}{(ab^2)^3}$

5. $\dfrac{(-p)^4}{-p^4}$

6. $\dfrac{28h^4}{21h^4}$

Express in simplest form without negative or zero exponents.

7. $(x^{-2}y^{-3})^{-1}$

8. $\left(\dfrac{u^{-1}}{v}\right)^{-2}$

9. $\dfrac{a^{-3}b^{-2}}{a^{-4}b^5}$

NAME ______________________ DATE ____________

5–4 Rational Algebraic Expressions

Objective: To simplify algebraic expressions.

Vocabulary

Rational algebraic expression (or **rational expression**) An algebraic expression that can be expressed as a quotient of polynomials.

Examples: $\frac{1}{x}$ $\quad \frac{5y + 15}{5}$ $\quad \frac{z^2 - 4}{2z(z + 2)}$

Simplified rational expression A quotient of polynomials whose greatest common factor is 1.

Examples: $\frac{2a}{3b}$ $\quad \frac{y + 3}{1}$ $\quad \frac{z - 2}{2z}$

Zero of a function A number r is a zero of a function f if $f(r) = 0$.

Example: 2 is a *zero* of $f(x) = x^2 + x - 6$ because $f(2) = 2^2 + 2 - 6 = 0$.

Example 1 Simplify: **a.** $\frac{8x^2 + 16x}{4x}$ **b.** $\frac{x^2 - 16}{x^2 + 7x + 12}$

Solution **a.** $\frac{8x^2 + 16x}{4x} = \frac{{}^{2}\cancel{8x}(x + 2)}{\cancel{4x}^{1}}$ Factor and then simplify.

$= \frac{2(x + 2)}{1}$

$= 2(x + 2)$

b. $\frac{x^2 - 16}{x^2 + 7x + 12} = \frac{\cancel{(x + 4)}(x - 4)}{(x + 3)\cancel{(x + 4)}}$ Factor and then simplify.

$= \frac{x - 4}{x + 3}$

Example 2 Simplify $(2 - x - 3x^2)(9x^2 - 4)^{-1}$.

Solution $(2 - x - 3x^2)(9x^2 - 4)^{-1} = \frac{2 - x - 3x^2}{9x^2 - 4}$ Definition: $a^{-n} = \frac{1}{a^n}$

$= \frac{-(3x^2 + x - 2)}{(3x + 2)(3x - 2)}$ Factor.

$= \frac{-\cancel{(3x - 2)}(x + 1)}{(3x + 2)\cancel{(3x - 2)}}$ $2 - 3x = -(3x - 2)$

$= \frac{-(x + 1)}{3x + 2}$

$= -\frac{x + 1}{3x + 2}$

NAME ______________________ DATE ______________

5-4 Rational Algebraic Expressions *(continued)*

Simplify.

1. $\dfrac{5x + 15}{10}$
2. $\dfrac{3y - 9}{6}$
3. $\dfrac{9s^3 + 27s^2}{6s}$
4. $\dfrac{r^2 + 9r + 18}{(r + 6)^2}$
5. $\dfrac{(p - 4)^2}{2p^2 - 9p + 4}$
6. $\dfrac{k^2 - 5k + 4}{k^2 + 2k - 3}$
7. $\dfrac{m^2 + 5m - 14}{m^2 - 4}$
8. $\dfrac{a^2 + 4a - 5}{a^3 - a}$
9. $\dfrac{t^2 - 9}{t^3 - 6t^2 + 9t}$
10. $(x - y)(y - x)^{-1}$
11. $(a^2 + ab)(a^2 - b^2)^{-1}$
12. $\dfrac{2x^2 + 5x - 3}{1 - 4x^2}$
13. $\dfrac{16 - 9b^2}{3b^2 + 11b - 20}$
14. $\dfrac{c^2 - d^2}{c^2 + 2cd + d^2}$
15. $\dfrac{y^4 - 16}{(y + 2)^2(y^2 + 4)}$

Example 3 Let $f(x) = \dfrac{2x^2 + x - 1}{x^3 - 9x}$.

a. Find the domain of f. **b.** Find the zeros of f, if there are any.

Solution $f(x) = \dfrac{2x^2 + x - 1}{x^3 - 9x} = \dfrac{(2x - 1)(x + 1)}{x(x - 3)(x + 3)}$

a. The function f will be undefined at any value for which the denominator equals 0. Find those values.

$x(x - 3)(x + 3) = 0$

$x = 0$ or $x - 3 = 0$ or $x + 3 = 0$ Use the zero-product property.

$x = 0$ $x = 3$ $x = -3$

∴ the domain of f consists of all real numbers except 0, 3, and -3.

b. $f(x) = 0$ if and only if $(2x - 1)(x + 1) = 0$. That is, if the numerator of a fraction equals 0, then the fraction equals 0.

$(2x - 1)(x + 1) = 0$

$2x - 1 = 0$ or $x + 1 = 0$ Use the zero-product property.

$x = \dfrac{1}{2}$ $x = -1$

∴ the zeros of f are $\dfrac{1}{2}$ and -1.

Find (a) the domain and (b) the zeros, if any, of each function.

16. $f(x) = \dfrac{x^2 - 4}{x^2 - 4x}$
17. $g(x) = \dfrac{x^2 + 5x - 24}{x^2 - 36}$
18. $F(y) = \dfrac{y^2 - y - 12}{2y^2 - 5y + 2}$
19. $h(t) = \dfrac{4t^2 - 11t + 6}{(3t + 1)^2}$

NAME ______________________________ DATE ______________

5–5 Products and Quotients of Rational Expressions

Objective: To multiply and divide rational expressions.

Vocabulary

Division rule for fractions Let p, q, r, and s be real numbers with $q \neq 0$, $r \neq 0$, and $s \neq 0$. Then $\frac{p}{q} \div \frac{r}{s} = \frac{p}{q} \cdot \frac{s}{r} = \frac{ps}{qr}$. (To divide by a fraction, you multiply by its *reciprocal*.)

Example: $\frac{3}{8} \div \frac{2}{5} = \frac{3}{8} \cdot \frac{5}{2} = \frac{3 \cdot 5}{8 \cdot 2} = \frac{15}{16}$

Example 1 Simplify: **a.** $\frac{14}{9} \cdot \frac{15}{28}$ **b.** $\frac{x^2 + x - 2}{x^2 - 4} \cdot \frac{x^2 - 5x - 6}{x^2 - 2x + 1}$

Solution Factor where possible, then multiply. Divide out common factors.

a. $\frac{14}{9} \cdot \frac{15}{28} = \frac{14 \cdot 15}{9 \cdot 28} = \frac{5}{6}$

b. $\frac{x^2 + x - 2}{x^2 - 4} \cdot \frac{x^2 - 5x - 6}{x^2 - 2x + 1} = \frac{(x + 2)(x - 1)}{(x + 2)(x - 2)} \cdot \frac{(x - 6)(x + 1)}{(x - 1)^2}$

$$= \frac{\cancel{(x + 2)}\cancel{(x - 1)}(x - 6)(x + 1)}{\cancel{(x + 2)}(x - 2)\cancel{(x - 1)}(x - 1)}$$

$$= \frac{(x - 6)(x + 1)}{(x - 2)(x - 1)}$$

Simplify.

1. $\frac{18}{5} \cdot \frac{10}{27}$
2. $\frac{4}{3} \cdot 12$
3. $\frac{x}{y^2} \cdot \frac{y}{x}$
4. $\frac{2p}{q} \cdot \frac{p}{8q^2}$
5. $\frac{5a}{2b^3} \cdot \frac{4b^2}{a}$
6. $\frac{3}{z + 2} \cdot \frac{z^2 - 4}{6}$
7. $\frac{2y^2 - 50}{2y - 10} \cdot \frac{4y - 2}{6y + 30}$
8. $\frac{t^2 + 3t}{t^2 + 2t - 3} \cdot \frac{t + 1}{t}$
9. $\frac{k^2 - 2k - 8}{2k^2 + 5k + 3} \cdot \frac{2k + 3}{k - 4}$

Example 2 Simplify: **a.** $\frac{-18}{25} \div \frac{6}{5}$ **b.** $\frac{6a^2b}{5c^3} \div \frac{ab^2}{10c}$

Solution

a. Multiply by the reciprocal.

$$\frac{-18}{25} \div \frac{6}{5} = \frac{-18}{25} \cdot \frac{5}{6}$$

$$= \frac{-18 \cdot 5}{25 \cdot 6} = -\frac{3}{5}$$

b. You can divide out factors common to the numerator and denominator *before* you write the product as a single fraction.

$$\frac{6a^2b}{5c^3} \div \frac{ab^2}{10c} = \frac{\overset{a}{\cancel{6a^2b}}}{\underset{c^2}{\cancel{5c^3}}} \cdot \frac{\overset{2}{\cancel{10c}}}{\underset{b}{\cancel{ab^2}}} = \frac{12a}{bc^2}$$

NAME ______________________ DATE ____________

5–5 Products and Quotients of Rational Expressions (continued)

Example 3 Simplify $\frac{2x-14}{x^2-2x-35} \div \frac{6x^3}{x^2-25}$.

Solution

$$\frac{2x-14}{x^2-2x-35} \div \frac{6x^3}{x^2-25} = \frac{2x-14}{x^2-2x-35} \cdot \frac{x^2-25}{6x^3}$$
$$= \frac{2(x-7)}{(x-7)(x+5)} \cdot \frac{(x+5)(x-5)}{6x^3}$$
$$= \frac{x-5}{3x^3}$$

Simplify.

10. $\frac{20}{9} \div \frac{5}{12}$

11. $\frac{-3}{8} \div \frac{4}{9}$

12. $\frac{45a^2}{8} \div \frac{35a}{4}$

13. $\frac{21t^3}{5} \div \frac{7t^2}{-5}$

14. $\frac{6bc^2}{15a^2} \div \frac{2bc}{5a}$

15. $\frac{u+1}{16} \div \frac{u+1}{4}$

16. $\frac{r-3}{2r^2} \div \frac{r^2-4r+3}{2r}$

17. $\frac{x^2-9}{x^2+6x+9} \div \frac{x-3}{x+3}$

18. $\frac{y^2-4y}{y^2+2y} \div \frac{y^2-9y+20}{2y^2-9y-5}$

Example 4 Simplify:

$$\frac{3m^2n}{2} \div \frac{n}{4m} \div \frac{15m^4}{2}$$

Solution

$$\frac{3m^2n}{2} \div \frac{n}{4m} \div \frac{15m^4}{2} = \frac{3m^2n}{2} \cdot \frac{4m}{n} \cdot \frac{2}{15m^4}$$
$$= \frac{\cancel{3m^2n} \cdot \cancel{4m} \cdot \cancel{2}}{\cancel{2} \cdot \cancel{n} \cdot \cancel{15m^4}} = \frac{4}{5m}$$

(with $5m$ written below the cancelled denominator)

Simplify.

19. $5xy \div \frac{10x^2}{y^2} \div \frac{y^3}{x}$

20. $\frac{12a^2}{b} \div \frac{2}{3ab} \div \frac{54a^3}{b}$

21. $\frac{24c^3}{d} \div \frac{40c}{de^2} \cdot \frac{5}{9e}$

22. $\frac{3p+6}{9p} \cdot \frac{12p}{p^2-4} \div \frac{18p^3}{2p-4}$

Mixed Review Exercises

Express in scientific notation. Assume that the zeros at the end of any integer are not significant.

1. 0.00276
2. 0.5
3. 7634
4. 72,000,000
5. 38.20

Simplify.

6. $\frac{20r^5s^2}{32r^3s^5}$

7. $\frac{3k+9}{2k+6}$

8. $\frac{(6x^2)^2}{(4x^3)^2}$

9. $\frac{4z^2-4}{8z^2+16z+8}$

NAME ______________________ DATE ______________

5–6 Sums and Differences of Rational Expressions

Objective: To add and subtract rational expressions.

Vocabulary

Least common multiple (LCM) of two or more polynomials The common multiple having least degree and least positive factors.

Examples: The LCM of $2x$, $6x^2$, and $3x$ is $6x^2$.
The LCM of $x(x + 1)$ and $(x + 1)(x - 1)$ is $x(x + 1)(x - 1)$.
The LCM of $x + 2$ and $x + 4$ is $(x + 2)(x + 4)$.

Example 1 Simplify: **a.** $\frac{5}{6} + \frac{13}{6} - \frac{7}{6}$ **b.** $\frac{2x - 5}{x - 3} - \frac{x - 1}{x - 3}$

Solution With fractions having the same denominator, add or subtract the numerators and write the result over the common denominator.

a. $\frac{5}{6} + \frac{13}{6} - \frac{7}{6} = \frac{5 + 13 - 7}{6}$

$= \frac{11}{6}$

b. $\frac{2x - 5}{x - 3} - \frac{x - 1}{x - 3} = \frac{2x - 5 - (x - 1)}{x - 3}$

$= \frac{2x - 5 - x + 1}{x - 3}$

$= \frac{x - 4}{x - 3}$

Example 2 Simplify: **a.** $\frac{5}{6} + \frac{1}{8} - \frac{1}{3}$ **b.** $\frac{1}{3a} - \frac{1}{4a} + \frac{2}{a^2}$

Solution With fractions having different denominators, rewrite the fractions using their *least common denominator* (LCD), which is the LCM of the denominators.

a. $6 = 2 \cdot 3$ and $8 = 2^3$. So the LCD $= 2^3 \cdot 3 = 24$.

$$\frac{5}{6} + \frac{1}{8} - \frac{1}{3} = \frac{5 \cdot 4}{6 \cdot 4} + \frac{1 \cdot 3}{8 \cdot 3} - \frac{1 \cdot 8}{3 \cdot 8}$$

$$= \frac{20}{24} + \frac{3}{24} - \frac{8}{24}$$

$$= \frac{20 + 3 - 8}{24} = \frac{15}{24} = \frac{5}{8}$$

b. The LCD for $3a$, $4a$, and a^2 is $12a^2$.

$$\frac{1}{3a} - \frac{1}{4a} + \frac{2}{a^2} = \frac{1 \cdot 4a}{3a \cdot 4a} - \frac{1 \cdot 3a}{4a \cdot 3a} + \frac{2 \cdot 12}{a^2 \cdot 12}$$

$$= \frac{4a}{12a^2} - \frac{3a}{12a^2} + \frac{24}{12a^2}$$

$$= \frac{4a - 3a + 24}{12a^2}$$

$$= \frac{a + 24}{12a^2}$$

NAME ______________________ DATE ______________

5–6 *Sums and Differences of Rational Expressions* *(continued)*

Simplify.

1. $\frac{7}{8} - \frac{3}{8} + \frac{1}{8}$
2. $\frac{1}{2} + \frac{1}{3} + \frac{3}{5}$
3. $\frac{5}{6} + \frac{2}{5} - \frac{8}{15}$
4. $\frac{3}{4} + \frac{5}{18} - \frac{7}{9}$
5. $\frac{5}{2x} - \frac{3}{2x}$
6. $\frac{3}{5x^3y} - \frac{2}{xy^2}$
7. $\frac{x}{x+1} + \frac{1}{x+1}$
8. $\frac{8t+4}{t-2} - \frac{6t-1}{t-2}$
9. $\frac{2}{3z} + \frac{7}{12z}$
10. $\frac{3}{rs} - \frac{4}{rs^2}$
11. $\frac{3m-2}{6} - \frac{m-3}{9}$
12. $\frac{2n+1}{3n} + \frac{2-3n}{4n}$

Example 3 Simplify $\frac{3}{x^2+x-2} - \frac{5}{x^2-x-6}$.

Solution

$x^2 + x - 2 = (x+2)(x-1)$
$x^2 - x - 6 = (x-3)(x+2)$
{ Factor the denominators to find the LCD.

So the LCD is $(x+2)(x-1)(x-3)$.

$$\frac{3}{x^2+x-2} - \frac{5}{x^2-x-6} = \frac{3}{(x+2)(x-1)} - \frac{5}{(x-3)(x+2)}$$
$$= \frac{3(x-3)}{(x+2)(x-1)(x-3)} - \frac{5(x-1)}{(x-3)(x+2)(x-1)}$$
$$= \frac{3(x-3) - 5(x-1)}{(x+2)(x-1)(x-3)}$$
$$= \frac{3x - 9 - 5x + 5}{(x+2)(x-1)(x-3)}$$
$$= \frac{-2x-4}{(x+2)(x-1)(x-3)}$$
$$= \frac{-2\cancel{(x+2)}}{\cancel{(x+2)}(x-1)(x-3)}$$
$$= \frac{-2}{(x-1)(x-3)}, \text{ or } -\frac{2}{(x-1)(x-3)}$$

Simplify.

13. $\frac{2}{k-3} + \frac{4}{k+3}$
14. $\frac{c+1}{c} - \frac{c}{c+1}$
15. $\frac{y}{y-1} + \frac{4}{y+1}$
16. $\frac{5m+1}{2m^2-2m} - \frac{3}{2m-2}$
17. $\frac{1}{x^2-3x} - \frac{1}{x^2-9}$
18. $\frac{1}{z^2-4} + \frac{1}{(z-2)^2}$
19. $\frac{3}{p^2-3p+2} - \frac{2}{p^2-1}$
20. $\frac{1}{x^2+x-2} + \frac{1}{x^2-5x+4}$

NAME ______________________ DATE ______________

5-7 Complex Fractions

Objective: To simplify complex fractions.

Vocabulary

Complex fraction A fraction that has a fraction or powers with negative exponents in its numerator or denominator (or both).

Symbol $\frac{a}{b}$ means $a \div b$

Example 1 Simplify $\dfrac{\frac{7}{15} + \frac{1}{5}}{2 + \frac{2}{9}}$.

Solution *Method 1:* Simplify the numerator and denominator separately.

$$\frac{\frac{7}{15} + \frac{1}{5}}{2 + \frac{2}{9}} = \frac{\frac{7}{15} + \frac{3}{15}}{\frac{18}{9} + \frac{2}{9}} = \frac{\frac{10}{15}}{\frac{20}{9}} = \frac{10}{15} \div \frac{20}{9} = \frac{10}{15} \cdot \frac{9}{20} = \frac{3}{10}$$

Method 2: Multiply the numerator and the denominator by the LCD.

The LCD for $\frac{7}{15}$, $\frac{1}{5}$, and $\frac{2}{9}$ is 45.

$$\frac{\frac{7}{15} + \frac{1}{5}}{2 + \frac{2}{9}} = \frac{\left(\frac{7}{15} + \frac{1}{5}\right)45}{\left(2 + \frac{2}{9}\right)45} = \frac{\left(\frac{7}{15}\right)45 + \left(\frac{1}{5}\right)45}{(2)45 + \left(\frac{2}{9}\right)45} = \frac{21 + 9}{90 + 10} = \frac{30}{100} = \frac{3}{10}$$

Simplify.

1. $\dfrac{\frac{3}{2} - 1}{\frac{5}{6} - \frac{2}{3}}$

2. $\dfrac{1 + \frac{2}{5}}{\frac{5}{2} - \frac{2}{5}}$

3. $\dfrac{\frac{1}{2} + \frac{1}{5}}{\frac{1}{6} + \frac{1}{8}}$

4. $\dfrac{\frac{4}{9} + \frac{1}{4}}{2 - \frac{1}{3}}$

Example 2 Simplify $\dfrac{\frac{1}{x} - \frac{1}{y}}{\frac{1}{x} + \frac{1}{y}}$.

Solution

Method 1:

$$\frac{\frac{1}{x} - \frac{1}{y}}{\frac{1}{x} + \frac{1}{y}} = \frac{\frac{y - x}{xy}}{\frac{y + x}{xy}} = \frac{y - x}{xy} \div \frac{y + x}{xy} = \frac{y - x}{xy} \cdot \frac{xy}{y + x} = \frac{y - x}{y + x}$$

Method 2:

$$\frac{\frac{1}{x} - \frac{1}{y}}{\frac{1}{x} + \frac{1}{y}} = \frac{\left(\frac{1}{x} - \frac{1}{y}\right)xy}{\left(\frac{1}{x} + \frac{1}{y}\right)xy} = \frac{\left(\frac{1}{x}\right)xy - \left(\frac{1}{y}\right)xy}{\left(\frac{1}{x}\right)xy + \left(\frac{1}{y}\right)xy} = \frac{y - x}{y + x}$$

NAME ______________________ DATE ____________

5–7 Complex Fractions *(continued)*

Example 3 Simplify $\dfrac{1 - m^{-1}}{1 - m^{-2}}$.

Solution Use the definition $a^{-n} = \dfrac{1}{a^n}$.

$$\frac{1 - m^{-1}}{1 - m^{-2}} = \frac{1 - \frac{1}{m}}{1 - \frac{1}{m^2}} = \frac{\frac{m - 1}{m}}{\frac{m^2 - 1}{m^2}}$$

$$= \frac{m - 1}{m} \div \frac{m^2 - 1}{m^2}$$

$$= \frac{(m - 1)}{m} \cdot \frac{m^2}{(m + 1)(m - 1)}$$

$$= \frac{m}{m + 1}$$

You can also use Method 2 to simplify the complex fraction above.

Simplify.

5. $\dfrac{\frac{1}{x}}{\frac{1}{y}}$

6. $\dfrac{\frac{m}{2} + \frac{2}{m}}{\frac{m + 2}{2m}}$

7. $\dfrac{a - 1}{1 - \frac{1}{a}}$

8. $\dfrac{c - \frac{1}{c}}{1 + \frac{1}{c}}$

9. $\dfrac{s + \frac{s}{t}}{1 + \frac{1}{t}}$

10. $\dfrac{1 + \frac{1}{z + 1}}{1 + \frac{3}{z - 1}}$

11. $\dfrac{r^{-2} + 1}{r - 1}$

12. $\dfrac{y^{-1} + x^{-1}}{y^{-2} - x^{-2}}$

13. $\dfrac{9 - k^{-2}}{3k^{-1} - k^{-2}}$

Mixed Review Exercises

Simplify.

1. $\dfrac{2}{3t} - \dfrac{t}{6t^2}$

2. $\dfrac{6}{a^2 - 4a} \cdot \dfrac{a^2 - 16}{10a}$

3. $\dfrac{49c^2}{25d} \div \dfrac{42c^4}{15d^2}$

4. $\dfrac{r}{r - 3} - \dfrac{1}{r + 3}$

5. $-\dfrac{10xy^4z}{15x^2yz}$

6. $\dfrac{5p - 10}{p^2 - 4p + 4}$

Find the unique solution of each system. Check your answer by using substitution.

7. $x + 2y = 6$
   $x - 2y = 4$

8. $3x + 7y = -4$
   $2x + 5y = -3$

9. $4x + 5y = -8$
   $3x - 4y = -6$

NAME ______________________ DATE ______________

5–8 Fractional Coefficients

Objective: To solve equations and inequalities having fractional coefficients.

Example 1 Solve $\frac{x^2}{5} = \frac{3x}{10} + \frac{1}{2}$.

Solution

$\frac{x^2}{5} = \frac{3x}{10} + \frac{1}{2}$ The LCD is 10.

$10\left(\frac{x^2}{5}\right) = 10\left(\frac{3x}{10} + \frac{1}{2}\right)$ Multiply *both* sides of the equation by the LCD to clear the denominators.

$2x^2 = 3x + 5$

$2x^2 - 3x - 5 = 0$ Make one side 0.

$(2x - 5)(x + 1) = 0$ Factor the polynomial.

$2x - 5 = 0$ or $x + 1 = 0$ Use the zero-product property.

$x = \frac{5}{2}$ or $x = -1$

$\therefore$ the solution set is $\left\{\frac{5}{2}, -1\right\}$.

Example 2 Solve $\frac{x}{6} - \frac{x - 3}{4} \le \frac{x + 1}{8}$.

Solution

$\frac{x}{6} - \frac{x - 3}{4} \le \frac{x + 1}{8}$ The LCD is 24.

$24\left(\frac{x}{6} - \frac{x - 3}{4}\right) \le 24\left(\frac{x + 1}{8}\right)$ Multiply *both* sides of the inequality by the LCD.

$4x - 6(x - 3) \le 3(x + 1)$ Use the distributive property and combine similar terms.

$4x - 6x + 18 \le 3x + 3$

$-2x + 18 \le 3x + 3$

$-5x \le -15$ Divide both sides by -5 and reverse the inequality sign.

$x \ge 3$

$\therefore$ the solution set is $\{x: x \ge 3\}$.

Solve each open sentence.

1. $\frac{2x}{3} - \frac{x}{6} = -1$
2. $\frac{4t}{3} + \frac{3t}{10} = \frac{7}{5}$
3. $\frac{h}{4} \le \frac{3}{2} - \frac{h}{5}$
4. $\frac{d}{6} - \frac{2}{3} \ge \frac{d}{4}$
5. $\frac{3u}{14} - \frac{5 - u}{21} = \frac{2}{7}$
6. $\frac{3r - 4}{5} - \frac{2r + 1}{4} = -\frac{1}{2}$
7. $\frac{y + 3}{2} + \frac{3}{5} \ge \frac{y + 1}{10}$
8. $\frac{c + 8}{12} - \frac{3c - 5}{15} < \frac{c}{20}$
9. $\frac{p^2}{6} - \frac{3p}{4} = \frac{15}{4}$
10. $\frac{2a}{5} - \frac{a^2}{3} = \frac{1}{15}$
11. $\frac{x^2}{4} + \frac{x - 1}{3} = 0$
12. $\frac{u(u + 6)}{5} = \frac{u - 1}{2}$

5–8 *Fractional Coefficients* *(continued)*

Example 3 How much pure alcohol must be added to 15 oz of a 60% solution of rubbing alcohol to change it to a 70% solution?

Solution Percents can be thought of as fractions since percent means *divided by 100.*

Step 1 The problem asks for the number of ounces of alcohol added to the 60% solution.

Step 2 Let x = number of ounces of alcohol to be added. Then $15 + x$ = number of oz in the 70% solution. } Show the known facts in a table.

	oz of solution ×	% alcohol =	oz of alcohol
60% solution	15	60%	0.60(15)
Alcohol added	x	100%	$1x$
70% solution	$15 + x$	70%	$0.70(15 + x)$

Step 3 alcohol in 60% solution + alcohol added = alcohol in 70% solution

$$0.60(15) + 1x = 0.70(15 + x)$$
or
$$60(15) + 100x = 70(15 + x)$$
{ To clear decimals, multiply both sides by 100.

Step 4
$$900 + 100x = 1050 + 70x$$
$$30x = 150$$
$$x = 5$$

Step 5 *Check:*
$$0.60(15) + 1(5) \stackrel{?}{=} 0.70(15 + 5)$$
$$9 + 5 \stackrel{?}{=} 0.70(20)$$
$$14 = 14 \ \checkmark$$
∴ 5 oz of alcohol must be added.

Solve.

13. How many liters of pure acid must be added to 5 L of a solution that is 20% acid to make a solution that is 60% acid?

14. A nurse has 6 L of a 3% boric acid solution. How much of a 10% boric acid solution must he add to produce a 4% solution?

15. How many gallons of cream that is 23% butterfat and milk that is 3% butterfat must a dairy farmer mix to make 30 gallons of milk that is 4% butterfat?

Complete the chart and solve.

16. John Gordon invested $1000, part at 5% and the rest at 6.5%. The income from the 5% investment exceeded the income from the 6.5% investment by $1.70. How much did he invest at each rate?

	Amount invested ×	Rate =	Interest earned
Investment at 5%	x	?	?
Investment at 6.5%	?	?	?

NAME ______________________ DATE ____________

5–9 Fractional Equations

Objective: To solve and use fractional equations.

Vocabulary

Fractional equation An equation in which a variable occurs in a denominator.

Extraneous root A root of a transformed equation that is not a root of the original equation.

CAUTION Since multiplying an equation by a polynomial may produce extraneous roots, you must *always check* each root of the new equation in the *original* equation.

Example Solve $\frac{12}{t^2 - 4} - \frac{3}{t - 2} = -1$.

Solution

$$\frac{12}{(t + 2)(t - 2)} - \frac{3}{t - 2} = -1$$ The LCD is $(t + 2)(t - 2)$.

$$(t + 2)(t - 2)\left[\frac{12}{(t - 2)(t + 2)} - \frac{3}{t - 2}\right] = (t + 2)(t - 2)(-1)$$ Multiply both sides by the LCD.

$$12 - 3(t + 2) = -1(t^2 - 4)$$
$$12 - 3t - 6 = -t^2 + 4$$
$$t^2 - 3t + 2 = 0$$
$$(t - 2)(t - 1) = 0 \longrightarrow t - 2 = 0 \text{ or } t - 1 = 0$$
$$t = 2 \text{ or } t = 1$$

Check the possible solutions in the *original* equation.

When $t = 2$: $\frac{12}{2^2 - 4} - \frac{3}{2 - 2} \stackrel{?}{=} -1$

$\frac{12}{0} - \frac{3}{0} \stackrel{?}{=} -1$

not defined

2 is an extraneous root.

When $t = 1$: $\frac{12}{1^2 - 4} - \frac{3}{1 - 2} \stackrel{?}{=} -1$

$\frac{12}{-3} - \frac{3}{-1} \stackrel{?}{=} -1$

$-4 + 3 = -1$ ✓

1 is a root of the original equation.

∴ the solution set is $\{1\}$.

Solve and check. If an equation has no solution, say so.

1. $\frac{3}{y} - \frac{1}{2y} = \frac{5}{4}$
2. $\frac{4}{3z} + \frac{2}{z} = \frac{5}{6}$
3. $\frac{6}{x + 1} = \frac{3}{x - 2}$
4. $\frac{12}{a} = \frac{4}{a - 4}$
5. $\frac{3}{r - 3} + 9 = \frac{r}{r - 3}$
6. $\frac{12}{n} = \frac{12}{n + 1} + 1$
7. $\frac{6p}{2p - 1} - 3 = \frac{3}{p}$
8. $\frac{7}{k - 3} - \frac{3}{k - 4} = \frac{1}{2}$
9. $\frac{9}{m + 5} - \frac{1}{m - 5} = \frac{3m}{m^2 - 25}$
10. $\frac{60}{d^2 - 36} + 1 = \frac{5}{d - 6}$
11. $\frac{2}{b^2 - 2b} - \frac{1}{b} = \frac{1}{3}$
12. $\frac{5}{x - 2} + \frac{x^2 - 4}{x^2 + 3x - 10} = \frac{x}{x + 5}$

NAME ______________________ DATE ______________

5-9 *Fractional Equations* *(continued)*

Vocabulary

Work rate The fractional part of a job done in a given unit of time.
Example: Lenny can paint a room in 3 h. His work rate is $\frac{1}{3}$ job per hour.

Special rate formulas work rate × time = work done rate × time = distance

Complete each table and solve.

13. Stan can load his truck in 24 min. If Chris helps him, it takes 15 min to load the truck. How long does it take Chris alone?

Let x = the time it takes Chris alone.

	Work rate × Time = Work done		
Stan	?	15	?
Chris	$\frac{1}{x}$	15	?

Stan's part of job + Chris' part of job = Whole job
? + ? = 1

14. Bonnie can complete her paper route in 45 min. When her sister Jean helps her it takes them 18 min to complete the route. How long would it take Jean alone?

Let x = the time it takes Jean alone.

	Work rate × Time = Work done		
Bonnie	?	18	?
Jean	$\frac{1}{x}$	18	?

15. An express train travels 150 km in the same time that a freight train travels 100 km. The average speed of the freight train is 20 km/h less than that of the express train. Find the speed of each train.

Use the fact that time $= \frac{\text{distance}}{\text{rate}}$.

	Distance	Rate	Time
Express	?	r	?
Freight	?	$r - 20$	?

time for express train = time for freight train

16. Helen can ride 15 km on her bicycle in the same time it takes her to walk 6 km. If her rate riding is 6 km/h faster than her rate walking, how fast does she walk?

	Distance	Rate	Time
Riding	?	?	?
Walking	?	r	?

Mixed Review Exercises

Simplify.

1. $\frac{x^2 - 4}{2 - x}$ **2.** $\frac{72m^2n^3}{27mn^4}$ **3.** $\frac{1 + a^{-1}}{a^{-2} - 1}$ **4.** $\frac{k^2 - k - 6}{k^2 - 2k - 8}$

NAME ______________________ DATE ____________

6 Irrational and Complex Numbers

6–1 Roots of Real Numbers

Objective: To find roots of real numbers.

Vocabulary

Square root A square root of a number b is a solution of the equation $x^2 = b$.
Example: The square roots of 36 are $\sqrt{36} = 6$ and $-\sqrt{36} = -6$, since $6^2 = 36$ and $(-6)^2 = 36$.

Cube root A cube root of a number b is a solution of the equation $x^3 = b$.
Example: The cube root of -125 is $\sqrt[3]{-125} = -5$, since $(-5)^3 = -125$.

***n*th root** An nth root of b is a solution of the equation $x^n = b$, where n is a positive integer.

1. a. If n is even and $b > 0$, there are two real nth roots of b, and they are opposites.
   b. If n is even and $b = 0$, there is one nth root of b, namely zero.
   c. If n is even and $b < 0$, there is no real nth root of b.
2. If n is odd, there is exactly one real nth root of b.

Radical symbol The symbol $\sqrt[n]{b}$, where n is the *index* (a positive integer), b is the *radicand*, and $\sqrt{}$ is the *radical sign*. It denotes the positive or *principal* nth root of b.

Properties of radicals	Examples	
1. $(\sqrt[n]{b})^n = b$, because $\sqrt[n]{b}$ satisfies the equation $x^n = b$.	$(\sqrt{11})^2 = 11$	$(\sqrt[5]{-2})^5 = -2$
2. $\sqrt[n]{b^n} = b$ if n is odd.	$\sqrt[5]{7^5} = 7$	$\sqrt[3]{y^3} = y$
3. $\sqrt[n]{b^n} = \lvert b\rvert$ if n is even, because the principal nth root is always nonnegative for even values of n.	$\sqrt{(-5)^2} = \lvert -5\rvert = 5$	$\sqrt[4]{x^4} = \lvert x\rvert$

CAUTION When *solving* an equation such as $x^2 = 4$, you must be sure to include both solutions: $x = \pm\sqrt{4} = \pm 2$ (read "plus-or-minus 2"). When *simplifying* an expression like $\sqrt{4}$, you must only give the principal square root: $\sqrt{4} = 2$.

Example 1 Simplify.

a. $\sqrt{49} = 7$ **b.** $-\sqrt{49} = -7$ **c.** $\sqrt{\dfrac{1}{49}} = \dfrac{1}{7}$ **d.** $\sqrt{0.49} = 0.7$

Example 2 Simplify: **a.** $\sqrt[5]{-243}$ **b.** $\sqrt{7^{-2}}$ **c.** $\sqrt[10]{a^{10}}$ **d.** $\sqrt[4]{a^8}$

Solution **a.** Since $-243 = (-3)^5$, $\sqrt[5]{-243} = -3$ **b.** Since $7^{-2} = \dfrac{1}{7^2} = \dfrac{1}{49}$, $\sqrt{7^{-2}} = \dfrac{1}{7}$

c. By property 3 above, $\sqrt[10]{a^{10}} = \lvert a\rvert$ **d.** Since $a^8 = (a^2)^4$, $\sqrt[4]{a^8} = a^2$

NAME ______________________ DATE ______________

6–1 Roots of Real Numbers *(continued)*

Simplify each expression that has a real root. If the expression does not represent a real number, say so.

1. a. $\sqrt{25}$ b. $-\sqrt{25}$ c. $\sqrt{-25}$ d. $\sqrt{0.25}$
2. a. $\sqrt{100}$ b. $\sqrt{-100}$ c. $-\sqrt{100}$ d. $\sqrt[4]{-100}$
3. a. $\sqrt{0.81}$ b. $-\sqrt{0.81}$ c. $\sqrt{-0.81}$ d. $\sqrt[4]{0.0081}$
4. a. $\sqrt{8^2}$ b. $\sqrt{-8^2}$ c. $\sqrt[4]{(-8)^4}$ d. $\sqrt[5]{(-8)^5}$
5. a. $\sqrt{\frac{1}{81}}$ b. $\sqrt{\frac{16}{81}}$ c. $\sqrt[4]{\frac{1}{81}}$ d. $\sqrt[4]{\frac{16}{81}}$
6. a. $\sqrt{6^2}$ b. $\sqrt{6^4}$ c. $\sqrt{6^{12}}$ d. $\sqrt{6^{24}}$
7. a. $\sqrt[5]{6^{-5}}$ b. $\sqrt[5]{6^{-10}}$ c. $\sqrt[5]{6^{-15}}$ d. $\sqrt[5]{6^{-25}}$
8. a. $\sqrt{a^{12}}$ b. $\sqrt[4]{a^{12}}$ c. $\sqrt[3]{a^{12}}$ d. $\sqrt[12]{a^{12}}$
9. a. $\sqrt{-a^4}$ b. $\sqrt{(-a)^4}$ c. $\sqrt[3]{(-a)^3}$ d. $\sqrt[8]{a^8}$

Example 3 Find the real roots of each equation. If there are none, say so.

a. $x^2 = 16$ b. $x^2 + 25 = 0$ c. $4x^2 = 12$

Solution

a. $x^2 = 16$
$x = \pm\sqrt{16} = \pm 4$
The roots are 4 and -4.

b. $x^2 + 25 = 0$
$x^2 = -25$
There are no real roots.

c. $4x^2 = 12$
$x^2 = 3$
$x = \pm\sqrt{3}$
The roots are $\sqrt{3}$ and $-\sqrt{3}$.

Find the real roots of each equation. If there are none, say so.

10. $x^2 = 121$ 11. $x^2 = 1$ 12. $x^2 + 16 = 0$ 13. $x^2 - 11 = 0$
14. $16x^2 = 9$ 15. $36x^2 = -49$ 16. $-9 = -25x^2$ 17. $25x^2 - 100 = 0$
18. $1 - 4x^2 = 0$ 19. $0 = 1 + 4x^2$ 20. $25 - 100x^2 = 0$ 21. $16x^2 + 8 = 9$

Mixed Review Exercises

Solve. If an equation has no solution, say so.

1. $\frac{x}{x-1} = \frac{3x}{x+3}$ 2. $-5y = y^2$ 3. $\frac{2x+1}{2} + \frac{x-1}{6} = -2$
4. $|3 - 2x| = 7$ 5. $4x^2 = 5x + 6$ 6. $\frac{x^2 - x}{2} = \frac{x+1}{3}$
7. $5(2x - 3) = 6x + 5$ 8. $2x + 6 = 2(x + 3)$ 9. $\frac{1}{x+1} - \frac{1}{x} = \frac{1}{x^2 + x}$

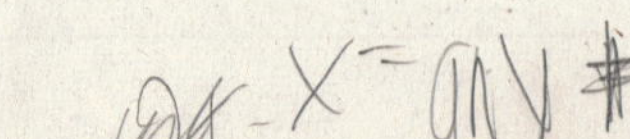

NAME ______________________________ DATE ______________

6–2 Properties of Radicals

Objective: To simplify expressions involving radicals.

Vocabulary

Rationalizing the denominator A process for eliminating radicals in denominators and fractions in radicands.

Perfect *n*th power A number whose *n*th root is a whole number.

Examples: 196 is a *perfect square,* since $\sqrt{196} = 14$.
8 is a *perfect cube,* since $\sqrt[3]{8} = 2$.
625 is a *perfect* 4th *power,* since $\sqrt[4]{625} = 5$.

Simplest radical form A radical expression is in simplest radical form if no radicand contains a factor (other than 1) that is a perfect *n*th power, and every denominator has been rationalized.

Product and quotient properties of radicals If $\sqrt[n]{a}$ and $\sqrt[n]{b}$ are real numbers, then:

1. $\sqrt[n]{ab} = \sqrt[n]{a} \cdot \sqrt[n]{b}$ 2. $\sqrt[n]{\frac{a}{b}} = \frac{\sqrt[n]{a}}{\sqrt[n]{b}}$ $(b \neq 0)$

Theorem 1 If each radical represents a real number, then $\sqrt[nq]{b} = \sqrt[n]{\sqrt[q]{b}}$.

Theorem 2 If $\sqrt[n]{b}$ represents a real number, then $\sqrt[n]{b^m} = (\sqrt[n]{b})^m$.

Example 1 Simplify: **a.** $\sqrt{50}$ **b.** $\sqrt{\frac{12}{25}}$ **c.** $\frac{10}{\sqrt{5}}$

Solution

a. $\sqrt{50} = \sqrt{25 \cdot 2}$
$= \sqrt{25} \cdot \sqrt{2}$
$= 5\sqrt{2}$

Find the *largest* perfect square that is a factor of the radicand. Then apply the product property of radicals.

b. $\sqrt{\frac{12}{25}} = \frac{\sqrt{12}}{\sqrt{25}}$
$= \frac{\sqrt{4 \cdot 3}}{5}$
$= \frac{2\sqrt{3}}{5}$

Apply the quotient property of radicals. Then find the square root of the numerator, and the square root of the denominator.

c. $\frac{10}{\sqrt{5}} = \frac{10}{\sqrt{5}} \cdot \frac{\sqrt{5}}{\sqrt{5}}$
$= \frac{10\sqrt{5}}{\sqrt{25}}$
$= \frac{10\sqrt{5}}{5} = 2\sqrt{5}$

Multiply the numerator and denominator by $\sqrt{5}$ so that the radicand in the denominator is a perfect square. This is an example of *rationalizing the denominator*.

NAME ______________________________ DATE ______________

6–2 Properties of Radicals *(continued)*

Example 2 Simplify: **a.** $\sqrt[3]{16} \cdot \sqrt[3]{12}$ **b.** $\dfrac{\sqrt[3]{140}}{\sqrt[3]{60}}$

Solution Use the strategies shown in Example 1, but look for perfect cubes.

a. $\sqrt[3]{16} \cdot \sqrt[3]{12} = \sqrt[3]{(4 \cdot 4)(4 \cdot 3)} = \sqrt[3]{4^3} \cdot \sqrt[3]{3} = 4\sqrt[3]{3}$

b. $\dfrac{\sqrt[3]{140}}{\sqrt[3]{60}} = \sqrt[3]{\dfrac{140}{60}} = \sqrt[3]{\dfrac{7}{3}} = \sqrt[3]{\dfrac{7}{3} \cdot \dfrac{3^2}{3^2}} = \sqrt[3]{\dfrac{63}{27}} = \dfrac{\sqrt[3]{63}}{\sqrt[3]{27}} = \dfrac{\sqrt[3]{63}}{3}$

Simplify.

1. $\sqrt{56}$
2. $\sqrt{\dfrac{20}{9}}$
3. $\sqrt{\dfrac{25}{11}}$
4. $\dfrac{8}{\sqrt{6}}$
5. $\dfrac{\sqrt{104}}{\sqrt{13}}$
6. $\sqrt{70} \cdot \sqrt{21}$
7. $\sqrt{125} \cdot \sqrt{10}$
8. $\sqrt{21} \cdot \sqrt{\dfrac{3}{7}}$
9. $\sqrt[3]{500}$
10. $\sqrt[3]{\dfrac{3}{16}}$
11. $\dfrac{10\sqrt{6}}{\sqrt{24}}$
12. $(5\sqrt{3})^2$
13. $\sqrt[3]{15} \cdot \sqrt[3]{18}$
14. $\dfrac{\sqrt[3]{120}}{\sqrt[3]{48}}$
15. $\sqrt[4]{81}$

Example 3 Simplify. Assume that each radical represents a real number.

a. $\sqrt{18x^9}$ **b.** $\sqrt[3]{\dfrac{8a^3}{25b^5}}$

Solution

a. $\sqrt{18x^9} = \sqrt{9x^8 \cdot 2x}$
$= \sqrt{9x^8} \cdot \sqrt{2x}$
$= 3x^4\sqrt{2x}$

Factor the radicand into two factors, one of which is a perfect square. Then, simplify by applying the product property of radicals.

b. $\sqrt[3]{\dfrac{8a^3}{25b^5}} = \dfrac{\sqrt[3]{8a^3}}{\sqrt[3]{25b^5}} = \dfrac{2a}{\sqrt[3]{25b^5}}$

Notice that the numerator is a perfect cube and simplify it.

$= \dfrac{2a}{\sqrt[3]{25b^5}} \cdot \dfrac{\sqrt[3]{5b}}{\sqrt[3]{5b}}$

$= \dfrac{2a\sqrt[3]{5b}}{\sqrt[3]{125b^6}}$

$= \dfrac{2a\sqrt[3]{5b}}{5b^2}$

Rationalize the denominator by multiplying the numerator and denominator by $\sqrt[3]{5b}$, making the radicand in the denominator a perfect cube.

Simplify. Assume that each radical represents a real number.

16. $\sqrt{48x^2}$
17. $\sqrt{54x^7}$

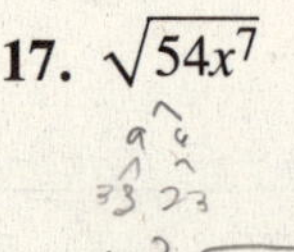

18. $\sqrt[3]{128a^5}$

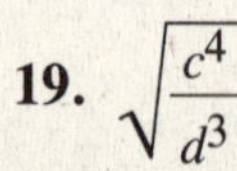

19. $\sqrt{\dfrac{c^4}{d^3}}$
20. $\sqrt[3]{\dfrac{125x}{16y^8}}$

NAME ________________ DATE ________

6–3 *Sums of Radicals*

Objective: To simplify expressions involving sums of radicals.

Vocabulary

Like radicals Two radicals with the same index and radicand.

Examples: $2\sqrt{3}$ and $5\sqrt{3}$ are like radicals, as are $\sqrt[3]{4}$ and $7\sqrt[3]{4}$; $\sqrt{6}$ and $\sqrt{3}$ are unlike radicals.

Example 1 Simplify $\sqrt{48} + \sqrt{27}$.

Solution

$$\sqrt{48} + \sqrt{27} = \sqrt{16 \cdot 3} + \sqrt{9 \cdot 3}$$ First simplify each radical.

$$= \sqrt{16} \cdot \sqrt{3} + \sqrt{9} \cdot \sqrt{3}$$

$$= 4\sqrt{3} + 3\sqrt{3}$$ Use the distributive property to add like radicals.

$$= (4 + 3)\sqrt{3} = 7\sqrt{3}$$

Example 2 Simplify $\sqrt[3]{135} - \sqrt[3]{40} + \sqrt[3]{2}$.

Solution

$$\sqrt[3]{135} - \sqrt[3]{40} + \sqrt[3]{2} = \sqrt[3]{27 \cdot 5} - \sqrt[3]{8 \cdot 5} + \sqrt[3]{2}$$ Simplify each radical.

$$= (\sqrt[3]{27} \cdot \sqrt[3]{5}) - (\sqrt[3]{8} \cdot \sqrt[3]{5}) + \sqrt[3]{2}$$

$$= 3\sqrt[3]{5} - 2\sqrt[3]{5} + \sqrt[3]{2}$$ Combine like radicals.

$$= \sqrt[3]{5} + \sqrt[3]{2}$$

Example 3 Simplify $\sqrt{\frac{7}{2}} + \sqrt{\frac{8}{7}}$.

Solution

$$\sqrt{\frac{7}{2}} + \sqrt{\frac{8}{7}} = \sqrt{\frac{7}{2} \cdot \frac{2}{2}} + \sqrt{\frac{8}{7} \cdot \frac{7}{7}}$$ Rationalize the denominator of each radical.

$$= \sqrt{\frac{14}{4}} + \sqrt{\frac{56}{49}}$$

$$= \frac{\sqrt{14}}{\sqrt{4}} + \frac{\sqrt{56}}{\sqrt{49}}$$ Simplify each radical.

$$= \frac{\sqrt{14}}{2} + \frac{\sqrt{4 \cdot 14}}{7}$$

$$= \frac{\sqrt{14}}{2} + \frac{2\sqrt{14}}{7}$$

$$= \frac{7\sqrt{14}}{14} + \frac{4\sqrt{14}}{14} = \frac{11\sqrt{14}}{14}$$ Express each fraction with a common denominator. Add.

NAME ______________________ DATE ______________

6–3 Sums of Radicals *(continued)*

Simplify. If no simplification is possible, say so.

1. $\sqrt{27} + \sqrt{12}$
2. $\sqrt{125} - 2\sqrt{80}$
3. $5\sqrt{6} - 6\sqrt{5}$
4. $\sqrt{28} + \sqrt{63}$
5. $\sqrt{32} - \sqrt{50} + \sqrt{98}$
6. $\sqrt[3]{54} - \sqrt[3]{16} + \sqrt[3]{27}$
7. $\sqrt[3]{32} + \sqrt[3]{64} + \sqrt[3]{108}$
8. $\sqrt{\frac{50}{3}} - \sqrt{\frac{2}{3}}$
9. $\sqrt{\frac{3}{5}} + \sqrt{\frac{5}{3}}$
10. $\sqrt[3]{9} + \sqrt[3]{\frac{1}{3}}$

Example 4 Simplify $\sqrt{15}(2\sqrt{3} - \sqrt{5})$.

Solution

$$\begin{aligned}\sqrt{15}\,(2\sqrt{3} - \sqrt{5}) &= (\sqrt{15}\cdot 2\sqrt{3}) - (\sqrt{15}\cdot\sqrt{5}) \\ &= 2\sqrt{45} - \sqrt{75} \\ &= 2\sqrt{9\cdot 5} - \sqrt{25\cdot 3} \\ &= 2\cdot 3\sqrt{5} - 5\sqrt{3} = 6\sqrt{5} - 5\sqrt{3}\end{aligned}$$

Multiply using the distributive property.

Simplify each radical.

Example 5 Simplify $\dfrac{\sqrt{12} + 3\sqrt{50}}{\sqrt{2}}$.

Solution

$$\begin{aligned}\frac{\sqrt{12} + 3\sqrt{50}}{\sqrt{2}} &= \frac{\sqrt{12}}{\sqrt{2}} + \frac{3\sqrt{50}}{\sqrt{2}} \\ &= \sqrt{\frac{12}{2}} + 3\sqrt{\frac{50}{2}} \\ &= \sqrt{6} + 3\sqrt{25} \\ &= \sqrt{6} + 3\cdot 5 = \sqrt{6} + 15\end{aligned}$$

Divide each term in the numerator by the radical in the denominator, and simplify.

Simplify.

11. $\sqrt{2}(\sqrt{32} + \sqrt{12})$
12. $\sqrt{11}(2\sqrt{66} - 3\sqrt{55})$
13. $3\sqrt{2}\,(\sqrt{18} - 5\sqrt{12})$
14. $\dfrac{\sqrt{10} - \sqrt{90}}{\sqrt{5}}$
15. $\dfrac{7\sqrt{72} - \sqrt{216}}{\sqrt{12}}$
16. $\sqrt{\frac{5}{3}}\left(\sqrt{\frac{5}{9}} + \frac{4}{\sqrt{5}}\right)$

Mixed Review Exercises

Simplify.

1. $\sqrt{44x^3}$
2. $\sqrt{\frac{64}{121}}$
3. $\dfrac{a^{[illegible]}}{[illegible]x^{[illegible]}y^{[illegible]}}$
4. $\sqrt{\frac{56x}{75y^2}}$
5. $\sqrt[3]{-72}$
6. $\dfrac{x^2 - 5x + 6}{3 - x}$
7. $(5x + 2)(5x - 2)$
8. $\sqrt{\frac{3}{x}}$
9. $\sqrt[3]{64x^5y^6}$

NAME ______________________ DATE ______________

6–4 Binomials Containing Radicals

Objective: To simplify products and quotients of binomials that contain radicals.

Vocabulary

Conjugates A pair of expressions of the form $a\sqrt{b} + c\sqrt{d}$ and $a\sqrt{b} - c\sqrt{d}$.

Examples: $3\sqrt{6} + 2\sqrt{5}$ is the conjugate of $3\sqrt{6} - 2\sqrt{5}$.

The product of conjugates is always an integer when a, b, c, and d are integers. For this reason conjugates are used to rationalize denominators that contain a binomial radical expression.

Example 1 Simplify: **a.** $(2 + \sqrt{5})(3 + 2\sqrt{5})$

b. $(5\sqrt{2} - \sqrt{6})^2$

c. $(5\sqrt{7} + 2\sqrt{3})(5\sqrt{7} - 2\sqrt{3})$

Solution **a.** Recall the FOIL (**F**irst, **O**uter, **I**nner, **L**ast) method used to multiply binomials. For example:

O, I, F, L

$$(2 + x)(3 + 2x) = 6 + 4x + 3x + 2x^2$$
$$= 6 + 7x + 2x^2$$

Use the FOIL method to simplify the given expression.

$$\therefore (2 + \sqrt{5})(3 + 2\sqrt{5}) = 6 + 4\sqrt{5} + 3\sqrt{5} + 2(\sqrt{5})^2$$
$$= 6 + 7\sqrt{5} + 10$$
$$= 16 + 7\sqrt{5}$$

b. Recall this pattern:

$$(a - b)^2 = a^2 - 2ab + b^2$$

$$\therefore (5\sqrt{2} - \sqrt{6})^2 = (5\sqrt{2})^2 - 2(5\sqrt{2})(\sqrt{6}) + (\sqrt{6})^2$$
$$= 25 \cdot 2 - 10\sqrt{12} + 6$$
$$= 50 - 10\sqrt{4 \cdot 3} + 6$$
$$= 50 - 20\sqrt{3} + 6$$
$$= 56 - 20\sqrt{3}$$

c. Recall this pattern:

$$(a + b)(a - b) = a^2 - b^2$$

$$\therefore (5\sqrt{7} + 2\sqrt{3})(5\sqrt{7} - 2\sqrt{3}) = (5\sqrt{7})^2 - (2\sqrt{3})^2$$
$$= 25 \cdot 7 - 4 \cdot 3$$
$$= 175 - 12$$
$$= 163$$

6–4 Binomials Containing Radicals *(continued)*

Example 2 Simplify: **a.** $\dfrac{3}{\sqrt{5} + \sqrt{3}}$ **b.** $\dfrac{4 + \sqrt{3}}{3 - 2\sqrt{2}}$

Solution Multiply the numerator and denominator of each fraction by the conjugate of the denominator to rationalize the denominator. Use the patterns illustrated in Example 1 to multiply binomials.

a.
$$\frac{3}{\sqrt{5} + \sqrt{3}} = \frac{3}{\sqrt{5} + \sqrt{3}} \cdot \frac{\sqrt{5} - \sqrt{3}}{\sqrt{5} - \sqrt{3}}$$
$$= \frac{3\sqrt{5} - 3\sqrt{3}}{(\sqrt{5})^2 - (\sqrt{3})^2}$$
$$= \frac{3\sqrt{5} - 3\sqrt{3}}{5 - 3}$$
$$= \frac{3\sqrt{5} - 3\sqrt{3}}{2}$$

b.
$$\frac{4 + \sqrt{3}}{3 - 2\sqrt{2}} = \frac{4 + \sqrt{3}}{3 - 2\sqrt{2}} \cdot \frac{3 + 2\sqrt{2}}{3 + 2\sqrt{2}}$$
$$= \frac{12 + 8\sqrt{2} + 3\sqrt{3} + 2\sqrt{6}}{3^2 - (2\sqrt{2})^2}$$
$$= \frac{12 + 8\sqrt{2} + 3\sqrt{3} + 2\sqrt{6}}{9 - 8}$$
$$= 12 + 8\sqrt{2} + 3\sqrt{3} + 2\sqrt{6}$$

Simplify.

1. $(4 + \sqrt{6})(4 - \sqrt{6})$

2. $(\sqrt{11} + 2)^2$

3. $(3 - \sqrt{5})(4 - \sqrt{5})$

4. $\dfrac{5}{4 + \sqrt{3}}$

5. $(\sqrt{7} - \sqrt{3})^2$

6. $(\sqrt{6} - 2\sqrt{5})^2$

7. $(2 + 3\sqrt{2})(3 + \sqrt{2})$

8. $(\sqrt{10} - \sqrt{3})(\sqrt{10} + \sqrt{3})$

9. $\dfrac{4}{\sqrt{7} - \sqrt{3}}$

10. $(5 + 2\sqrt{3})(4 - 3\sqrt{3})$

11. $\dfrac{\sqrt{14}}{\sqrt{2} + \sqrt{7}}$

12. $(2\sqrt{10} - \sqrt{5})^2$

13. $(3\sqrt{2} - 2\sqrt{5})(3\sqrt{2} + 2\sqrt{5})$

14. $(6\sqrt{7} + \sqrt{14})(\sqrt{7} - \sqrt{2})$

15. $(2\sqrt{10} + 3\sqrt{15})^2$

16. $\dfrac{\sqrt{6} + \sqrt{5}}{2} \cdot \dfrac{\sqrt{6} - \sqrt{5}}{2}$

17. $\dfrac{2\sqrt{11} - 2}{3} \cdot \dfrac{2\sqrt{11} + 2}{3}$

18. $(\sqrt{6} - \sqrt{21})^2$

19. $\dfrac{\sqrt{2} + 1}{\sqrt{3} - 4}$

20. $\dfrac{8\sqrt{3} - \sqrt{2}}{\sqrt{3} + \sqrt{2}}$

21. $\dfrac{2 + \sqrt{5}}{\sqrt{3} - \sqrt{5}}$

NAME ______________________ DATE ______________

6–5 *Equations Containing Radicals*

Objective: To solve equations containing radicals.

Vocabulary

Radical equation An equation containing a radical with a variable in the radicand.

CAUTION Since you may get an extraneous root when you solve a radical equation, make sure that you check all answers in the *original* equation.

Example 1 Solve.

a. $\sqrt{2x+1} = 5$ **b.** $3 + \sqrt[3]{x} = 5$ **c.** $x - \sqrt{x-1} = 3$

Solution Follow these steps: (1) transform the equation to isolate the radical term, (2) eliminate the radical by squaring or cubing both sides, and (3) solve for x.

a.

$$\sqrt{2x+1} = 5$$
$$(\sqrt{2x+1})^2 = 5^2$$
$$2x + 1 = 25$$
$$2x = 24$$
$$x = 12$$

Check: $\sqrt{2 \cdot 12 + 1} \stackrel{?}{=} 5$

$\sqrt{25} = 5$ ✓

$\therefore$ the solution set is $\{12\}$.

b.

$$3 + \sqrt[3]{x} = 5$$
$$\sqrt[3]{x} = 2$$
$$(\sqrt[3]{x})^3 = 2^3$$
$$x = 8$$

Check: $3 + \sqrt[3]{8} \stackrel{?}{=} 5$

$3 + 2 = 5$ ✓

$\therefore$ the solution set is $\{8\}$.

c.

$$x - \sqrt{x-1} = 3$$
$$x - 3 = \sqrt{x-1}$$
$$(x-3)^2 = (\sqrt{x-1})^2$$
$$x^2 - 6x + 9 = x - 1$$
$$x^2 - 7x + 10 = 0$$
$$(x-5)(x-2) = 0$$
$$x = 5 \text{ or } x = 2$$

Check: $5 - \sqrt{5-1} \stackrel{?}{=} 3$

$5 - 2 = 3$ ✓

$2 - \sqrt{2-1} \stackrel{?}{=} 3$

$2 - 1 \neq 3$

$\therefore$ 2 is an extraneous solution, and the solution set is $\{5\}$.

Solve. If an equation has no real solution, say so.

1. $\sqrt{x-5} = 3$ **2.** $\sqrt{3x-5} = 4$ **3.** $2\sqrt{x} - 3 = 5$ **4.** $10 + 3\sqrt{x} = 1$

5. $\sqrt{3x^2+4} = 4$ **6.** $\sqrt[3]{7x+1} = 2$ **7.** $\sqrt[3]{4x} + 7 = 5$ **8.** $2\sqrt[3]{5x} = \sqrt[3]{4x+72}$

9. $\sqrt{3x+4} = x$ **10.** $\sqrt{x+3} - x = 3$ **11.** $\sqrt{10x+15} - 4 = x$

6–5 Equations Containing Radicals *(continued)*

Example 2 Solve $5x = 1 + x\sqrt{3}$ without squaring both sides.

Solution

$$5x = 1 + x\sqrt{3}$$
$$5x - x\sqrt{3} = 1$$
$$(5 - \sqrt{3})x = 1$$
$$x = \frac{1}{5 - \sqrt{3}} \cdot \frac{5 + \sqrt{3}}{5 + \sqrt{3}}$$
$$= \frac{5 + \sqrt{3}}{22}$$

Note: The equation at the left is *not* a radical equation because there is no variable in the radicand. Therefore, it is not necessary to square both sides to solve the equation.

The solution set is $\left\{\frac{5 + \sqrt{3}}{22}\right\}$.

In Exercises 12–14, a radical and a linear equation are given. Solve the radical equation by squaring both sides. Solve the linear equation without squaring both sides.

12. a. $2\sqrt{x} = 6$
b. $x\sqrt{2} = 6$

13. a. $4 + 2\sqrt{x} = 14$
b. $4 + x\sqrt{2} = 14$

14. a. $x = 8 + 2\sqrt{x}$
b. $x = 8 + x\sqrt{2}$

Example 3 Solve $\sqrt{2x - 4} - \sqrt{x - 3} = 1$.

Solution

$\sqrt{2x - 4} = 1 + \sqrt{x - 3}$	Isolate one radical term.
$2x - 4 = 1 + 2\sqrt{x - 3} + x - 3$	Square both sides.
$x - 2 = 2\sqrt{x - 3}$	Isolate the other radical term.
$x^2 - 4x + 4 = 4x - 12$	Square both sides again.
$x^2 - 8x + 16 = 0$	Solve for x.
$(x - 4)^2 = 0$	The check is left for you.
$x = 4$	The solution set is $\{4\}$.

Solve. If an equation has no real solution, say so.

15. $\sqrt{2y - 7} + \sqrt{2y} = 7$ **16.** $\sqrt{4x - 7} = \sqrt{2x} + 1$ **17.** $\sqrt{3a + 5} - \sqrt{a - 4} = 2$

Mixed Review Exercises

Simplify.

1. $\frac{\sqrt{42} - \sqrt{21}}{\sqrt{7}}$ **2.** $\sqrt{15}(2\sqrt{3} - 3\sqrt{5})$ **3.** $\frac{a^3 + b^3}{a + b}$ **4.** $\frac{3}{5 - \sqrt{3}}$

5. $(x^2y^3)^2(-xy^2)^3$ **6.** $(\sqrt{3} + \sqrt{15})^2$ **7.** $\frac{5\sqrt{32}}{4\sqrt{125}}$ **8.** $\sqrt{45} - \sqrt{20}$

9. $(2 - x)^{-2}(x - 2)$ **10.** $(3y - 5)(y^2 - 2y + 4)$ **11.** $\sqrt[3]{54} + \sqrt[3]{24} + \sqrt[3]{375}$

NAME ______________________ DATE ____________

6–6 *Rational and Irrational Numbers*

Objective: To find and use decimal representations of real numbers.

Vocabulary

Rational number A number that can be written as a fraction with an integer in the numerator and a nonzero integer in the denominator. The decimal equivalent of such a number either terminates (ends) or repeats. Examples:

$\frac{3}{4} = 0.75$ (a *terminating*, or *finite*, *decimal*) $\frac{5}{3} = 1.\overline{6}$ (a *repeating*, or *infinite*, *decimal*)

Irrational number A real number that is not rational. The decimal equivalent of such a number is infinite and nonrepeating. Examples: π $6\sqrt{2}$ $0.525225222\ldots$

Completeness property of real numbers Every real number has a decimal representation, and every decimal represents a real number.

CAUTION Since calculators have finite displays, calculators can give only a *rational approximation* of the decimal representation of an irrational number.

Example 1 Classify each real number as either rational or irrational.

a. $\sqrt{3}$ **b.** $\sqrt{\frac{9}{16}}$ **c.** $3.712712\ldots$

Solution **a.** Since $\sqrt{3} = 1.7320508\ldots$, an infinite, nonrepeating decimal, $\sqrt{3}$ is *irrational*. (Note: $\sqrt{3}$ can't be written as the quotient of two integers.)

b. Since $\sqrt{\frac{9}{16}} = \frac{3}{4} = 0.75$, a terminating decimal, $\sqrt{\frac{9}{16}}$ is *rational*. (Note: $\frac{3}{4}$ is a fraction with integers in the numerator and denominator.)

c. Since $3.712712\ldots$ (or $3.\overline{712}$) is a repeating decimal, it is *rational*.

Classify each real number or expression as either rational or irrational.

1. a. $1.\overline{48}$ **b.** $1.481481148111\ldots$ **2. a.** $\sqrt{3} + \sqrt{27}$ **b.** $\sqrt{3} \cdot \sqrt{27}$

3. a. $\pi - \frac{1}{\pi}$ **b.** $\pi \div \pi$ **4. a.** $\sqrt{\frac{25}{64}}$ **b.** $\sqrt{\frac{1}{10}}$

Example 2 Write as a decimal: **a.** $\frac{9}{8}$ **b.** $\frac{3}{11}$

Solution **a.** $\frac{9}{8} = 9 \div 8 = 1.125$ **b.** $\frac{3}{11} = 3 \div 11 = 0.2727\ldots = 0.\overline{27}$

Write each fraction as a terminating or repeating decimal.

5. $\frac{2}{7}$ **6.** $\frac{7}{9}$ **7.** $\frac{15}{16}$ **8.** $\frac{32}{5}$

NAME ______________________________ DATE ______________

6–6 Rational and Irrational Numbers *(continued)*

Example 3 Write each terminating decimal as a common fraction in lowest terms.

a. 5.635 **b.** 0.0048

Solution **a.** The decimal place is thousandths. **b.** The decimal place is ten-thousandths.

$$5.635 = \frac{5635}{1000} = \frac{1127}{200} \qquad 0.0048 = \frac{48}{10{,}000} = \frac{3}{625}$$

Example 4 Write each repeating decimal as a common fraction in lowest terms.

a. $0.2\overline{13}$ **b.** $5.76\overline{576}$

Solution In each case let N be the number. Multiply the given number by 10^n where n is the number of digits in the block of repeating digits. Then subtract N. This will eliminate the repeating block. Finally, solve for N.

a.

$$\begin{aligned} N &= 0.2\overline{13} \\ 100N &= 21.3\overline{13} \\ -\quad N &= 0.2\overline{13} \\ \hline 99N &= 21.1 \end{aligned}$$

$$N = \frac{21.1}{99} = \frac{211}{990}$$

$$\therefore 0.2\overline{13} = \frac{211}{990}$$

b.

$$\begin{aligned} N &= 5.76\overline{576} \\ 1000N &= 5765.76\overline{576} \\ -\quad N &= 5.76\overline{576} \\ \hline 999N &= 5760 \end{aligned}$$

$$N = \frac{5760}{999} = \frac{640}{111}$$

$$\therefore 5.76\overline{576} = \frac{640}{111}$$

Write each decimal as a common fraction in lowest terms.

9. 2.005 **10.** 3.74 **11.** 5.0125 **12.** $0.8\overline{7}$ **13.** $3.\overline{72}$ **14.** $4.62\overline{462}$

Example 5 Find a rational number r and an irrational number s between $\sqrt{5}$ and $\sqrt{6}$.

Solution Begin by finding the decimal approximations for the radicals:

$$\sqrt{5} \approx 2.236068\ldots \quad \text{and} \quad \sqrt{6} \approx 2.4494897\ldots$$

Then find one finite or repeating decimal between them, and one infinite, nonrepeating decimal between them. Two possible answers are:

$$r = 2.\overline{37} \qquad \text{and} \qquad s = 2.301401501\ldots$$

Find (a) a rational and (b) an irrational number between each pair.

15. 0.7 and 0.8 **16.** $\sqrt{7}$ and $\sqrt{8}$ **17.** $\sqrt{8}$ and 3

18. 10^{-4} and 10^{-3} **19.** $2\frac{1}{12}$ and $2\frac{1}{11}$ **20.** $1.\overline{96}$ and $1.\overline{9}$

NAME ______________________ DATE ____________

6–7 The Imaginary Number i

Objective: To use the number i to simplify square roots of negative numbers.

Vocabulary

The number i The basic element of the set of *imaginary numbers.* It is defined as follows: $i = \sqrt{-1}$, and $i^2 = -1$. The number i is used to simplify square roots of negative numbers. For instance, if r is a positive real number, then $\sqrt{-r} = i\sqrt{r}$.

Examples: $\sqrt{-3} = i\sqrt{3}$ $\quad \sqrt{-36} = i\sqrt{36} = 6i$

Pure imaginary number Any number of the form bi, $b \neq 0$. Examples: $3i$ and $i\sqrt{7}$

CAUTION When a and b are negative, $\sqrt{a} \cdot \sqrt{b} \neq \sqrt{ab}$.

For example: $\sqrt{-4} \cdot \sqrt{-9} \neq \sqrt{36}$

Correct: $\sqrt{-4} \cdot \sqrt{-9} = 2i \cdot 3i = 6i^2 = 6(-1) = -6$

To avoid making mistakes, always express the square root of a negative number as a pure imaginary number before performing any other operation.

Example 1 Simplify: **a.** $\sqrt{-98}$ **b.** $\sqrt{-9} \cdot \sqrt{-25}$

Solution **a.** $\sqrt{-98} = i\sqrt{98}$
$= i\sqrt{49 \cdot 2}$
$= 7i\sqrt{2}$

b. $\sqrt{-9} \cdot \sqrt{-25} = i\sqrt{9} \cdot i\sqrt{25}$
$= 3i \cdot 5i$
$= 15i^2 = -15$

Simplify.

1. $\sqrt{-49}$ **2.** $\sqrt{-10}$ **3.** $-3\sqrt{-144}$ **4.** $\sqrt{-28}$

5. $3\sqrt{-12}$ **6.** $7i \cdot 5i$ **7.** $\sqrt{7} \cdot \sqrt{-14}$ **8.** $\sqrt{-6} \cdot \sqrt{-15}$

9. $(6i)^2$ **10.** $(-3i)^2$ **11.** $(-i\sqrt{5})^2$ **12.** $(2i\sqrt{3})^2$

Example 2 Simplify: **a.** $\frac{4}{5i}$ **b.** $\frac{10}{\sqrt{-5}}$

Solution To rationalize the denominator of a fraction, you must eliminate the imaginary number i from the denominator. Use the fact that $i^2 = -1$.

a. $\frac{4}{5i} = \frac{4}{5i} \cdot \frac{i}{i}$
$= \frac{4i}{5i^2}$
$= \frac{4i}{5(-1)}$
$= -\frac{4i}{5}$

b. $\frac{10}{\sqrt{-5}} = \frac{10}{i\sqrt{5}} = \frac{10}{i\sqrt{5}} \cdot \frac{i\sqrt{5}}{i\sqrt{5}}$
$= \frac{10i\sqrt{5}}{\sqrt{25}i^2}$
$= \frac{10i\sqrt{5}}{5(-1)}$
$= -2i\sqrt{5}$

NAME ______________________ DATE ______________

6–7 The Imaginary Number *i* *(continued)*

Simplify.

13. $\frac{5}{i}$ **14.** $\frac{6}{7i}$ **15.** $\frac{1}{\sqrt{-3}}$ **16.** $-\frac{9}{\sqrt{-9}}$

17. $\frac{\sqrt{24}}{3i\sqrt{8}}$ **18.** $\frac{\sqrt{48}}{2i\sqrt{3}}$ **19.** $\frac{\sqrt{56}}{\sqrt{-7}}$ **20.** $-\frac{\sqrt{21}}{\sqrt{-35}}$

Example 3 Solve $3x^2 + 23 = 5$.

Solution

$$3x^2 + 23 = 5$$
$$3x^2 = -18$$
$$x^2 = -6$$
$$x = \pm\sqrt{-6}$$
$$x = \pm i\sqrt{6}$$

$\therefore$ the solution set is $\{i\sqrt{6}, -i\sqrt{6}\}$.

Solve.

21. $x^2 + 100 = 0$

22. $y^2 + 81 = 0$

23. $2z^2 = -128$

24. $7a^2 = -28$

25. $3b^2 + 28 = 4$

26. $4x^2 + 78 = 6$

Example 4 Simplify: **a.** $\sqrt{-20} + \sqrt{-45}$ **b.** $\sqrt{-20} \cdot \sqrt{-45}$

Solution

a. $\sqrt{-20} + \sqrt{-45} = i\sqrt{20} + i\sqrt{45}$
$= 2i\sqrt{5} + 3i\sqrt{5}$
$= (2 + 3)i\sqrt{5}$
$= 5i\sqrt{5}$

b. $\sqrt{-20} \cdot \sqrt{-45} = i\sqrt{20} \cdot i\sqrt{45}$
$= 2i\sqrt{5} \cdot 3i\sqrt{5}$
$= [2 \cdot 3 \cdot (\sqrt{5})^2]i^2$
$= 30i^2 = -30$

Simplify.

27. a. $\sqrt{-5} + \sqrt{-80}$ **b.** $\sqrt{-5} \cdot \sqrt{-80}$ **28. a.** $4\sqrt{-3} - \sqrt{-75}$ **b.** $4\sqrt{-3} \cdot (-\sqrt{-75})$

29. a. $i\sqrt{27} + \sqrt{-12}$ **b.** $i\sqrt{27} \cdot \sqrt{-12}$ **30. a.** $2i\sqrt{32} - \sqrt{-72}$ **b.** $2i\sqrt{32} \cdot (-\sqrt{-72})$

Mixed Review Exercises

Solve. If an equation has no real solution, say so.

1. $\sqrt{3x + 1} = 5$ **2.** $n^2 = 9n - 20$ **3.** $\sqrt{z^2 + 75} = 2z$

4. $\frac{4y + 5}{7} = \frac{y - 1}{2}$ **5.** $\frac{1}{x} + \frac{3}{x - 3} = \frac{5}{x(x - 3)}$ **6.** $\sqrt[3]{t} + 12 = 7$

7. $x = \sqrt{x + 12}$ **8.** $2|m| - 5 = 1$ **9.** $w = 3 + \sqrt{w + 3}$

Classify each real number as either rational or irrational.

10. $\sqrt[3]{-8}$ **11.** $5.7\overline{62}$ **12.** $\sqrt{24}$ **13.** 0.10111213…

NAME ______________________ DATE ____________

6–8 The Complex Numbers

Objective: To add, subtract, multiply, and divide complex numbers.

Vocabulary

Imaginary number A number of the form $a + bi$, where a and b are real numbers and $b \neq 0$. Examples: $3 + 7i$ $\quad 2 + i\sqrt{5}$

Complex number A number of the form $a + bi$, where a and b are real numbers. The number a is called the *real part* of $a + bi$, and b (not bi) is called the *imaginary part*. Notice that when $b = 0$, the complex number $a + bi$ becomes the real number a. This shows that a complex number can be either real or imaginary.

Complex conjugates The numbers $a + bi$ and $a - bi$ are complex conjugates. Their product is the real number $a^2 + b^2$.

Equality of complex numbers $a + bi = c + di$ if and only if $a = c$ and $b = d$.

Sum of complex numbers $(a + bi) + (c + di) = (a + c) + (b + d)i$

Product of complex numbers $(a + bi)(c + di) = (ac - bd) + (ad + bc)i$
This results from using the FOIL method and the fact that $i^2 = -1$.

Example 1 Simplify: **a.** $(4 + 5i) + (8 - 7i)$ **b.** $3(2 + 4i) - 2(4 - 5i)$

Solution **a.** $(4 + 5i) + (8 - 7i) = (4 + 8) + (5 - 7)i = 12 - 2i$

b. $3(2 + 4i) - 2(4 - 5i) = (6 + 12i) - (8 - 10i)$
$= (6 - 8) + (12 + 10)i = -2 + 22i$

Example 2 Simplify: **a.** $(2 + 3i)(3 + 4i)$ **b.** $(2 + 5i)^2$ **c.** $(3 + 2i)(3 - 2i)$

Solution **a.** $(2 + 3i)(3 + 4i) = 6 + 8i + 9i + 12i^2$ Use the FOIL method.
$= 6 + 17i + 12(-1)$
$= -6 + 17i$

b. $(2 + 5i)^2 = 4 + 20i + 25i^2$ Use the pattern:
$= 4 + 20i + 25(-1)$ $(a + b)^2 = a^2 + 2ab + b^2$.
$= -21 + 20i$

c. $(3 + 2i)(3 - 2i) = 9 - 4i^2$ Use the pattern:
$= 9 - 4(-1)$ $(a + b)(a - b) = a^2 - b^2$.
$= 13$

Simplify.

1. $(7 + 3i) + (2 - 5i)$ **2.** $(4 - 7i) - (5 - 3i)$ **3.** $2(-3 + i) - 5(2 - 2i)$

4. $2i(4 + 7i)$ **5.** $-4i(-5 - i)$ **6.** $(2 + i)(2 - i)$

7. $(4 + 9i)(4 - 9i)$ **8.** $(-3 + 2i)(4 + 5i)$ **9.** $(-3 + i)(4 + 3i)$

10. $(2 - 5i)(3 + 4i)$ **11.** $(4 - i\sqrt{7})(4 + i\sqrt{7})$ **12.** $(3 - 5i)^2$

13. $(-2 + i\sqrt{5})^2$ **14.** $(1 + 4i)^2(1 - 4i)^2$ **15.** $(\sqrt{5} - \sqrt{-8})(\sqrt{5} + \sqrt{-8})$

NAME ______________________ DATE ____________

6-8 *The Complex Numbers* (continued)

Example 3 Simplify $\frac{6}{1 + 3i}$.

Solution $\frac{6}{1+3i} = \frac{6}{1+3i} \cdot \frac{1-3i}{1-3i}$ Multiply the numerator and denominator by the conjugate of the denominator. This will rationalize the denominator.

$= \frac{6 - 18i}{1 - 9i^2}$

$= \frac{6 - 18i}{1 - 9(-1)}$

$= \frac{6 - 18i}{10} = \frac{3}{5} - \frac{9}{5}i$ Give your answer in the form $a + bi$.

Simplify each quotient.

16. $\frac{3}{2 + 3i}$ **17.** $\frac{12}{1 - i}$ **18.** $\frac{4 + i}{4 - i}$ **19.** $\frac{-2 + i\sqrt{5}}{-2 - i\sqrt{5}}$

Example 4 Find the reciprocal of $2 - i$.

Solution The reciprocal of $2 - i$ is $\frac{1}{2 - i}$. The denominator must be rationalized.

$$\frac{1}{2-i} = \frac{1}{2-i} \cdot \frac{2+i}{2+i} = \frac{2+i}{4-i^2} = \frac{2+i}{4-(-1)} = \frac{2}{5} + \frac{1}{5}i$$

Find the reciprocal of each complex number.

20. $3 + 2i$ **21.** $2 - 5i$ **22.** $4 - i\sqrt{6}$ **23.** $-\sqrt{3} + i\sqrt{2}$

Example 5 If $f(x) = 2x - \frac{1}{x}$, find $f(3 + 2i)$.

Solution $f(3 + 2i) = 2(3 + 2i) - \frac{1}{3 + 2i}$ Substitute $3 + 2i$ for x.

$= 2(3 + 2i) - \frac{1}{3 + 2i} \cdot \frac{3 - 2i}{3 - 2i}$ Rationalize the denominator.

$= 6 + 4i - \frac{3 - 2i}{9 - (-4)}$ Multiply.

$= 6 + 4i - \frac{3 - 2i}{13}$

$= \frac{78 + 52i}{13} - \frac{3 - 2i}{13}$ Find a common denominator.

$= \frac{75}{13} + \frac{54}{13}i$ Simplify.

24. If $f(x) = \frac{1}{2x} + 2x$, find $f(1 + i)$. **25.** If $f(x) = \frac{x + 3}{x - 3}$, find $f(4 - i\sqrt{3})$.

NAME ______________________________ DATE ______________

7 Quadratic Equations and Functions

7–1 Completing the Square

Objective: To solve quadratic equations by completing the square.

Vocabulary

Quadratic equation An equation that can be written in the form $ax^2 + bx + c = 0$ $(a \neq 0)$.

Completing the square Changing the form of a quadratic equation into the form $(x + q)^2 = r$.

Solving $ax^2 + bx + c = 0$ by completing the square (See Examples 2 and 3.)

Step 1 Transform the equation so that the constant term c is alone on the right side.

Step 2 If a, the coefficient of the second-degree term, is not equal to 1, then divide both sides by a.

Step 3 Complete the square by adding half the coefficient of the first-degree term, $\left(\frac{b}{2a}\right)^2$, to *both sides*.

Step 4 Factor the left side as the square of a binomial.

Step 5 Complete the solution using the fact that $(x + q)^2 = r$ is equivalent to $x + q = \pm\sqrt{r}$.

Example 1 Solve.

a. $(x - 2)^2 = 5$

b. $(5x + 4)^2 = -36$

Solution Recall that if $x^2 = r$, then $x = \pm\sqrt{r}$.

a.
$$\begin{aligned}(x - 2)^2 &= 5\\ x - 2 &= \pm\sqrt{5}\\ x &= 2 \pm \sqrt{5}\end{aligned}$$

∴ the solution set is $\{2 + \sqrt{5}, 2 - \sqrt{5}\}$.

b.
$$\begin{aligned}(5x + 4)^2 &= -36\\ 5x + 4 &= \pm\sqrt{-36}\\ 5x + 4 &= \pm 6i\\ 5x &= -4 \pm 6i\\ x &= \frac{-4 \pm 6i}{5}\end{aligned}$$

∴ the solution set is $\left\{\frac{-4 + 6i}{5}, \frac{-4 - 6i}{5}\right\}$.

Solve.

1. a. $x^2 = 5$ b. $(x - 1)^2 = 5$ c. $(2x - 1)^2 = 5$
2. a. $y^2 = 64$ b. $(y + 4)^2 = 64$ c. $(5y + 4)^2 = 64$
3. a. $z^2 = -9$ b. $(z + 3)^2 = -9$ c. $(2z + 3)^2 = -9$
4. $(x - 5)^2 = 20$
5. $\left(\frac{1}{2}z + 1\right)^2 = -18$
6. $3(y - 2)^2 = -12$
7. $\frac{(x + 3)^2}{4} = 8$

NAME ______________________ DATE ______________

7–1 Completing the Square (continued)

Example 2 Solve $x^2 - 8x + 2 = 0$ by completing the square.

Solution $x^2 - 8x + 2 = 0$

Step 1 Add -2 to both sides to get -2 alone on one side. $x^2 - 8x = -2$

Step 2 Check the coefficient of x^2. It is 1, so you are ready to complete the square. $1x^2 - 8x = -2$

Step 3 Add the square of half the coefficient of x to both sides. Thus, add $\left(\frac{-8}{2}\right)^2$, or 16, to both sides. $x^2 - 8x + 16 = -2 + 16$

Step 4 Factor the left side as the square of a binomial. $(x - 4)^2 = 14$

Step 5 Solve for x as in Example 1. $x - 4 = \pm\sqrt{14}$

$\therefore$ the solution set is $\{4 + \sqrt{14}, 4 - \sqrt{14}\}$. $x = 4 \pm \sqrt{14}$

Example 3 Solve $3y^2 + 5y + 9 = 0$.

Solution $3y^2 + 5y + 9 = 0$

Step 1 $3y^2 + 5y = -9$ Add -9 to both sides.

Step 2 $y^2 + \frac{5}{3}y = -3$ Divide both sides by 3 so that $a = 1$.

Step 3 $y^2 + \frac{5}{3}y + \frac{25}{36} = -3 + \frac{25}{36}$ Complete the square by adding $\left(\frac{5}{6}\right)^2$.

Step 4 $\left(y + \frac{5}{6}\right)^2 = -\frac{108}{36} + \frac{25}{36}$ Factor the left side.

Step 5 $y + \frac{5}{6} = \pm\sqrt{-\frac{83}{36}}$ Solve. The solution set is

$y = -\frac{5}{6} \pm i\frac{\sqrt{83}}{6}$ $\left\{-\frac{5}{6} + i\frac{\sqrt{83}}{6}, -\frac{5}{6} - i\frac{\sqrt{83}}{6}\right\}$.

Solve by completing the square.

8. $x^2 + 4x - 6 = 0$ **9.** $y^2 + 2y + 15 = 0$ **10.** $t^2 - 5 = 3t$

11. $2k^2 - 8k + 5 = 0$ **12.** $5n^2 - 10n + 7 = 0$ **13.** $3x^2 - 10 = 12x$

14. $y^2 - y - 3 = 0$ **15.** $k^2 + 5k - 10 = 0$ **16.** $3v^2 + 7v + 4 = 0$

Mixed Review Exercises

Simplify.

1. $(2 + 3i)(4 - i)$ **2.** $(2\sqrt{5})^3$ **3.** $\sqrt{75} + \sqrt{48}$ **4.** $(3i\sqrt{7})^2$

5. $3(6 - 2i) + (-4 + 7i)$ **6.** $\sqrt{\frac{2}{3}} + \sqrt{\frac{3}{2}}$ **7.** $\frac{12xy^5}{-4x^3y^2}$ **8.** $\frac{x^2 - 8x + 7}{x^2 - 49}$

NAME ______________________ DATE ______________

7–2 The Quadratic Formula

Objective: To solve quadratic equations by using the quadratic formula.

Vocabulary

The quadratic formula The solutions of the quadratic equation $ax^2 + bx + c = 0$ $(a \neq 0)$ are given by the formula

$$x = \frac{-b \pm \sqrt{b^2 - 4ac}}{2a}.$$

Example 1 Solve $5x^2 - 4x - 2 = 0$.

Solution For the equation $5x^2 - 4x - 2 = 0$, $a = 5$, $b = -4$, and $c = -2$. Substitute these values in the quadratic formula. Then simplify.

$$x = \frac{-b \pm \sqrt{b^2 - 4ac}}{2a}$$

$$x = \frac{-(-4) \pm \sqrt{(-4)^2 - 4(5)(-2)}}{2(5)}$$

$$= \frac{4 \pm \sqrt{16 + 40}}{10}$$

$$= \frac{4 \pm \sqrt{56}}{10}$$

$$= \frac{4 \pm 2\sqrt{14}}{10} = \frac{2 \pm \sqrt{14}}{5}$$

$\therefore$ the solution set is $\left\{\frac{2 + \sqrt{14}}{5}, \frac{2 - \sqrt{14}}{5}\right\}$.

The solutions just obtained are *exact* and expressed in simplest radical form. In applications you may want to approximate solutions to the nearest hundredth. Since $\sqrt{14} \approx 3.7417$ (from a calculator or a square root table),

$$x \approx \frac{2 + 3.7417}{5} \quad \text{or} \quad x \approx \frac{2 - 3.7417}{5}$$

$$\approx 1.1483 \quad \text{or} \quad \approx -0.3483$$

$\therefore$ the solution set is approximately $\{1.15, -0.35\}$.

Solve each equation. Give answers involving radicals in simplest radical form.

1. $x^2 + 2x - 4 = 0$

2. $y^2 + 3y - 1 = 0$

3. $p^2 - 5p + 3 = 0$

4. $x^2 + 4x + 2 = 0$

5. $x^2 - 2x - 2 = 0$

6. $k^2 - 6k - 1 = 0$

7. $2y^2 - 3y - 5 = 0$

8. $2t^2 + 7t + 3 = 0$

NAME ______________________ DATE ______________

7-2 The Quadratic Formula (continued)

Example 2 Solve $2y(4y - 5) = -5$.

Solution First rewrite the equation in the form $ax^2 + bx + c = 0$:

$$2y(4y - 5) = -5$$
$$8y^2 - 10y = -5$$
$$8y^2 - 10y + 5 = 0$$

Then substitute 8 for a, -10 for b, and 5 for c in the quadratic formula.

$$y = \frac{-b \pm \sqrt{b^2 - 4ac}}{2a}$$

$$y = \frac{-(-10) \pm \sqrt{(-10)^2 - 4(8)(5)}}{2(8)}$$

$$= \frac{10 \pm \sqrt{-60}}{16}$$

$$= \frac{10 \pm 2i\sqrt{15}}{16}$$

$$= \frac{5 \pm i\sqrt{15}}{8}$$

$\therefore$ the solution set is $\left\{\frac{5 + i\sqrt{15}}{8}, \frac{5 - i\sqrt{15}}{8}\right\}$.

Solve each equation after rewriting it in the form $ax^2 + bx + c = 0$. Give answers involving radicals in simplest radical form.

9. $3x^2 = -2x - 1$

10. $y^2 - 2y = 5$

11. $7n^2 + 3 = 2n$

12. $4x(x + 3) = 15$

13. $9 = 2w(w - 3)$

14. $(3m - 1)(m + 4) = -9$

15. $\frac{x^2}{2} + 1 = \frac{x}{3}$

16. $\frac{4n^2 + 1}{3} = 2n$

Solve each equation and approximate solutions to the nearest hundredth. A calculator may be helpful.

17. $3x^2 + 4x - 2 = 0$

18. $5n^2 - 2n = 6$

19. $2t(2t + 5) = -5$

20. $3x(x - 2) = -1.5$

Solve each equation (a) by factoring and (b) by using the quadratic formula.

21. $2x^2 - 50 = 0$

22. $3y^2 - 12 = 0$

23. $2y^2 - 14y + 20 = 0$

24. $6x^2 + x - 12 = 0$

NAME ______________________________ DATE ______________

7–3 The Discriminant

Objective: To determine the nature of the roots of a quadratic equation by using its discriminant.

Vocabulary

Discriminant (D) The expression $b^2 - 4ac$ is called the discriminant of the quadratic equation $ax^2 + bx + c = 0$. It appears under the radical symbol in the quadratic formula.

Roots of a quadratic equation The quadratic formula gives the two roots (or solutions) of the quadratic equation $ax^2 + bx + c = 0$:

$$r_1 = \frac{-b + \sqrt{D}}{2a} \quad \text{and} \quad r_2 = \frac{-b - \sqrt{D}}{2a}.$$

Example 1 Solve.

a. $x^2 + 7x - 3 = 0$ **b.** $4x^2 - 4\sqrt{5}\,x + 5 = 0$ **c.** $x^2 - 5x + 9 = 0$

Solution First evaluate the discriminant. Then use it to find the roots.

a. $x^2 + 7x - 3 = 0.$

$D = 7^2 - 4(1)(-3) = 49 + 12 = 61$ D is positive.

$r_1 = \frac{-7 + \sqrt{61}}{2}, \quad r_2 = \frac{-7 - \sqrt{61}}{2}$ The roots are real and unequal.

b. $4x^2 - 4\sqrt{5}\,x + 5 = 0$

$D = (-4\sqrt{5})^2 - 4(4)(5) = 80 - 80 = 0$ D is zero.

$r_1 = \frac{-(-4\sqrt{5}) + \sqrt{0}}{2(4)} = \frac{4\sqrt{5}}{8} = \frac{\sqrt{5}}{2}$ The roots are real and equal. We say there is a *double root.*

$r_2 = \frac{-(-4\sqrt{5}) - \sqrt{0}}{2(4)} = \frac{4\sqrt{5}}{8} = \frac{\sqrt{5}}{2}$

c. $x^2 - 5x + 9 = 0$

$D = (-5)^2 - 4(1)(9) = 25 - 36 = -11$ D is negative.

$r_1 = \frac{-(-5) + \sqrt{-11}}{2} = \frac{5 + i\sqrt{11}}{2}$ The roots are imaginary conjugates.

$r_2 = \frac{-(-5) - \sqrt{-11}}{2} = \frac{5 - i\sqrt{11}}{2}$

Solve each equation using whichever method seems easiest to you.

1. $x^2 - 13x + 42 = 0$ **2.** $n^2 - 2n - 35 = 0$ **3.** $4p^2 + 20p + 25 = 0$

4. $4(y - 2)^2 = 16$ **5.** $(x + 11)^2 = 36$ **6.** $3y - y^2 = 6$

7. $16x^2 + 25 = 40x$ **8.** $(3n - 8)(2n + 7) = 0$ **9.** $(3x - 1)(x - 3) = -1$

7–3 The Discriminant *(continued)*

Vocabulary

The nature of the roots of a quadratic equation
Given a quadratic equation with *real* coefficients and discriminant D:

1. If D is positive, then the equation has two unequal real roots.
2. If D is zero, then the equation has a real double root.
3. If D is negative, then the equation has two conjugate imaginary roots.

Test for rational roots If a quadratic equation has *integral* coefficients and its discriminant is a perfect square, or if it can be transformed into an equivalent equation that meets these conditions, then the equation has rational roots.

Example 2 Without solving each equation, determine the nature of its roots.

a. $4x^2 - 19x + 12 = 0$ **b.** $x^2 + 2\sqrt{5}\,x - 4 = 0$

c. $x^2 - 6x + 9 = 0$ **d.** $5x^2 - 3x + \sqrt{6} = 0$

Solution Evaluate the discriminant, and determine whether it is positive, zero, or negative. If it is positive or zero, apply the test for rational roots.

a. $D = (-19)^2 - 4(4)(12) = 361 - 192 = 169 = 13^2$ (positive)
The coefficients are integers and the discriminant is a perfect square. It passes the test for rational roots. The roots are unequal and rational.

b. $D = (2\sqrt{5})^2 - 4(1)(-4) = 20 + 16 = 36 = 6^2$ (positive)
The discriminant is a perfect square, but the equation doesn't pass the other conditions of the test for rational roots. The roots are unequal and irrational.

c. $D = (-6)^2 - 4(1)(9) = 36 - 36 = 0 = 0^2$ (zero)
The coefficients are integers and the discriminant is a perfect square. The roots are equal and rational, that is, there is a rational double root.

d. $D = (-3)^2 - 4(5)(\sqrt{6}) = 9 - 20\sqrt{6} \approx 9 - 49$, or -40 (negative)
The roots are imaginary conjugates.

Without solving each equation, determine the nature of its roots. A calculator may be helpful.

10. $x^2 + 10x + 25 = 0$ **11.** $y^2 + 9y - 6 = 0$ **12.** $5x^2 - 6x + 2 = 0$

13. $3m^2 - 4m - 7 = 0$ **14.** $y^2 - \frac{3}{2}y = 5$ **15.** $\frac{u^2}{6} + 1 = u$

16. $\sqrt{2}x^2 - 9\sqrt{2} = 0$ **17.** $5z^2 - 2\sqrt{10}z + 2 = 0$ **18.** $\sqrt{5}x^2 + 2\sqrt{5}x = -4$

Mixed Review Exercises

Express in simplest form without negative exponents. Assume all radicals represent real numbers.

1. $(x^4)^3$ **2.** $\sqrt{98x^3}$ **3.** $(x^3)^{-1}$ **4.** $(x^3y^2)(x^4y)$

5. $(\sqrt{x})^{-1}$ **6.** $\frac{x^3y^{-1}}{(xy)^{-2}}$ **7.** $(\sqrt{7x^2})^2$ **8.** $\left(\frac{x^3}{y^2}\right)^{-4}$

NAME ______________________ DATE ______________

7–4 Equations in Quadratic Form

Objective: To recognize and solve equations in quadratic form.

Vocabulary

Equation in quadratic form An equation that can be written as $a[f(x)]^2 + b[f(x)] + c = 0$ where $a \neq 0$ and $f(x)$ is some function of x.

Example 1 Solve $(2x + 5)^2 - 3(2x + 5) - 4 = 0$.

Solution To transform the equation into a simpler equivalent equation, replace $2x + 5$ with a single variable.

Let $y = 2x + 5$. Then $y^2 = (2x + 5)^2$ and the equation becomes

$$y^2 - 3y - 4 = 0$$
$$(y - 4)(y + 1) = 0 \qquad \text{Solve by factoring.}$$
$$y = 4 \quad \text{or} \quad y = -1$$

Next, substitute $2x + 5$ back in for y and solve for the original variable, x.

$$y = 4 \quad \text{or} \quad y = -1$$
$$2x + 5 = 4 \quad \text{or} \quad 2x + 5 = -1 \qquad \text{Substitute } 2x + 5 \text{ for } y.$$
$$2x = -1 \quad \text{or} \quad 2x = -6 \qquad \text{Solve for } x.$$
$$x = -\frac{1}{2} \quad \text{or} \quad x = -3$$

$\therefore$ the solution set is $\left\{-\frac{1}{2}, -3\right\}$.

Solve each equation.

1. a. $(x + 2)^2 - 7(x + 2) + 12 = 0$ b. $(x - 4)^2 - 7(x - 4) + 12 = 0$
2. a. $(2x - 1)^2 - 5(2x - 1) + 6 = 0$ b. $(3x + 2)^2 - 5(3x + 2) + 6 = 0$
3. a. $6(x - 3)^2 + 7(x - 3) - 5 = 0$ b. $6(1 - 2x)^2 + 7(1 - 2x) - 5 = 0$

Example 2 Solve $6x^{-2} - 14x^{-1} - 12 = 0$.

Solution Let $y = x^{-1}$. Then $y^2 = (x^{-1})^2 = x^{-2}$ and the equation becomes

$$6y^2 - 14y - 12 = 0$$
$$(3y + 2)(2y - 6) = 0 \qquad \text{Factor.}$$
$$3y + 2 = 0 \quad \text{or} \quad 2y - 6 = 0$$
$$y = -\frac{2}{3} \quad \text{or} \quad y = 3 \qquad \text{Solve for } y.$$
$$x^{-1} = \frac{1}{x} = -\frac{2}{3} \quad \text{or} \quad x^{-1} = \frac{1}{x} = 3 \qquad \text{Substitute } x^{-1} \text{ for } y.$$
$$x = -\frac{3}{2} \quad \text{or} \quad x = \frac{1}{3} \qquad \text{Solve for } x.$$

$\therefore$ the solution set is $\left\{-\frac{3}{2}, \frac{1}{3}\right\}$.

NAME ______________________ DATE ______________

7-4 Equations in Quadratic Form (continued)

CAUTION Keep in mind that the principal square root of a real number, denoted by the symbol $\sqrt{\ }$, must be *nonnegative*. Be on the lookout for extraneous roots.

Example 3 Solve $6x - 14\sqrt{x} - 12 = 0$.

Solution Let $y = \sqrt{x}$. Then $y^2 = x$ and the equation becomes

$$6y^2 - 14y - 12 = 0$$
$$(3y + 2)(2y - 6) = 0 \quad \text{Factor.}$$

$3y + 2 = 0$ or $2y - 6 = 0$

$y = -\frac{2}{3}$ or $y = 3$ Solve for y.

$\sqrt{x} = -\frac{2}{3}$ or $\sqrt{x} = 3$ Replace y by $\sqrt{x}$.

$-\frac{2}{3}$ is *not* a solution. The principal square root cannot be negative. $x = 9$ Square both sides.

$\therefore$ the solution set is $\{9\}$.

Solve each equation.

4. a. $x - 6\sqrt{x} + 8 = 0$ **b.** $(x^2 - 4)^2 - 6(x^2 - 4) + 8 = 0$

5. a. $5\left(\frac{x}{5}\right)^2 - 8\left(\frac{x}{5}\right) - 4 = 0$ **b.** $5x^{-2} - 8x^{-1} - 4 = 0$

Example 4 Solve $x^4 + 3x^2 - 18 = 0$.

Solution

Method 1

Do all of your work in terms of x.

$$(x^2)^2 + 3(x^2) - 18 = 0$$
$$(x^2 + 6)(x^2 - 3) = 0$$

$x^2 = -6$ or $x^2 = 3$

$x = \pm\sqrt{-6}$ or $x = \pm\sqrt{3}$

$= \pm i\sqrt{6}$

Method 2

Let $z = x^2$. Then $z^2 = x^4$ and the equation becomes

$$z^2 + 3z - 18 = 0$$
$$(z + 6)(z - 3) = 0$$

$z = -6$ or $z = 3$

$x^2 = -6$ or $x^2 = 3$

$x = \pm\sqrt{-6}$ or $x = \pm\sqrt{3}$

$= \pm i\sqrt{6}$

$\therefore$ the solution set is $\{i\sqrt{6}, -i\sqrt{6}, \sqrt{3}, -\sqrt{3}\}$.

Solve each equation.

6. a. $x^4 + 2x^2 - 15 = 0$ **7. a.** $2(3x - 4)^2 - 3(3x - 4) + 1 = 0$

b. $x + 2\sqrt{x} - 15 = 0$ **b.** $2x^4 - 3x^2 + 1 = 0$

8. a. $x^{-4} - 7x^{-2} - 18 = 0$ **9. a.** $2x^4 + x^2 - 3 = 0$

b. $x - 7\sqrt{x} - 18 = 0$ **b.** $2(x + 4)^4 + (x + 4)^2 - 3 = 0$

NAME ______________________________ DATE ______________________

7–5 Graphing $y - k = a(x - h)^2$

Objective: To graph parabolas whose equations have the form $y - k = a(x - h)^2$, and to find the vertices and axes of symmetry.

Vocabulary

Parabola The graph of an equation in the form $y - k = a(x - h)^2$, $a \neq 0$.

Axis of symmetry (or axis) The line about which the two "halves" of a parabola are *mirror images*. The line $x = h$ is the axis of the graph of $y - k = a(x - h)^2$.

Vertex The point where a parabola crosses its axis. This is its highest or lowest point. The point (h, k) is the vertex of the graph of $y - k = a(x - h)^2$.

Example 1 Graph $y - 1 = 2(x - 3)^2$. Label the vertex and axis.

Solution First, use the definitions to find the vertex and the axis.

$h = 3$, and $k = 1$.

$\therefore$ the vertex (h, k) is $(3, 1)$, and the axis $x = h$ is the line $x = 3$.

Next, make a table of values.

$y = 2(x - 3)^2 + 1$

x	y
1	9
2	3

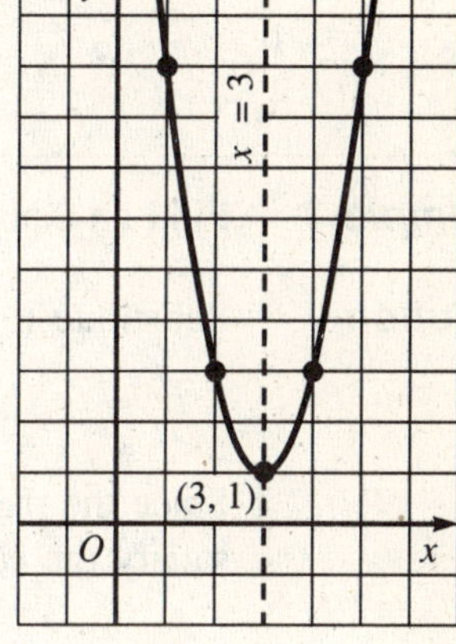

Finally, construct the graph by following these steps:

1. Plot the vertex.
2. Draw the axis using a dotted line.
3. Plot the points from the table.
4. Using the axis, plot the mirror images of the points in the table.
5. Draw a smooth curve connecting the points.

Note: A parabola opens upward when a is positive, and downward when a is negative.

Example 2 Graph $y - 3 = -(x + 2)^2$. Label the vertex and axis. Find all intercepts.

Solution

1. Since $a = -1$, the parabola opens downward. Since $h = -2$ and $k = 3$, the vertex is $(-2, 3)$. The axis of symmetry is the line $x = -2$.
2. To find the y-intercept, set $x = 0$ and solve for y.

$$y - 3 = -(0 + 2)^2$$
$$y - 3 = -4$$
$$y = -1 \quad \leftarrow \quad y\text{-intercept}$$

Therefore, the graph crosses the y-axis at $(0, -1)$. Since $(0, -1)$ is on the graph, its mirror image across the axis of symmetry, $(-4, -1)$, is on the graph also.

(Solution continues on the next page.)

NAME ______________________ DATE ______________________

7–5 Graphing $y - k = a(x - h)^2$ *(continued)*

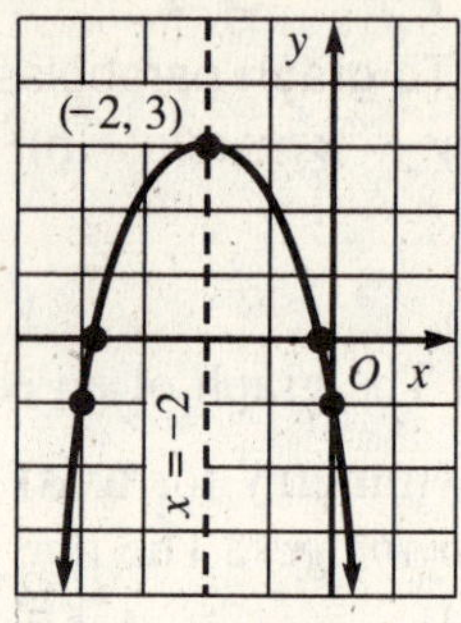

3. To find any x-intercepts, set $y = 0$.

$$0 - 3 = -(x + 2)^2$$
$$3 = (x + 2)^2$$
$$\pm\sqrt{3} = x + 2$$
$$x = -2 \pm \sqrt{3}$$

Since $\sqrt{3} \approx 1.7$, the x-intercepts are:

$x \approx -2 + 1.7$ or $x \approx -2 - 1.7$

$x \approx -0.3$ or $x \approx -3.7$

4. Plot the vertex $(-2, 3)$ and the intercepts. Then complete the curve using symmetry.

Note: There are no x-intercepts when the graph is all above or all below the x-axis.

Graph each equation. Label the vertex and axis of symmetry. Find all intercepts.

1. $y = -4x^2$ **2.** $y = \frac{1}{4}x^2$ **3.** $y - 2 = -x^2$ **4.** $y + 2 = \frac{1}{3}x^2$

5. $y = (x - 1)^2$ **6.** $y = -(x + 2)^2$ **7.** $y - 4 = -3(x + 2)^2$ **8.** $y - 2 = -(x - 3)^2$

9. $y - 1 = \frac{1}{4}(x - 2)^2$ **10.** $y - 3 = -\frac{1}{3}(x - 2)^2$ **11.** $y + 2 = \frac{1}{2}(x + 1)^2$

Example 3 Find an equation for the parabola with vertex $(-3, 2)$ and containing $(1, 8)$.

Solution Substitute $(-3, 2)$ for (h, k) in the equation $y - k = a(x - h)^2$:

$$y - 2 = a(x - (-3))^2$$
$$y - 2 = a(x + 3)^2$$

Since the parabola contains the point $(1, 8)$, the coordinates of this point must satisfy the equation. Substitute $(1, 8)$ for (x, y) and solve for a:

$$8 - 2 = a(1 + 3)^2$$
$$6 = 16a$$
$$a = \frac{3}{8}$$

$\therefore$ an equation of the parabola is $y - 2 = \frac{3}{8}(x + 3)^2$.

Find an equation $y - k = a(x - h)^2$ for each parabola described.

12. Vertex $(0, 6)$; contains $(1, 3)$ **13.** Vertex $(2, 3)$; contains $(1, -2)$ **14.** Vertex $(4, -7)$; contains $(8, 1)$

$y-6 = a(x-0)^2$ $y-6=ax^2$ $3-6 = a(1)^2$ $-3=a$ $y-6=-3x^2$

Mixed Review Exercises

Without solving each equation, determine the nature of its roots.

1. $3x^2 - 5x + 4 = 0$ **2.** $4x^2 + 12x + 9 = 0$ **3.** $x^2 + x - 7 = 0$

Solve each equation over the complex numbers.

4. $x + \sqrt{x} - 12 = 0$ **5.** $4x^2 + 8x + 5 = 0$ **6.** $x^4 - 6x^2 + 8 = 0$

NAME ______________________________ DATE ____________

7–6 Quadratic Functions

Objective: To analyze and graph a quadratic function, and find its minimum or maximum value.

Vocabulary

Quadratic function A function that can be written in either of two forms.

General form: $f(x) = ax^2 + bx + c \ (a \neq 0)$

Completed-square form: $f(x) = a(x - h)^2 + k \ (a \neq 0)$

The general form can be transformed into the following form.

$$f(x) - \left(-\frac{b^2 - 4ac}{4a}\right) = a\left(x - \frac{-b}{2a}\right)^2$$

This is the equation of a parabola with vertex $(h, k) = \left(-\frac{b}{2a}, -\frac{b^2 - 4ac}{4a}\right)$.

Maximum or minimum value (of a quadratic function) The y-coordinate at the vertex of the graph of that function, that is, when $x = -\frac{b}{2a}$. You can tell whether the y-coordinate is a maximum or a minimum value by applying the following test.

Let $f(x) = ax^2 + bx + c, a \neq 0$.

If $a < 0$, the graph of f opens downward, and f has a maximum value.
If $a > 0$, the graph of f opens upward, and f has a minimum value.

Example 1 Graph $f(x) = 3(x - 1)^2 + 2$.

Solution Replace $f(x)$ with y:

$$f(x) = 3(x - 1)^2 + 2$$
$$y = 3(x - 1)^2 + 2$$
$$y - 2 = 3(x - 1)^2$$

The last equation is the equation of a parabola with vertex (1, 2) and axis $x = 1$, and it opens upward.

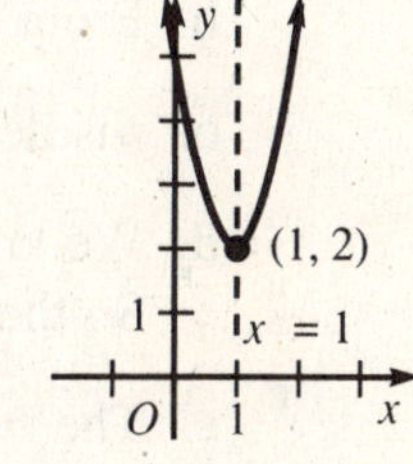

Example 2 **a.** Show that the graph of $f(x) = 2x^2 - 8x + 3$ is a parabola.
b. Find the vertex and graph the parabola.

Solution **a.** You must show that f can be written in the form $y - k = a(x - h)^2$.

Step 1	Replace $f(x)$ with y.	$y = 2x^2 - 8x + 3$
Step 2	Subtract the constant term from both sides.	$y - 3 = 2x^2 - 8x$
Step 3	Factor so that the coefficient of x^2 is 1.	$y - 3 = 2(x^2 - 4x)$
Step 4	Complete the square in x.	$y - 3 + \mathbf{2 \cdot 4} = \mathbf{2}(x^2 - 4x + \mathbf{4})$
		$y + 5 = 2(x - 2)^2$

∴ the graph is a parabola, since the last equation is in the form $y - k = a(x - h)^2$.

b. Since $h = 2$ and $k = -5$, the vertex is $(2, -5)$. Since $a = 2$, a is positive and the parabola opens upward.

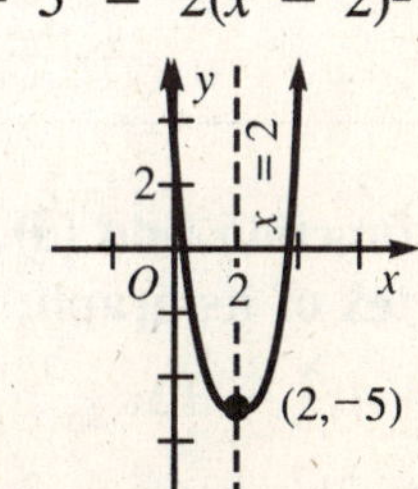

NAME ______________________________ DATE ______________

7–6 Quadratic Functions *(continued)*

Graph each function. Follow the method of Example 1.

1. $f(x) = x^2 - 2$
2. $f(x) = 2 - x^2$
3. $f(x) = (x + 2)^2 + 3$
4. $f(x) = 2(x - 3)^2 - 5$
5. $f(x) = 1 - 2(x + 1)^2$
6. $f(x) = 5 - \frac{1}{2}(x + 3)^2$

Graph each function by finding its vertex. Follow the method of Example 2.

7. $g(x) = x^2 + 2x - 5$
8. $h(x) = 2x^2 + 4x - 2$
9. $f(x) = -x^2 - 4x + 1$
10. $h(x) = 12x - 2x^2$
11. $f(x) = 3x^2 - 12x + 12$
12. $g(x) = \frac{1}{3}x^2 + 2x + 1$

Example 3 Given the function $f(x) = -3x^2 - 6x + 3$, find (a) its maximum value or minimum value, (b) the vertex of its graph, (c) its domain, (d) its range, and (e) its zeros.

Solution

a. Since $a = -3$, $a < 0$ and f has a maximum value. This maximum is the value of f when $x = -\frac{b}{2a} = -\frac{-6}{2(-3)} = -1$.

Find the maximum value by substituting -1 for x in the original function.

$$f(-1) = -3(-1)^2 - 6(-1) + 3 = -3 + 6 + 3 = 6$$

$\therefore$ the maximum value of f is 6.

b. From part (a), we can conclude that the vertex is $(-1, 6)$.

c. f is defined for all real values of x. So the domain $D = \{\text{real numbers}\}$.

d. We know from part (a) that the maximum value of this function is 6. So the range $R = \{y: y \le 6\}$.

e. The zeros of f are the same as the roots of $f(x) = 0$, that is, the zeros are the x-intercepts of the graph of f. Let $f(x) = 0$, and use the quadratic formula to find the roots.

$$f(x) = -3x^2 - 6x + 3$$
$$0 = -3x^2 - 6x + 3$$
$$x = \frac{-(-6) \pm \sqrt{36 - 4(-3)(3)}}{2(-3)}$$
$$x = \frac{6 \pm \sqrt{72}}{-6}$$
$$x = -1 \pm \sqrt{2}$$

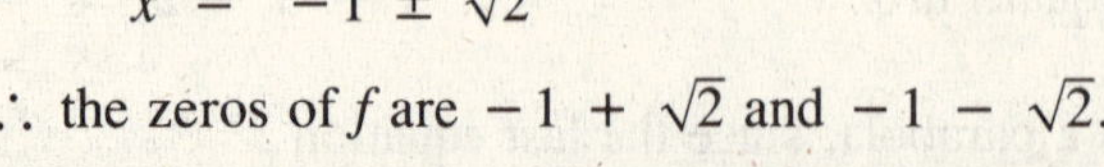

$\therefore$ the zeros of f are $-1 + \sqrt{2}$ and $-1 - \sqrt{2}$.

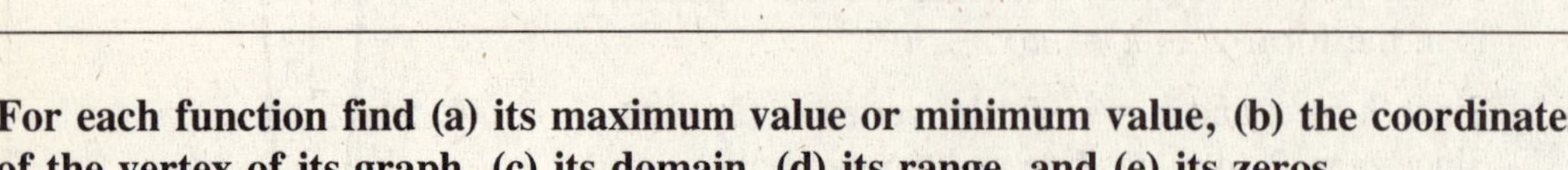

For each function find (a) its maximum value or minimum value, (b) the coordinates of the vertex of its graph, (c) its domain, (d) its range, and (e) its zeros.

13. $f(x) = 4x^2 - 12x$
14. $h(x) = x^2 - 6x + 3$
15. $g(x) = -3x^2 - 2x + 1$
16. $g(x) = 2x^2 + 4x - 3$
17. $h(x) = (3 - x)(2 + x)$
18. $f(x) = (2x + 1)(2x + 3)$

NAME ______________________ DATE ______________

7–7 Writing Quadratic Equations and Functions

Objective: To learn the relationship between the roots and coefficients of a quadratic equation. To write a quadratic equation or function using information about the roots or the graph.

Vocabulary

Theorem 1 A quadratic equation with roots r_1 and r_2 will have either of the forms

$$x^2 - (r_1 + r_2)x + r_1r_2 = 0 \quad \text{or} \quad a[x^2 - (r_1 + r_2)x + r_1r_2] = 0.$$

The second form is equivalent to $a[x^2 - (\text{sum of roots})x + (\text{product of roots})] = 0$.

Theorem 2 If r_1 and r_2 are the roots of a quadratic equation $ax^2 + bx + c = 0$, then

$$r_1 + r_2 = \text{sum of roots} = -\frac{b}{a} \quad \text{and} \quad r_1r_2 = \text{product of roots} = \frac{c}{a}.$$

Example 1 Find the roots of $3x^2 + 8x + 6 = 0$. Check your answer by using the theorem about the sum and product of the roots.

Solution

$$x = \frac{-8 \pm \sqrt{8^2 - 4(3)(6)}}{2(3)} = \frac{-8 \pm \sqrt{-8}}{6} = \frac{-4 \pm i\sqrt{2}}{3}$$

$\therefore$ the roots are $r_1 = \dfrac{-4 + i\sqrt{2}}{3}$ and $r_2 = \dfrac{-4 - i\sqrt{2}}{3}$.

Check:

$$r_1 + r_2 = \frac{-4 + i\sqrt{2}}{3} + \frac{-4 - i\sqrt{2}}{3} = -\frac{8}{3} = -\frac{b}{a} \quad \checkmark$$

$$r_1r_2 = \frac{-4 + i\sqrt{2}}{3} \cdot \frac{-4 - i\sqrt{2}}{3} = \frac{16 + 2}{9} = \frac{6}{3} = \frac{c}{a} \quad \checkmark$$

Solve the following equations. Check each answer as in Example 1.

1. $x^2 - 5x - 6 = 0$
2. $3u^2 + u - 4 = 0$
3. $3y^2 + 5 = -2y$
4. $4x^2 - 7 = 0$

Example 2 Find a quadratic equation with roots $\dfrac{2 - \sqrt{3}}{3}$ and $\dfrac{2 + \sqrt{3}}{3}$.

Solution Find the sum of the roots and the product of the roots. Then use these values in the formula for a quadratic equation given in Theorem 1.

$$\text{sum of roots} = \frac{2 - \sqrt{3}}{3} + \frac{2 + \sqrt{3}}{3} = \frac{4}{3}$$

$$\text{product of roots} = \frac{2 - \sqrt{3}}{3} \cdot \frac{2 + \sqrt{3}}{3} = \frac{4 - 3}{9} = \frac{1}{9}$$

$x^2 - (\text{sum of roots})x + (\text{product of roots}) = 0;\ x^2 - \dfrac{4}{3}x + \dfrac{1}{9} = 0.$

To clear fractions, multiply both sides of the equation by 9.

$\therefore$ the quadratic equation is $9x^2 - 12x + 1 = 0$.

NAME ______________________________ DATE ________________

7–7 Writing Quadratic Equations and Functions (continued)

Find a quadratic equation with *integral coefficients* having the given roots.

5. 3, 5 **6.** $-1, 2$ **7.** $\frac{2}{3}, -\frac{4}{3}$ **8.** $\frac{\sqrt{6}}{2}, -\frac{\sqrt{6}}{2}$

9. $1 + \sqrt{5}, 1 - \sqrt{5}$ **10.** $\frac{-2 + \sqrt{5}}{3}, \frac{-2 - \sqrt{5}}{3}$ **11.** $3i\sqrt{3}, -3i\sqrt{3}$

12. $5 - 2i, 5 + 2i$ **13.** $-2 - i\sqrt{5}, -2 + i\sqrt{5}$ **14.** $\frac{-2 + i\sqrt{5}}{3}, \frac{-2 - i\sqrt{5}}{3}$

Example 3 Find a quadratic function $f(x) = ax^2 + bx + c$ such that the minimum value of f is -4 and the graph of f has x-intercepts -1 and 7.

Solution The x-intercepts of the graph of f are also the roots of the equation $f(x) = 0$. You can use them to write the equation. The sum of the roots is $-1 + 7 = 6$, and the product of the roots is $-1(7) = -7$, so $f(x) = a(x^2 - 6x - 7)$.

Since the minimum value of f is -4, the y-coordinate of the vertex of the graph is -4. To find the x-coordinate of the vertex, note that the axis of symmetry intersects the x-axis at a point midway between the x-intercepts. This means the x-coordinate of the vertex is the *average* of the x-intercepts:

$$x = \frac{-1 + 7}{2} = 3$$

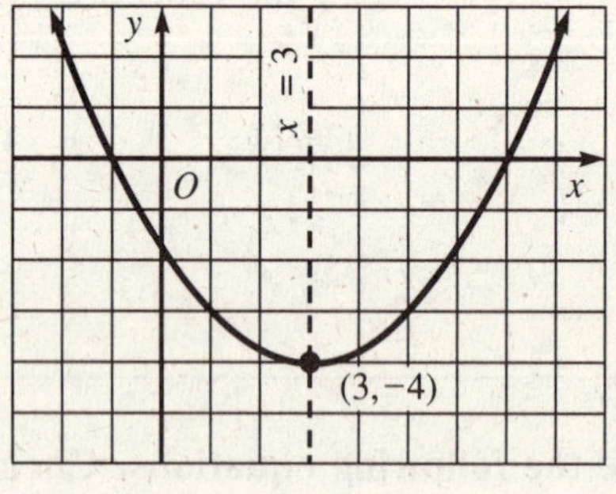

Therefore, the vertex is $(3, -4)$. Since the vertex lies on the graph of f, you can substitute its coordinates in the equation of f to find a.

$$\begin{aligned} f(x) &= a(x^2 - 6x - 7) \\ -4 &= a(3^2 - 6 \cdot 3 - 7) \\ -4 &= -16a \\ \frac{1}{4} &= a \end{aligned}$$

$\therefore$ the equation is $f(x) = \frac{1}{4}(x^2 - 6x - 7)$, or $f(x) = \frac{1}{4}x^2 - \frac{3}{2}x - \frac{7}{4}$.

Find a quadratic function $f(x) = ax^2 + bx + c$ for each parabola described.

15. Maximum value 2; x-intercepts -3 and 1

16. Minimum value -18; x-intercepts -1 and 5

17. Maximum value 4; x-intercepts 0 and 6

Mixed Review Exercises

Sketch the graph of each equation.

1. $y - 1 = (x - 3)^2$ **2.** $3x + 5y = 6$ **3.** $y = -x^2 + 4x + 1$ **4.** $y = 2$

Find the domain and zeros of each function. If there are no zeros, say so.

5. $g(x) = 5x + 3$ **6.** $h(x) = x^2 - 6x + 3$ **7.** $F(x) = -1$ **8.** $H(x) = \frac{2x - 5}{x + 3}$

NAME ______________________ DATE ______________

8 Variation and Polynomial Equations

8–1 *Direct Variation and Proportion*

Objective: To solve problems involving direct variation.

Vocabulary

Direct variation A linear function defined by an equation of the form $y = mx$ $(m \neq 0)$. The constant m in the equation is called the *constant of variation* or *constant of proportionality*. We say that *y varies directly as x* because if x increases, y also increases, and if x decreases, y also decreases.

Proportion An equality of ratios. A proportion can be written in the form $\frac{y_1}{x_1} = \frac{y_2}{x_2}$, or $y_1 : x_1 = y_2 : x_2$, where the ordered pairs (x_1, y_1) and (x_2, y_2) are solutions of a direct variation, and $\frac{y_1}{x_1} = \frac{y_2}{x_2} = m$ (x_1 and $x_2 \neq 0$).

In a direct variation, y is often said to be *directly proportional* to x.

Means and extremes In the proportion $y_1 : x_1 = y_2 : x_2$, the means are the numbers x_1 and y_2, and the extremes are y_1 and x_2. In a proportion, *the product of the extremes equals the product of the means*, or $y_1 x_2 = x_1 y_2$.

Example 1 If y varies directly as x, and $y = 12$ when $x = 20$, find y when $x = 50$.

Solution First find m and write an equation of the direct variation.

$y = mx$	Start with the general equation.
$12 = m(20)$	Substitute $y = 12$ and $x = 20$.
$m = \frac{12}{20} = \frac{3}{5}$	Solve for m.

$\therefore y = \frac{3}{5}x$ is an equation of the direct variation.

You can find y when $x = 50$ by substituting 50 for x in this equation.

$$y = \frac{3}{5}(50) = 30$$

Example 2 If a is directly proportional to $b + 3$, and $a = 6$ when $b = 15$, find b when $a = 7$.

Solution Since a is directly proportional to $b + 3$, you can write a proportion.

$\frac{a_1}{b_1 + 3} = \frac{a_2}{b_2 + 3}$	Set up a proportion.
$\frac{6}{15 + 3} = \frac{7}{b_2 + 3}$	Substitute values for variables.
$6(b_2 + 3) = 18(7)$	Multiply the extremes, and the means.
$b_2 + 3 = 126 \div 6 = 21$	Solve for b_2.
$b_2 = 18$	

NAME ____________________ DATE ____________

8–1 Direct Variation and Proportion *(continued)*

CAUTION It doesn't matter whether you solve direct variation problems by finding the equation first or by using the proportion method. However, if you decide to use the proportion method, be sure that the values in the numerators represent the same variable.

Solve.

1. If y varies directly as x, and $y = 6$ when $x = 4$, find y when $x = 12$.
2. If a is directly proportional to b, and $a = 25$ when $b = 35$, find b when $a = 40$.
3. If w varies directly as z, and $w = 4.5$ when $z = 3$, find z when $w = 1.5$.
4. If p is directly proportional to q^3, and $p = 3$ when $q = 2$, find p when $q = 4$.
5. If r varies directly as $s + 1$, and $r = 4$ when $s = 5$, find r when $s = 8$.
6. If a varies directly as $3b + 2$, and $a = 10$ when $b = 6$, find b when $a = 7$.

Example 3 If a car travels 70 km in 2 hours, how far can it travel in 4.5 hours, traveling at the same rate of speed?

Solution Let d be the required distance in kilometers.

Since the ratio $\frac{\textit{distance}}{\textit{time}} = \textit{rate}$ is constant, a proportion can be written.

$$\frac{d_1}{t_1} = \frac{d_2}{t_2} \longrightarrow \frac{70}{2} = \frac{d}{4.5}$$

$$2d = 70(4.5)$$

$$d = 157.5$$

∴ the distance the car will travel is 157.5 km.

Solve.

7. If the sales tax on a \$38 purchase is \$2.85, what will the tax be on an \$84 purchase?
8. A survey showed that 52 out of 234 people questioned preferred hot cereal to cold cereal. In a school population of 1800, how many people are likely to prefer hot cereal?
9. A real estate agent received a commission of \$2232 on a piece of land that sold for \$124,000. At this rate, what commission will the agent receive for a piece of land that sold for \$160,000?

Mixed Review Exercises

Solve each equation over the real numbers.

1. $4x^2 - 7x + 2 = 0$
2. $\frac{a-1}{a+5} = 1 - \frac{3}{a}$
3. $\frac{3y^2}{8} + \frac{y}{4} = 1$
4. $|2a - 8| = 6$
5. $\sqrt{2m + 15} = m$
6. $4q^{-2} + 7q^{-1} = 2$
7. $6n^2 = 7n$
8. $(2x - 5)^2 = 18$

NAME ______________________________ DATE ______________

8–2 *Inverse and Joint Variation*

Objective: To solve problems involving inverse and joint variation.

Vocabulary

Inverse variation A function defined by an equation of the form $xy = k$ or $y = \frac{k}{x}$ ($x \neq 0$, $k \neq 0$). The constant k is called the *constant of variation* or *constant of proportionality*. We say that *y varies inversely as x*, or *y is inversely proportional to x*, because if x increases, y decreases, and if x decreases, y increases.

Joint variation When a quantity varies directly as the product of two or more other quantities, the variation is called joint variation. Example: If m varies jointly as n and p, then $m = knp$ for some nonzero constant k. Another way to state this relationship is "m is jointly proportional to n and p."

Example 1 If y is inversely proportional to x, and $y = 8$ when $x = 6$, find x when $y = 15$.

Solution First find k and write an equation of the inverse variation.

$xy = k$	Start with the general equation.
$(6)(8) = k$	Substitute $x = 6$ and $y = 8$.
$k = 48$	Solve for k.

$\therefore$ an equation of the inverse variation is $xy = 48$.

To find x when $y = 15$, substitute this value in $xy = 48$.

$$x(15) = 48$$
$$x = 3.2$$

Solve.

1. If y varies inversely as x, and $y = 5$ when $x = 4$, find x when $y = 10$.
2. If p is inversely proportional to q, and $p = 10$ when $q = 5$, find q when $p = 2$.
3. If a is inversely proportional to b, and $b = 12$ when $a = 8$, find b when $a = 3$.
4. If x varies inversely as the square of y, and $x = 2$ when $y = 12$, find y when $x = 8$.

Example 2 If z varies jointly as x and the square of y, and $z = 12$ when $x = 6$ and $y = 2$, find z when $x = 10$ and $y = 3$.

Solution First find k and write an equation of the joint variation. Remember it's a *direct* variation.

$$z = kxy^2$$
$$12 = k(6)(2)^2$$
$$12 = 24k$$
$$k = \frac{1}{2}$$

$\therefore$ an equation of the joint variation is $z = \frac{1}{2}xy^2$.

Then, substitute $x = 10$ and $y = 3$ in this equation.

$$z = \frac{1}{2}xy^2$$
$$z = \frac{1}{2}(10)(3)^2$$
$$z = 45$$

NAME ______________________ DATE ______________

8–2 *Inverse and Joint Variation* *(continued)*

Solve.

5. If x varies jointly as y and z, and $x = 100$ when $y = 20$ and $z = 10$, find x when $y = 60$ and $z = 30$.

6. If a is jointly proportional to b and c, and $a = 48$ when $b = 6$ and $c = 4$, find c when $a = 540$ and $b = 18$.

7. If I varies jointly as p and r, and $I = 14$ when $p = 100$ and $r = 0.07$, find p when $I = 48$ and $r = 0.08$.

8. If x varies jointly as y and the square root of z, and $x = 20$ when $y = 5$ and $z = 9$, find x when $y = 6$ and $z = 25$.

Example 3 The intensity of light, measured in *lux*, is inversely proportional to the square of the distance between the light source and the object illuminated. A light meter 6.4 m from a light source registers 30 lux. What intensity would it register 16 m from the light source?

Solution Let I = the intensity of light in lux, and d = the distance in meters of the illuminated object from the light source. I varies inversely as the square of d. That is,

$$Id^2 = k.$$

Find k when $I = 30$ lux and $d = 6.4$ m.

$$30(6.4)^2 = k$$
$$k = 1228.8$$

$\therefore$ an equation of the variation is $Id^2 = 1228.8$.

Then, to find I when $d = 16$ m, substitute this value in $Id^2 = 1228.8$.

$$I(16)^2 = 1228.8$$
$$I = 4.8$$

$\therefore$ the intensity is 4.8 lux.

Solve.

9. A light meter 5.4 m from a light source registers 20 lux. What intensity would it measure 1.8 m from the light source?

10. The frequency of a radio signal varies inversely as the wavelength. A signal of frequency 1250 kilohertz (kHz) has a wavelength of 240 m. What frequency has a signal of wavelength 300 m?

11. The volume of a cylinder is jointly proportional to the height and the square of the radius of a base. A cylinder of height 12 cm and base radius 3 cm has volume 108π cm^3. Find the radius of the base of a cylinder if the height is 2 cm and the volume is 72π cm^3.

12. The stretch in a wire under a given tension varies directly as the length of the wire and inversely as the square of its diameter. If the length and the diameter of a wire are both doubled, what is the effect on the stretch of the wire?

NAME ________________ DATE ________________

8–3 Dividing Polynomials

Objective: To divide one polynomial by another polynomial.

Vocabulary

Dividend A quantity to be divided. In long division, the number *under* the division symbol is the dividend. In a fraction, the numerator is the dividend.

Divisor The quantity by which another quantity is divided. In long division, the number *outside* the division symbol is the divisor. In a fraction, the denominator is the divisor.

Division algorithm (rule) $\dfrac{\text{Dividend}}{\text{Divisor}} = \text{Quotient} + \dfrac{\text{Remainder}}{\text{Divisor}}$

or $\text{Dividend} = \text{Quotient} \times \text{Divisor} + \text{Remainder}$

CAUTION Before using long division, always arrange the terms of both dividend and divisor in order of decreasing degree. Also, insert any "missing" terms by using zero as a coefficient. Example: $x - 3x^3 - 5$ becomes $-3x^3 + 0x^2 + x - 5$.

Example 1 Divide: $\dfrac{3x + 2x^3 - 5}{x + 2}$

Solution

1. Rewrite the problem in the form you would use to do long division.
2. Divide the first term of the dividend by the first term of the divisor. Write this monomial above the line as the first term of the quotient.
3. Multiply by the new quotient term.
4. Subtract.
5. Repeat Steps 2–4 until the remainder is 0 or the degree of the remainder is less than the degree of the divisor.

	$2x^2 - 4x + 11$ ←	**quotient**
divisor → $x + 2$	$\overline{)2x^3 + 0x^2 + 3x - 5}$ ←	**dividend**
	$\underline{2x^3 + 4x^2}$ ←	Multiply $x + 2$ by $2x^2$.
	$-4x^2 + 3x$ ←	Subtract.
	$\underline{-4x^2 - 8x}$ ←	Multiply $x + 2$ by $-4x$.
	$11x - 5$ ←	Subtract.
	$\underline{11x + 22}$ ←	Multiply $x + 2$ by 11.
remainder →	-27 ←	Subtract.

$$\therefore \underbrace{\frac{2x^3 + 3x - 5}{x + 2}}_{\frac{\text{Dividend}}{\text{Divisor}}} = \underbrace{2x^2 - 4x + 11}_{\text{Quotient}} + \underbrace{\frac{-27}{x + 2}}_{\frac{\text{Remainder}}{\text{Divisor}}}$$

Check: To check the result, use the following form of the division algorithm:

$$\text{Quotient} \times \text{Divisor} + \text{Remainder} = \text{Dividend}$$
$$(2x^2 - 4x + 11)(x + 2) + (-27) \stackrel{?}{=} 2x^3 + 3x - 5$$
$$2x^3 - 4x^2 + 11x + 4x^2 - 8x + 22 - 27 \stackrel{?}{=} 2x^3 + 3x - 5$$
$$2x^3 + 3x - 5 = 2x^3 + 3x - 5 \ \checkmark$$

NAME ______________________________ DATE ______________

8-3 Dividing Polynomials *(continued)*

Divide. Additional answers are given at the back of this Answer Key.

1. $\dfrac{x^2 + 8x + 12}{x + 4}$
2. $\dfrac{-68 + 17x - x^2}{6 - x}$
3. $\dfrac{x^2 - 16x}{x - 4}$
4. $\dfrac{9x^2 - 6x + 1}{3x + 1}$
5. $\dfrac{6x^2 - 13x + 6}{2x - 3}$
6. $\dfrac{10x^2 + x - 3}{5x + 3}$
7. $\dfrac{2x^3 - 15x + 8}{x - 2}$
8. $\dfrac{x^3 - 4x - 8}{x - 2}$
9. $\dfrac{6x^3 - 19x^2 + 15}{3x - 5}$

Example 2 Divide: $\dfrac{x^4 + x^2 + 4}{x^2 + x + 1}$

Solution Insert the "missing" terms, $0x^3$ and $0x$, in the dividend.

$$\begin{array}{r|l}
 & x^2 - x + 1 \\
x^2 + x + 1 & x^4 + 0x^3 + x^2 + 0x + 4 \\
 & \underline{x^4 + 1x^3 + x^2} \\
 & -x^3 + 0x^2 + 0x \\
 & \underline{-x^3 - x^2 - x} \\
 & x^2 + x + 4 \\
 & \underline{x^2 + x + 1} \\
 & 3
\end{array}$$

$\therefore$ the quotient is $x^2 - x + 1$, and the remainder is 3.

$$\therefore \frac{x^4 + x^2 + 4}{x^2 + x + 1} = x^2 - x + 1 + \frac{3}{x^2 + x + 1}$$

The check is left for you.

Divide.

10. $\dfrac{x^3 - 6x^2 + 12x - 8}{x^2 - 4x + 4}$
11. $\dfrac{x^4 - x^3 + 7x + 5}{x^2 + 2x + 1}$
12. $\dfrac{3x^4 + x^3 - 2x + 7}{x^2 - x + 1}$
13. $\dfrac{4x^4 + 5x^2 + 12x + 1}{2x^2 - x - 3}$
14. $\dfrac{6x^4 + 5x^3 - 9x^2 + 5}{3x^2 + x - 2}$
15. $\dfrac{10x^4 - 2x^3 + x^2 + x - 3}{2x^2 - 1}$

Mixed Review Exercises

In Exercises 1–4, assume that y varies directly as x.

1. If $y = 10$ when $x = 4$, find y when $x = 5$.
2. If $y = 2$ when $x = 0.5$, find y when $x = 6$.
3. If $y = 7$ when $x = \sqrt{3}$, find y when $x = \sqrt{15}$.
4. If $y = 4.2$ when $x = 0.4$, find y when $x = 2.5$.

5–8. Solve 1–4, assuming that y varies inversely as x.

NAME ______________________ DATE ______________

8–4 Synthetic Division

Objective: To use synthetic division to divide a polynomial by a first-degree binomial.

Vocabulary

Synthetic division An efficient way to divide a polynomial in x by a binomial of the form $x - c$. It uses only the coefficients of the polynomials involved.

Example 1 Use synthetic division to divide $x^4 - 3x^3 + 2x^2 - 4$ by $x + 2$.

Solution

1. Arrange the terms of the polynomial in order from the largest to the smallest exponent: $x^4 - 3x^3 + 2x^2 + 0x - 4$. Then write the coefficients of the terms in a row. The coefficient of x^4 is 1. Use a 0 as the coefficient of any power of x that is "missing."
2. Write the constant term c of the binomial divisor $x - c$ to the left of the coefficients. *Note:* Since $x + 2 = x - (-2)$, you use -2 for c.
3. Bring down the first coefficient.
4. Multiply what you've brought down by c and write the result below the next coefficient.
5. Add.
6. Repeat Steps 4 and 5 for every coefficient.

Step 2 ↓

-2	1	-3	2	0	-4	← *Step 1*
	↓	-2	10	-24	48	
Step 3 →	1	-5	12	-24	44	
		↑ *Steps 4, 5*			↑ *Remainder*	

The resulting numbers 1, -5, 12, and -24 are the coefficients of the quotient. The degree of the quotient will be 1 less than the degree of the dividend, or 3. The quotient is $1x^3 - 5x^2 + 12x - 24$, and the remainder is 44.

$$\therefore \frac{x^4 - 3x^3 + 2x^2 - 4}{x + 2} = x^3 - 5x^2 + 12x - 24 + \frac{44}{x + 2}$$

Example 2 Express as a polynomial (a) the divisor, (b) the dividend, (c) the quotient, and (d) the remainder for the synthetic division shown below.

3	1	2	-10	0	-41
		3	15	15	45
	1	5	5	15	4

(*Solution is on the next page.*)

NAME ______________________________ DATE ______________

8–4 Synthetic Division *(continued)*

Solution You need to put the variables back in place.

a. The divisor has the form $x - c$, and $c = 3$. Thus, the divisor is $x - 3$.

b. There are 5 coefficients in the top row. The coefficient -41 is a constant. The dividend is the *fourth*-degree polynomial $x^4 + 2x^3 - 10x^2 + 0x - 41$, or $x^4 + 2x^3 - 10x^2 - 41$.

c. The first four coefficients in the bottom row form the quotient. The coefficient 15 is a constant. Thus, the quotient is the *third*-degree polynomial $x^3 + 5x^2 + 5x + 15$.

d. The remainder is the last number in the bottom row, or 4.

For each synthetic division shown below, express as a polynomial (a) the divisor, (b) the dividend, (c) the quotient, and (d) the remainder. Use x as the variable.

1.

-3	1	5	7	3
		-3	-6	-3
	1	2	1	0

2.

4	1	-6	0	18	-10
		4	-8	-32	-56
	1	-2	-8	-14	-66

Example 3 Divide using synthetic division: $\dfrac{x^4 - 16}{x - 2}$

Solution Use 2 for c. Insert zeros for the missing terms x^3, x^2, and x in the dividend.

2	1	0	0	0	-16
		2	4	8	16
	1	2	4	8	0

The quotient is $1x^3 + 2x^2 + 4x + 8$, and the remainder is 0.

$$\therefore \frac{x^4 - 16}{x - 2} = x^3 + 2x^2 + 4x + 8$$

Divide using synthetic division.

3. $\dfrac{x^3 + 2x^2 - 2x - 4}{x + 1}$

4. $\dfrac{x^3 + 2x^2 - 2x - 1}{x - 1}$

5. $\dfrac{x^3 + x^2 - 8x - 12}{x + 3}$

6. $\dfrac{2x^3 - 5x^2 + 4x - 3}{x - 2}$

7. $\dfrac{x^4 + 2x^3 + 3x^2 - x - 3}{x + 1}$

8. $\dfrac{3x^4 + 10x^3 - 7x^2 + 5x + 6}{x + 4}$

9. $\dfrac{x^4 + 3x^2 - x - 5}{x + 2}$

10. $\dfrac{2x^4 - 7x^3 - x + 10}{x - 3}$

11. $\dfrac{x^5 - 32}{x - 2}$

12. $\dfrac{x^7 - 1}{x - 1}$

13. $\dfrac{x^6 + 2x^3 - 7x^2 + 8x + 4}{x + 2}$

14. $\dfrac{x^6 - 3x^5 + 14x + 4}{x - 2}$

NAME ______________________________ DATE ______________

8–5 The Remainder and Factor Theorems

Objective: To use the remainder and factor theorems to find factors of polynomials and to solve polynomial equations.

Vocabulary

Synthetic substitution Another name for synthetic division.

Remainder theorem Let $P(x)$ be a polynomial of positive degree n, with real or complex coefficients. For any number c, $P(x) = Q(x) \cdot (x - c) + P(c)$, where $Q(x)$ is a polynomial of degree $n - 1$. In other words, the *remainder* when the polynomial $P(x)$ is divided by the binomial $x - c$ is equal to $P(c)$.

Factor theorem The polynomial $P(x)$ has $x - r$ as a factor if and only if r is a root of the equation $P(x) = 0$.

Depressed equation The equation $Q(x) = 0$, where $Q(x)$ is the quotient when $P(x)$ is divided by one of its factors. The roots of the depressed equation are also roots of the equation $P(x) = 0$.

Example 1 Use synthetic substitution to find the value $P(-3)$ for the polynomial $P(x) = x^4 + 4x^3 - x^2 - 10x + 12$.

Solution By the remainder theorem, $P(-3)$ is equal to the remainder when $P(x)$ is divided by $x - (-3)$. Using synthetic substitution, divide by -3.

$c \to -3$	1	4	−1	−10	12	
		−3	−3	12	−6	
	1	1	−4	2	6 ← $P(c)$	$\therefore P(-3) = 6$

Check: You can check the answer by directly calculating $P(-3)$.

$$P(-3) = (-3)^4 + 4(-3)^3 - (-3)^2 - 10(-3) + 12$$
$$= 81 - 108 - 9 + 30 + 12 = 6 \ \checkmark$$

Use synthetic substitution to find $P(c)$ for the given polynomial $P(x)$ and the given number c.

1. $P(x) = x^3 - 3x^2 - 4x + 6;\ c = 3$
2. $P(x) = x^3 + 2x^2 - 6x - 4;\ c = -2$
3. $P(x) = 12 - 17x - 10x^2 - x^3;\ c = -3$
4. $P(x) = x^4 - 4x^2 + x + 4;\ c = -2$
5. $P(x) = 10 - 16x^2 + 6x^3;\ c = -\frac{1}{3}$
6. $P(x) = 4x^3 + 2x^2 - 4x + 6;\ c = -\frac{3}{2}$

Example 2 Determine whether the binomial is a factor of the given polynomial.

a. $x - 3;\ P(x) = 2x^3 - 4x^2 - 7x + 3$ **b.** $z - i;\ P(z) = z^3 + z^2 + z + 1$

Solution

a. By the factor theorem, if $P(3) = 0$, then $x - 3$ is a factor. Find $P(3)$.

$P(3) = 2(3)^3 - 4(3)^2 - 7(3) + 3$
$= 54 - 36 - 21 + 3 = 0$

$\therefore x - 3$ is a factor of $P(x)$.

b. By the factor theorem, if $P(i) = 0$, then $z - i$ is a factor. Find $P(i)$.

$P(i) = i^3 + i^2 + i + 1$
$= -i + -1 + i + 1 = 0$

$\therefore z - i$ is a factor of $P(z)$.

NAME ______________________ DATE ______________

8–5 The Remainder and Factor Theorems *(continued)*

Use the factor theorem to determine whether the binomial is a factor of the given polynomial.

7. $x - 1$; $P(x) = x^6 - x^4 + x^2 - 1$ **8.** $y + 1$; $P(y) = y^7 + y^4 + y$

9. $n - 2$; $P(n) = n^4 - 2n^3 - 3n + 6$ **10.** $z + \sqrt{5}$; $P(z) = z^4 - 2z^3 - 3z^2 + 10z - 10$

11. $y + 2$; $P(y) = y^4 - y^2 + 4y + 2$ **12.** $x - i$; $P(x) = x^5 - x^4 + x^3 - x^2 + x - 1$

Example 3 Solve $x^3 + x^2 - 11x - 3 = 0$ given that 3 is a root.

Solution Find the depressed equation by dividing $P(x) = x^3 + x^2 - 11x - 3$ by $x - 3$.

$$\begin{array}{r|rrrr} 3 & 1 & 1 & -11 & -3 \\ & & 3 & 12 & 3 \\ \hline & 1 & 4 & 1 & 0 \end{array}$$

So $x^3 + x^2 - 11x - 3 = (x - 3)(x^2 + 4x + 1)$.

The *depressed equation* is $x^2 + 4x + 1 = 0$. Solve the depressed equation.

$$x = \frac{-4 \pm \sqrt{16 - 4(1)(1)}}{2(1)} = \frac{-4 \pm 2\sqrt{3}}{2} = -2 \pm \sqrt{3}$$

$\therefore$ the solution set is $\{3, -2 + \sqrt{3}, -2 - \sqrt{3}\}$.

In Exercises 13–15, a root of the equation is given. Solve the equation.

13. $x^3 + 3x^2 - 18x - 40 = 0$; 4 **14.** $x^3 - 5x - 2 = 0$; -2 **15.** $x^3 - x^2 + 3x + 5 = 0$; -1

Example 4 Find a polynomial equation that has 3, $3i$, and $-3i$ as roots.

Solution By the factor theorem, the required polynomial must have factors $(x - 3)$, $(x - 3i)$, and $(x - (-3i))$. Therefore, a solution is

$$(x - 3)(x - 3i)(x + 3i) = 0, \text{ or}$$
$$x^3 - 3x^2 + 9x - 27 = 0.$$

Find a polynomial equation that has the given numbers as roots.

16. 1, -2, 3 **17.** -1, 1, -2 **18.** 0, -2, 4 **19.** -4, $2i$, $-2i$

Mixed Review Exercises

Divide.

1. $\dfrac{x^2 + 5x - 2}{x + 1}$ **2.** $\dfrac{y^3 + 1}{y^2 - y + 1}$ **3.** $\dfrac{6a^2 - 9a + 5}{2a - 1}$ **4.** $\dfrac{4c^3 + 7c + 4}{2c^2 - c + 4}$

Without solving, determine whether the roots of each equation are rational, irrational, or imaginary.

5. $2x^2 + 7x + 8 = 0$ **6.** $2x^2 - 7x + 6 = 0$ **7.** $2x^2 - 7x - 6 = 0$

NAME ________________________________ DATE ______________

8–6 Some Useful Theorems

Objective: To find or solve a polynomial equation with real coefficients and positive degree *n* by using these facts:

1. There are exactly *n* roots.
2. The imaginary roots occur in conjugate pairs.
3. Descartes' rule of signs gives information about the number of positive and negative real roots.

Vocabulary

Degree of a polynomial equation The degree of a polynomial equation of the form $P(x) = 0$ is the degree of $P(x)$. Example: The degree of $x^3 - 2x + 7 = 0$ is 3.

Theorem Every polynomial equation with complex coefficients and positive degree n has exactly n roots (provided that a multiple root is counted as many times as it appears). Examples: $4x^3 + x + 5$ has exactly 3 roots; $x^5 - 5ix^4 + 9x^2 - 2ix + 12$ has exactly 5 roots.

Conjugate root theorem If a polynomial equation with *real* coefficients has $a + bi$ as a root (a and b real, $b \neq 0$), then $a - bi$ is also a root.

Variation in sign A change of sign from one term of a polynomial to the next. Missing terms are ignored. Example: $3 + x^5 + 2x - x^2 = x^5 - x^2 + 2x + 3$ has two variations in sign.

Descartes' rule of signs Let $P(x)$ be a simplified polynomial with real coefficients and terms arranged in decreasing degree of x.

1. The number of *positive* real roots of $P(x) = 0$ equals the number of variations of sign of $P(x)$ or is fewer than this number by an even integer. Example: $P(x) = x^5 - x^2 + 2x + 3$ has two variations in sign, so the equation $x^5 - x^2 + 2x + 3 = 0$ has either two or zero positive real roots.
2. The number of *negative* real roots of $P(x) = 0$ equals the number of variations in sign of $P(-x)$ or is fewer than this number by an even integer. Example: $P(-x) = (-x)^5 - (-x)^2 + 2(-x) + 3 = -x^5 - x^2 - 2x + 3$ has one variation in sign. Thus, the equation $-x^5 - x^2 + 2x + 3 = 0$ has only one negative real root.

Example 1 Find a cubic equation with integral coefficients that has 3 and $2 - i$ as roots.

Solution The third root must be the conjugate of $2 - i$, which is $2 + i$. An equation with these three roots is:

$$\begin{aligned}
(x - 3)[x - (2 - i)][x - (2 + i)] &= 0\\
(x - 3)[x^2 - (2 - i)x - (2 + i)x + (2 - i)(2 + i)] &= 0\\
(x - 3)(x^2 - 4x + 5) &= 0\\
x^3 - 3x^2 - 4x^2 + 12x + 5x - 15 &= 0\\
x^3 - 7x^2 + 17x - 15 &= 0
\end{aligned}$$

Find a cubic equation with integral coefficients that has the given numbers as roots.

1. $2, i$ **2.** $5, i\sqrt{3}$ **3.** $-1, 1 - i$ **4.** $-3, 2 + 3i$

8–6 Some Useful Theorems *(continued)*

Example 2 Solve $x^3 - 5x^2 + 11x - 15 = 0$ given that $1 - 2i$ is a root.

Solution Let $P(x) = x^3 - 5x^2 + 11x - 15$.

1. If $1 - 2i$ is a root of $P(x) = 0$, then $1 + 2i$ is also a root, by the conjugate root theorem. Therefore, $x - (1 - 2i)$ and $x - (1 + 2i)$ are factors of $P(x)$. Their product, $[x - (1 - 2i)][x - (1 + 2i)] = x^2 - 2x + 5$, is also a factor of $P(x)$; that is $P(x) = (x^2 - 2x + 5) \cdot Q(x)$.
2. Long division will show that $Q(x) = P(x) \div (x^2 - 2x + 5) = x - 3$.
3. The depressed equation, $x - 3 = 0$, has root 3.

$\therefore$ the solution set of the given equation is $\{1 - 2i, 1 + 2i, 3\}$.

In each exercise all but one of the equation's roots are given. Find the remaining root. Check your answer by substituting it for x in the equation.

5. $x^3 + 2x^2 + 2x + 40 = 0$; -4 and $1 + 3i$ **6.** $x^4 - 8x^3 + 22x^2 - 56x + 105 = 0$; 3, 5, $i\sqrt{7}$

In each exercise a root of the equation is given. Solve the equation.

7. $x^3 - 10x^2 + 35x - 38 = 0$; $4 + i\sqrt{3}$ **8.** $x^3 - 5x^2 + 2x - 10 = 0$; $-i\sqrt{2}$

9. $x^4 - 2x^3 - 4x^2 + 14x - 5 = 0$; $2 + i$ **10.** $x^4 + 8x^3 + 19x^2 + 2x - 30 = 0$; $-3 - i$

Example 3 List the possibilities for the nature of the roots (positive real, negative real, and imaginary) for the equation $P(x) = 0$, where

$$P(x) = -3x^4 + x^3 - 6x^2 + 3x + 2.$$

Solution

1. $P(x)$ has three variations in sign, so by the first part of Descartes' rule, the number of positive real roots of $P(x) = 0$ is 3 or 1.
2. $P(-x) = -3(-x)^4 + (-x)^3 - 6(-x)^2 + 3(-x) + 2$
   $= -3x^4 - x^3 - 6x^2 - 3x + 2$
   $P(-x)$ has one variation in sign, so by the second part of Descartes' rule the number of negative roots of $P(x) = 0$ is 1.
3. Since $P(x)$ has four roots in all, by the first theorem on the previous page, there are only two possibilities. These are listed in the chart below.

Number of positive real roots	Number of negative real roots	Number of imaginary roots
3	1	0
1	1	2

List the possibilities for the nature of the roots of each equation.

11. $x^3 - 2x^2 + 3x + 3 = 0$ **12.** $x^4 + x^3 - 16 = 0$

13. $x^4 - 5x^3 - 2x^2 + x + 1 = 0$ **14.** $x^6 - x^4 + x - 1 = 0$

NAME ______________________ DATE ____________

8–7 *Finding Rational Roots*

Objective: To find rational roots of polynomial equations with integral coefficients.

Vocabulary

Rational root theorem Suppose that a polynomial equation with integral coefficients has the root $\frac{h}{k}$, where h and k are relatively prime integers (that is, integers whose GCF is 1). Then h must be a factor of the constant term of the polynomial and k must be a factor of the coefficient of the highest-degree term.

Example 1 Solve the equation $2x^4 + 5x^3 + 3x^2 + 15x - 9 = 0$ by first using the rational root theorem to find any rational roots.

Solution Since h is ± 1, ± 3, or ± 9 (the factors of 9), and k is ± 1 or ± 2 (the factors of 2), the only possible rational roots are

$$\pm 1, \pm 3, \pm 9, \pm\frac{1}{2}, \pm\frac{3}{2}, \text{ and } \pm\frac{9}{2}.$$

You should try the integers first. Check 1 and -1 by substitution. You will find that neither is a root. Use synthetic substitution to try the other possibilities.

3	2	5	3	15	−9
		6	33	108	369
	2	11	36	123	360

3 is *not* a root.

−3	2	5	3	15	−9
		−6	3	−18	9
	2	−1	6	−3	0

Since the remainder is 0, -3 *is* a root.

Once you have a root, use the depressed equation $(2x^3 - x^2 + 6x - 3 = 0)$ to find the remaining roots. The depressed equation has ± 1, ± 3, $\pm\frac{1}{2}$, and $\pm\frac{3}{2}$ as possible rational roots. You eliminated 1, -1, and 3 using the original equation. Since -3 may be a double root, it is still a possibility, along with $\pm\frac{1}{2}$ and $\pm\frac{3}{2}$.

−3	2	−1	6	−3
		−6	21	−81
	2	−7	27	−84

-3 is *not* a double root.

$\frac{1}{2}$	2	−1	6	−3
		1	0	3
	2	0	6	0

$\frac{1}{2}$ *is* a root.

The second depressed equation is $2x^2 + 6 = 0$ (or $x^2 + 3 = 0$). If you rewrite this quadratic equation as $x^2 = -3$, and take the square root of both sides, you obtain the solutions $x = \pm i\sqrt{3}$.

$\therefore$ the original equation has solution set $\left\{-3, \frac{1}{2}, i\sqrt{3}, -i\sqrt{3}\right\}$.

Find any rational roots. If the equation has at least one rational root, solve it completely.

1. $x^3 + 2x^2 - 13x + 10 = 0$

2. $x^3 - 4x^2 - x + 4 = 0$

3. $x^3 - 3x^2 - 5x + 15 = 0$

4. $x^4 + x^2 + 9 = 0$

NAME ______________________ DATE ______________

8–7 Finding Rational Roots *(continued)*

Find any rational roots. If the equation has at least one rational root, solve it completely.

5. $x^4 - 4x^3 + 6x^2 - 12x + 9 = 0$
6. $x^4 - x^3 - 6x^2 + 14x - 12 = 0$
7. $4x^4 - 8x^3 - 9x^2 + 2x + 2 = 0$
8. $3x^4 - 5x^3 + 10x^2 - 20x - 8 = 0$

Example 2 Show that $\sqrt{8}$ is irrational.

Solution $\sqrt{8}$ is a root of the equation $x^2 - 8 = 0$. The only possible rational roots of this equation are ± 1, ± 2, ± 4, and ± 8. If you substitute these numbers into the equation, you will find that none of them satisfy it. Since $x^2 - 8 = 0$ has no rational roots, $\sqrt{8}$ is irrational.

Use the method of Example 2 to show that the following numbers are irrational.

9. $\sqrt{2}$
10. $\sqrt{10}$
11. $\sqrt[3]{-9}$
12. $\sqrt[3]{5}$
13. $\sqrt[4]{9}$
14. $\sqrt[5]{-12}$

Example 3 Verify that $\sqrt{3} - \sqrt{2}$ is a root of $x^4 - 10x^2 + 1 = 0$ and use this fact to show that the number is irrational.

Solution To verify that $\sqrt{3} - \sqrt{2}$ is a root, substitute it into the equation.

$$(\sqrt{3} - \sqrt{2})^4 - 10(\sqrt{3} - \sqrt{2})^2 + 1 \stackrel{?}{=} 0$$
$$49 - 20\sqrt{6} - 10(5 - 2\sqrt{6}) + 1 \stackrel{?}{=} 0$$
$$49 - 20\sqrt{6} - 50 + 20\sqrt{6} + 1 \stackrel{?}{=} 0$$
$$0 = 0 \ \checkmark$$

$\therefore \sqrt{3} - \sqrt{2}$ is a root, or solution, of the equation $x^4 - 10x^2 + 1 = 0$.

By the rational root theorem, the only possible rational roots of the equation are ± 1. If you substitute these numbers into the equation, you will find that neither of them is a solution. Since the equation has no rational roots, $\sqrt{3} - \sqrt{2}$ is an irrational number.

Verify that the given number is a root of the equation and use this fact to show that the number is irrational.

15. $\sqrt{2} + \sqrt{5}$; $x^4 - 14x^2 + 9 = 0$
16. $\sqrt{3} - \sqrt{6}$; $x^4 - 18x^2 + 9 = 0$

Mixed Review Exercises

Solve each equation completely. In Exercises 7 and 8, one root is given.

1. $5(2 - x) + 3 = 7 - 4x$
2. $2x^2 - 2x + 3 = 0$
3. $\sqrt{2x + 7} = x + 2$
4. $\frac{1}{x} + \frac{x}{x - 1} + \frac{1}{x - x^2} = 0$
5. $8x^{-2} + 4x^{-1} - 12 = 0$
6. $x^3 - 6x^2 + 13x - 10 = 0$
7. $2x^3 - 11x^2 + 12x + 9 = 0$; 3
8. $x^4 - x^2 - 6 = 0$; $i\sqrt{2}$

NAME ______________________ DATE ______________

8–8 Approximating Irrational Roots

Objective: To approximate the real roots of a polynomial equation $P(x) = 0$ by using the graph of $y = P(x)$.

Vocabulary

Intermediate-value theorem
If P is a polynomial function with real coefficients, and m is any number between $P(a)$ and $P(b)$, then there is at least one number c between a and b for which $P(c) = m$.

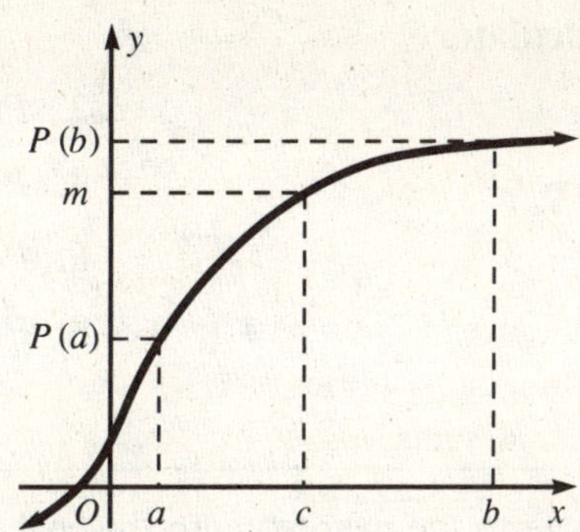

CAUTION When using a graph to approximate the roots of a function, keep in mind that you can only approximate *real* roots by this method, not imaginary roots.

Example 1 **a.** Graph the equation $y = x^3 - 4x^2 - 4x + 17$.

b. Use the graph to estimate to the nearest unit the roots of the equation $x^3 - 4x^2 - 4x + 17 = 0$.

Solution **a.** You may wish to use a computer or graphing calculator if one is available. Otherwise make a table of values, plot the corresponding points, and join the points with a smooth unbroken curve. When making your table, apply Descartes' rule of signs to choose an appropriate range of values for x. When making your graph, use a smaller scale on the y-axis than on the x-axis to accommodate the greater range of y-values.

$y = x^3 - 4x^2 - 4x + 17$

x	y
−3	−34
−2	1
−1	16
0	17
1	10
2	1
3	−4
4	1
5	22

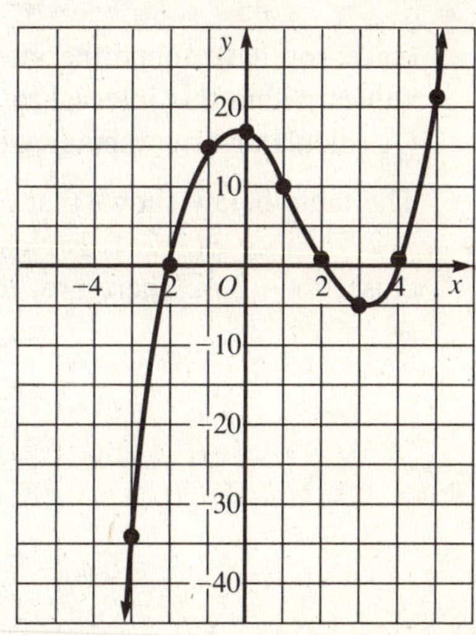

b. The roots of the equation are the x-coordinates of the points where the graph intersects the x-axis. To the nearest unit, the roots are −2, 2, and 4.
Note: Roots occur in intervals where the sign of the y-value changes.

NAME ______________________ DATE ______________

8–8 *Approximating Irrational Roots* (continued)

For each polynomial P, draw a graph to approximate to the nearest unit the real root(s) of the equation $P(x) = 0$. You may wish to use a computer or a graphing calculator.

1. $P(x) = x^3 - 6$
2. $P(x) = x^3 + 9$
3. $P(x) = x^3 - x^2 + 5x + 6$
4. $P(x) = -x^3 - x + 8$
5. $P(x) = x^3 - 2x^2 + x - 3$
6. $P(x) = 2x^3 - 3x^2 - 3x + 3$
7. $P(x) = x^4 + x^3 - 14$
8. $P(x) = x^4 - 2x^3 - 4x^2 + 4x$

Example 2 Approximate to the nearest tenth the real zero of the function $P(x) = x^3 + 4x^2 + 10x + 15$.

Solution Recall that the zeros of P are the roots of the polynomial equation $P(x) = 0$. By Descartes' rule, $P(x) = 0$ has no positive roots, so make a table using nonpositive values of x. Begin with integers, and look for the interval where the value of $P(x)$ changes sign.

The table below shows that $P(x)$ changes sign when x is in the interval between -3 and -2. By the intermediate-value theorem, there is a value r between -3 and -2 such that $P(r) = 0$.

	x	$P(x)$	
	0	15	
	−1	8	
	−2	3	
$r \rightarrow$	−3	−6	$\leftarrow 0$

Once you have found the interval that contains the root, make a table using values within this interval to find the answer to the nearest tenth. (A calculator or computer will be very helpful.)

The table below shows that $P(x)$ changes sign when x is between -2.4 and -2.5. Since $P(-2.4)$ is closer to 0 than $P(-2.5)$, you may assume that r is closer to -2.4. Therefore, to the nearest tenth, the real zero of P is -2.4.

	x	$P(x)$	
	−2.1	2.379	
	−2.2	1.712	
	−2.3	0.993	
	−2.4	0.216	
$r \rightarrow$	−2.5	−0.625	$\leftarrow 0$

9–13. In Exercises 1–5 above, each polynomial equation $P(x) = 0$ has only one real root. Approximate it to the nearest tenth.

NAME ______________________________ DATE ______________

8–9 Linear Interpolation

Objective: To use linear interpolation to find values not listed in a given table of data.

Vocabulary

Linear interpolation A process used to approximate values not given in a table of data. The method is based upon the assumption that within a given interval, the change in one value is proportional to the change in the other value.

Example 1 Use linear interpolation and the table to find these values to the nearest integer:

a. the approximate number of bachelor's degrees awarded in computer science in 1971

b. the approximate year that 10,000 bachelor's degrees were awarded in computer science

Year	Bachelor's degrees awarded in computer science (U.S.)
1968	459
1972	3,402
1976	5,700
1980	11,154
1984	32,172

Solution **a.** 1971 is three quarters of the way from 1968 to 1972. Assume that the value you're looking for is similarly three quarters of the way from 459 to 3402. Write a proportion to find three quarters of the difference between 459 and 3402, and then add this amount to 459 to find d, the approximate number of degrees awarded.

Year	Bachelor's degrees awarded in computer science (U.S.)
1968	459
1971	d
1972	3402

(Year differences: 1968 to 1971 is 3; 1968 to 1972 is 4. Degree differences: 459 to d is x; 459 to 3402 is 2943.)

$$\frac{x}{2943} = \frac{3}{4}$$

$$x = \frac{3}{4}(2943) \approx 2207$$

$$d = 459 + x$$

$$\approx 459 + 2207 = 2666$$

∴ the number of bachelor's degrees awarded in computer science in 1971 was about 2666.

b. Finding a value in the first column to correspond with a given value in the second column is called *inverse interpolation*. While the name of the process is different, the method is basically the same.

Year	Bachelor's degrees awarded in computer science (U.S.)
1976	5,700
y	10,000
1980	11,154

(Year differences: 1976 to y is x; 1976 to 1980 is 4. Degree differences: 5,700 to 10,000 is 4300; 5,700 to 11,154 is 5454.)

$$\frac{x}{4} = \frac{4300}{5454}$$

$$x = 4\left(\frac{4300}{5454}\right) \approx 3$$

$$y = 1976 + x$$

$$\approx 1976 + 3 = 1979$$

∴ there were 10,000 bachelor's degrees awarded in computer science in 1979.

NAME ______________________________ DATE ______________

8–9 Linear Interpolation *(continued)*

Use linear interpolation and the table in Example 1 to find the approximate number of bachelor's degrees awarded in computer science in each of the following years.

1. 1975 **2.** 1981 **3.** 1978 **4.** 1969

Use inverse interpolation to approximate the year in which the given number of bachelor's degrees were awarded in computer science.

5. 1930 **6.** 7080 **7.** 27,000 **8.** 3970

Example 2 The table gives the temperature in degrees Fahrenheit on a spring day in Boston, Massachusetts.

a. Approximate the temperature at 3:40 P.M.

b. At about what time was the temperature 40°?

Time (P.M.)	Temperature
1:00	68°
2:00	66°
3:00	63°
4:00	59°
5:00	53°
6:00	45°
7:00	39°

Solution **a.** The temperature values in the table *decrease* as time *increases*. As a result, you will have to *subtract* the change in temperature at the last step.

		Time (P.M.)	Temperature		
60	40	3:00	63°	x	4
		3:40	t		
		4:00	59°		

$$\frac{x}{4} = \frac{40}{60}$$

$$x = 4\left(\frac{40}{60}\right) \approx 2.7°$$

$$t = 63 - x$$

$$\approx 63 - 2.7 = 60.3°$$

∴ the temperature at 3:40 was about 60.3°.

b. In this part of the problem, since time is increasing, you will have to *add* the change in time at the last step. When you do this, be sure not to confuse minutes and hours.

		Time (P.M.)	Temperature		
60	x	6:00	45°	5	6
		t	40°		
		7:00	39°		

$$\frac{x}{60} = \frac{5}{6}$$

$$x = 60\left(\frac{5}{6}\right) = 50 \text{ min}$$

∴ the temperature was 40° at about 6:50 P.M.

Use linear interpolation and the table in Example 2 to approximate the temperature at each given time.

9. 4:20 P.M. **10.** 5:45 P.M. **11.** 1:15 P.M. **12.** 6:24 P.M.

Use inverse interpolation and the table in Example 2 to find the approximate time associated with the given temperature.

13. 55° **14.** 44° **15.** 62° **16.** 49°

NAME ______________________ DATE ____________

9 Analytic Geometry

9–1 Distance and Midpoint Formulas

Objective: To find the distance between any two points and the midpoint of the line segment joining them.

Vocabulary

Pythagorean theorem If the length of the hypotenuse of a right triangle is c, and the length of the other two sides (legs) are a and b, then $c^2 = a^2 + b^2$.

Distance formula The distance between the points $P_1(x_1, y_1)$ and $P_2(x_2, y_2)$ is

$$P_1P_2 = \sqrt{(x_2 - x_1)^2 + (y_2 - y_1)^2}.$$

Midpoint formula The midpoint of the line segment joining $P_1(x_1, y_1)$ and $P_2(x_2, y_2)$ is

$$M\left(\frac{x_1 + x_2}{2}, \frac{y_1 + y_2}{2}\right).$$

Symbol PQ (the distance between points P and Q)

Example 1 Find the distance between $P_1(-3, -5)$ and $P_2(-1, 6)$.

Solution

$$\begin{aligned} P_1P_2 &= \sqrt{(-1 - (-3))^2 + (6 - (-5))^2} \\ &= \sqrt{2^2 + 11^2} \\ &= \sqrt{4 + 121} \\ &= \sqrt{125} \\ &= 5\sqrt{5} \end{aligned}$$

Find the distance between each pair of points.

1. $(5, 3), (1, 4)$

2. $(0, 6), (4, 9)$

3. $(-4, 8), (-7, 2)$

4. $(2, 3), (6, 5)$

5. $(3, 4), (1, -2)$

6. $(0, 6), (-4, 0)$

7. $(-3, -4), (1, 2)$

8. $(2, 2), (9, 9)$

9. $(2, 2), \left(\frac{1}{2}, -1\right)$

Example 2 Find the midpoint of the line segment joining $(-3, -5)$ and $(-1, 6)$.

Solution $\left(\frac{x_1 + x_2}{2}, \frac{y_1 + y_2}{2}\right) = \left(\frac{-3 + -1}{2}, \frac{-5 + 6}{2}\right) = \left(-2, \frac{1}{2}\right)$

Find the midpoint of the line joining the points.

10. $(0, 8), (10, 0)$

11. $(-4, 2), (-9, -6)$

12. $(0, 5), (-2, -3)$

13. $(8, 1), (3, -2)$

14. $(0, 0), (7, 7)$

15. $\left(\frac{1}{2}, -2\right), \left(\frac{5}{2}, 1\right)$

NAME ____________________ DATE ____________

9–1 Distance and Midpoint Formulas *(continued)*

Example 3 Find (a) the distance between the points $P_1(3 - \sqrt{5}, \sqrt{7} - 2)$ and $P_2(3 + \sqrt{5}, \sqrt{7} + 2)$, and (b) the midpoint of the segment joining them.

Solution

a. $P_1P_2 = \sqrt{[(3 + \sqrt{5}) - (3 - \sqrt{5})]^2 + [(\sqrt{7} + 2) - (\sqrt{7} - 2)]^2}$

$= \sqrt{(2\sqrt{5})^2 + 4^2} = \sqrt{20 + 16} = \sqrt{36} = 6$

b. Find x: $\dfrac{x_1 + x_2}{2} = \dfrac{(3 - \sqrt{5}) + (3 + \sqrt{5})}{2} = \dfrac{6}{2} = 3$

Find y: $\dfrac{y_1 + y_2}{2} = \dfrac{(\sqrt{7} - 2) + (\sqrt{7} + 2)}{2} = \dfrac{2\sqrt{7}}{2} = \sqrt{7}$

$\therefore$ the midpoint of $\overline{P_1P_2}$ is the point $(3, \sqrt{7})$.

Find (a) the distance between each pair of points and (b) the midpoint of the line segment joining the points. Express all radicals in simplest form.

16. $(6, \sqrt{7}), (4, -\sqrt{7})$

17. $(\sqrt{6} + 2, \sqrt{2} - \sqrt{3}), (\sqrt{6} - 2, \sqrt{2} + \sqrt{3})$

18. $(-r, s), (3r, 5s)$

19. $(r - s, r + s), (s - r, s + r)$

Example 4 $M(3, -1)$ is the midpoint of $\overline{PQ}$. Given $P(5, 7)$, find the coordinates of Q.

Solution Let Q be the point (x, y). Then use the midpoint formula.

Find x:

$$\frac{x_1 + x_2}{2} = 3$$
$$\frac{5 + x}{2} = 3$$
$$5 + x = 6$$
$$x = 1$$

Find y:

$$\frac{y_1 + y_2}{2} = -1$$
$$\frac{7 + y}{2} = -1$$
$$7 + y = -2$$
$$y = -9$$

$\therefore Q$ is the point $(1, -9)$.

Find the coordinates of Q given that M is the midpoint of $\overline{PQ}$.

20. $P(0, 0), M(2, -4)$

21. $P(6, 5), M(3, 4)$

22. $P(0, 9), M(-2, -3)$

Mixed Review Exercises

Graph each equation.

1. $x - 3y = 3$

2. $y = x^2 + 2$

3. $y = x^2 + 6x + 3$

Find the value of each function if $x = -3$.

4. $F(x) = -2x + 8$

5. $g(x) = \dfrac{x + 3}{2x - 1}$

6. $h(x) = x^4 + 5x^3 - 2x + 12$

NAME ______________________ DATE ______________

9–2 Circles

Objective: To learn the relationship between the center and the radius of a circle and the equation of the circle.

Vocabulary

Conic sections or conics Curves in a plane having second-degree equations. They can be obtained by cutting a cone, or two cones connected at their tips, with a plane.

Circle The set of all points in a plane that are a fixed distance, called the *radius*, from a fixed point, called the *center*.

Equation of a circle The circle with the center (h, k) and radius r has the equation

$$(x - h)^2 + (y - k)^2 = r^2.$$

Translation Sliding a graph to a new position in the coordinate plane without changing its shape.

Example 1 Find an equation of the circle with center $(3, -1)$ and radius 2.

Solution Substitute $h = 3$, $k = -1$, and $r = 2$ in this equation:

$$(x - h)^2 + (y - k)^2 = r^2$$
$$(x - 3)^2 + [y - (-1)]^2 = 2^2$$
$$(x - 3)^2 + (y + 1)^2 = 4$$

Find an equation of the circle with the given center and radius.

1. $(2, 1)$; 4
2. $(0, -2)$; 1
3. $(4, -1)$; 7
4. $(-2, -1)$; 5
5. $(0, 0)$; 10
6. $(-5, -1)$; 11
7. $(3, 2)$; $\sqrt{3}$
8. $(-5, 2)$; $\frac{1}{5}$
9. $(4, -3)$; $2\sqrt{3}$ h k r

Example 2 Graph $(x - 2)^2 + (y + 4)^2 = 9$.

Solution 1 Use a translation. Begin with the graph of the equation $x^2 + y^2 = 9$, which is a circle with its center at the origin and radius 3. Slide this circle 2 units to the right and then 4 units down to get the graph of $(x - 2)^2 + [y - (-4)]^2 = 3^2$, or $(x - 2)^2 + (y + 4)^2 = 9$.

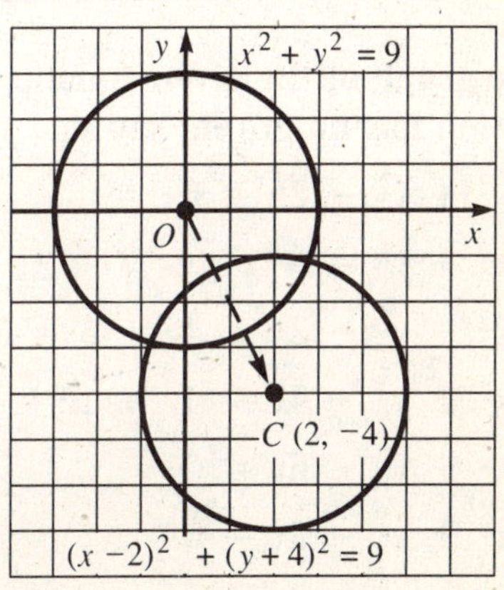

Solution 2 Rewrite the given equation in the form $(x - h)^2 + (y - k)^2 = r^2$. You should get $(x - 2)^2 + [y - (-4)]^2 = 3^2$. The graph of this equation is a circle. The center (h, k) is $(2, -4)$ and the radius r is 3.

9–2 Circles *(continued)*

Graph each equation. You may wish to check your graphs on a computer or a graphing calculator.

10. $x^2 + y^2 = 16$

11. $x^2 + y^2 = 9$

12. $(x - 2)^2 + (y - 3)^2 = 1$

13. $(x + 3)^2 + (y + 2)^2 = 36$

14. $(x - 4)^2 + y^2 = 49$

15. $x^2 + (y - 4)^2 = 9$

Example 3 If the graph of the given equation is a circle, find its center and radius. If the equation has no graph, say so.

a. $x^2 + y^2 + 6x - 2y + 5 = 0$ **b.** $x^2 + y^2 - 4x + 8y + 26 = 0$

Solution If the graph is a circle, the equation can be written in the form

$$(x - h)^2 + (y - k)^2 = r^2.$$

In order to write the equation in this form, you must complete the square *twice*, once using the terms in x, and once using the terms in y.

a.

$$x^2 + y^2 + 6x - 2y + 5 = 0$$

$$(x^2 + 6x + \underline{?}) + (y^2 - 2y + \underline{?}) = -5 + \underline{?} + \underline{?}$$ Rearrange terms.

$$(x^2 + 6x + 9) + (y^2 - 2y + 1) = -5 + 9 + 1$$ Add 9 and 1.

$$(x + 3)^2 + (y - 1)^2 = 5$$ Factor twice.

$\therefore$ the center is $(-3, 1)$ and the radius is $\sqrt{5}$.

b.

$$x^2 + y^2 - 4x + 8y + 26 = 0$$ Rearrange terms.

$$(x^2 - 4x + \underline{?}) + (y^2 + 8y + \underline{?}) = -26 + \underline{?} + \underline{?}$$ Add 4 and 16.

$$(x^2 - 4x + 4) + (y^2 + 8y + 16) = -26 + 4 + 16$$ Factor twice.

$$(x - 2)^2 + (y + 4)^2 = -6$$

Since the square of any number is positive and the sum of two positive numbers is positive, *no* ordered pair satisfies the equation. This equation has no graph.

If the graph of the given equation is a circle, find its center and radius. If the equation has no graph, say so.

16. $x^2 + y^2 - 25 = 0$

17. $x^2 + y^2 - 64 = 0$

18. $x^2 + y^2 = -6x$

19. $x^2 + y^2 = -4y$

20. $x^2 + y^2 - 2x + 4y + 3 = 0$

21. $x^2 + y^2 + 12x - 8y + 20 = 0$

22. $x^2 + y^2 - 4x + 4y + 12 = 0$

23. $x^2 + y^2 + 10x - 8y - 8 = 0$

24. $x^2 + y^2 - 6y = 0$

25. $x^2 + y^2 + 8x - 9 = 0$

26. $x^2 + y^2 - 5x = 0$

27. $x^2 + y^2 + 3x - 2y = 0$

NAME ______________________ DATE ______________

9–3 Parabolas

Objective: To learn the relationship between the focus, directrix, vertex, and axis of a parabola and the equation of the parabola.

Vocabulary

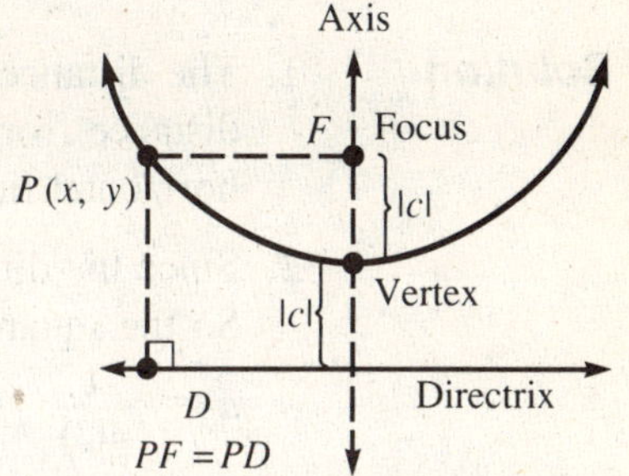

Parabola The set of all points equidistant from a fixed line, called the *directrix,* and a fixed point, called the *focus*. The parabola is symmetric about a line, called the *axis of symmetry*. The *vertex* is the point where the parabola crosses its axis. The vertex is midway between the focus and the directrix, and the distance between the focus and vertex is $|c|$.

If a parabola has an equation of the form

$y - k = a(x - h)^2$, where $a = \frac{1}{4c}$,

then the parabola:

1. opens up if $a > 0$; opens down if $a < 0$
2. has vertex (h, k)
3. has focus $(h, k + c)$
4. has horizontal directrix $y = k - c$
5. has axis of symmetry $x = h$

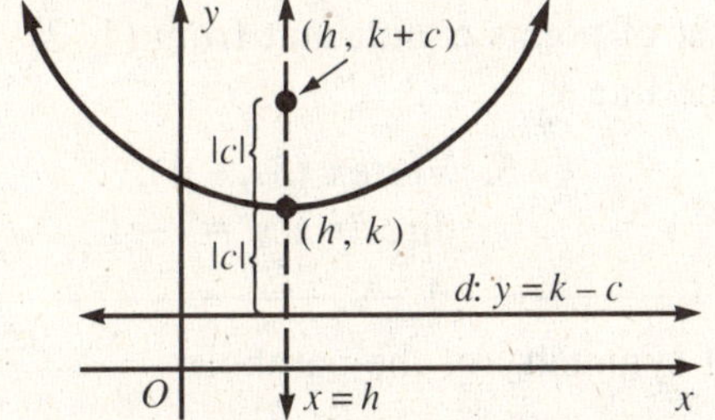

If a parabola has an equation of the form

$x - h = a(y - k)^2$, where $a = \frac{1}{4c}$,

then the parabola:

1. opens right if $a > 0$; opens left if $a < 0$
2. has vertex (h, k)
3. has focus $(h + c, k)$
4. has vertical directrix $x = h - c$
5. has axis of symmetry $y = k$

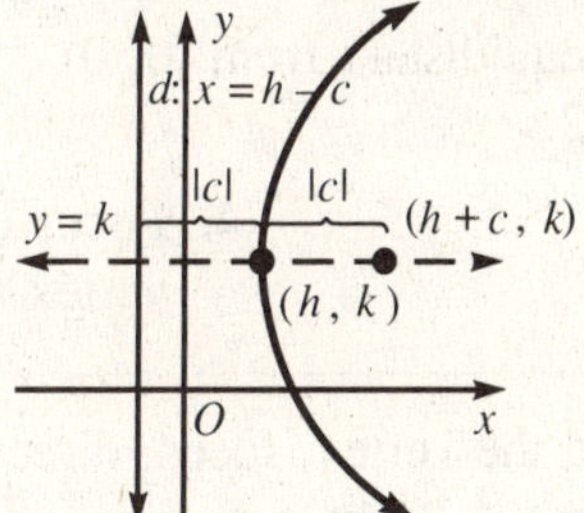

Example 1 Find an equation of the set of all points equidistant from the point $(-1, 2)$ and the line $x = 3$. Then graph the parabola.

Solution

1. The focus is $(-1, 2)$ and the directrix is the vertical line $x = 3$. The distance from the focus to directrix is 4 units. The distance from the vertex to the focus is then 2 units, so $|c| = 2$. The vertex is $(-1 + 2, 2) = (1, 2)$.
2. The directrix is to the right of the vertex, so the parabola opens to the left. So the squared term is the term with y, and c is negative. If $c = -2$, then $a = \frac{1}{4(-2)} = -\frac{1}{8}$.
3. The parabola has an equation of the form $x - h = a(y - k)^2$. Since $(h, k) = (1, 2)$ and $a = -\frac{1}{8}$, the equation is $x - 1 = -\frac{1}{8}(y - 2)^2$.
4. If $x = -1$, then $y = 6$ or -2. Use the points $(-1, 6)$ and $(-1, -2)$ as well as the vertex $(1, 2)$ to sketch the parabola. The graph is shown at the right.

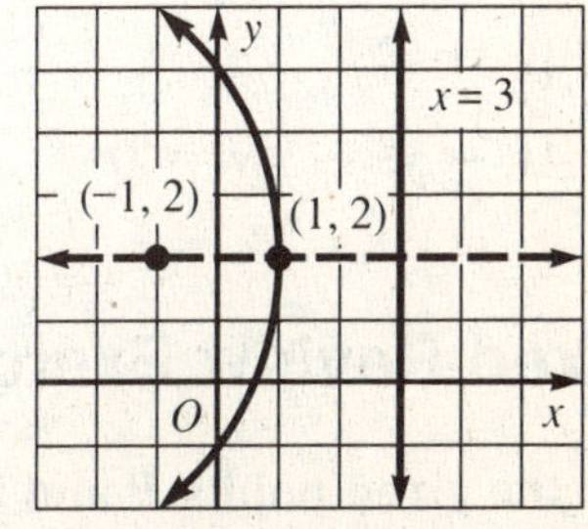

NAME ________________________________ DATE ______________

Example 2 Find an equation of the parabola that has focus (2, 5) and vertex (2, 2).

Solution

1. The distance from the focus to the vertex is 3 units, so $|c| = 3$. The distance from the vertex to the directrix is 3 units, so the directrix is the horizontal line $y = -1$.
2. Since the directrix is below the vertex, the parabola opens upward. So the squared term is the term with x, and c is positive. If $c = 3$, then $a = \frac{1}{4(3)} = \frac{1}{12}$.
3. The parabola has an equation of the form $y - k = a(x - h)^2$. Since $(h, k) = (2, 2)$ and $a = \frac{1}{12}$, the equation is $y - 2 = \frac{1}{12}(x - 2)^2$.
4. If $y = 5$, then $x = 8$ or -4. Use the points $(8, 5)$ and $(-4, 5)$ as well as the vertex $(2, 2)$ to sketch the parabola, shown at the right.

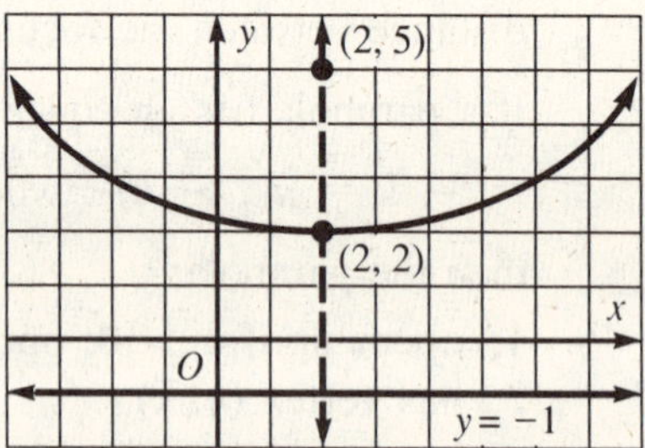

Find an equation of the parabola described. Then graph the parabola.

1. The set of points equidistant from (0, 0) and the line $y = 2$.
2. The set of points equidistant from (1, 2) and the line $x = -3$.
3. Focus (2, 0); vertex (2, 2)
4. Focus (2, −1); vertex (5, −1)
5. Vertex (3, −1); directrix $x = -1$

Example 3 Find the vertex, focus, directrix, and axis of symmetry of the parabola $4x = y^2 + 4y$.

Solution

$4x + \underline{?} = y^2 + 4y + \underline{?}$ Complete the square using the terms in y.

$4x + 4 = y^2 + 4y + 4$

$4(x + 1) = (y + 2)^2$

$x - (-1) = \frac{1}{4}[y - (-2)]^2$ Since $a = \frac{1}{4} = \frac{1}{4c}$, $c = 1$.

∴ vertex: $(-1, -2)$ focus: $(0, -2)$ directrix: $x = -2$ axis: $y = -2$

Find the vertex, focus, directrix, and axis of symmetry of each parabola. Then graph the parabola. You may wish to check your graphs on a computer or a graphing calculator.

6. $4x = y^2$
7. $8y + x^2 = 0$
8. $y^2 + 8y - 8x = 0$
9. $4x - y^2 - 4y = 0$
10. $x^2 = y + 2x$
11. $x^2 + 4x + 2y = 0$

Mixed Review Exercises

For the given points P and Q, find the midpoint of PQ.

1. $P(-1, 3)$, $Q(4, 5)$
2. $P(-1, 2)$, $Q(3, -2)$
3. $P(3, 2)$, $Q(1, 4)$
4. $P(2, -4)$, $Q(-3, -5)$

NAME ______________________ DATE ______________

9–4 Ellipses

Objective: To learn the relationship between the center, foci, and intercepts of an ellipse and the equation of the ellipse.

Vocabulary

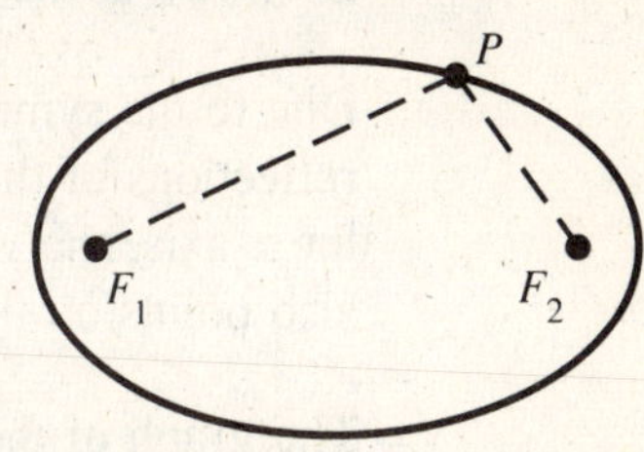

Ellipse The set of all points P in the plane such that the sum of the distances from P to two fixed points F_1 and F_2 is a given constant (that is, the sum $PF_1 + PF_2$ is the same for all positions of P). Each of the fixed points F_1 and F_2 is called a *focus* (plural: *foci*) of the ellipse. For each point P of the ellipse, PF_1 and PF_2 are the *focal radii* of P. The *center* of the ellipse is the midpoint of the line segment joining the foci. The *major axis* of the ellipse is the chord passing through the foci. The *minor axis* is the chord containing the center and perpendicular to the major axis. An ellipse is symmetric with respect to its major axis, its minor axis, and its center.

If an ellipse has an equation of the form

$$\frac{x^2}{a^2} + \frac{y^2}{b^2} = 1 \ (a > b > 0),$$

then the ellipse has:

1. center (0, 0)
2. x-intercepts $\pm a$
3. y-intercepts $\pm b$
4. horizontal major axis
5. foci $(-c, 0)$ and $(c, 0)$, where $c^2 = a^2 - b^2$
6. sum of focal radii $2a$

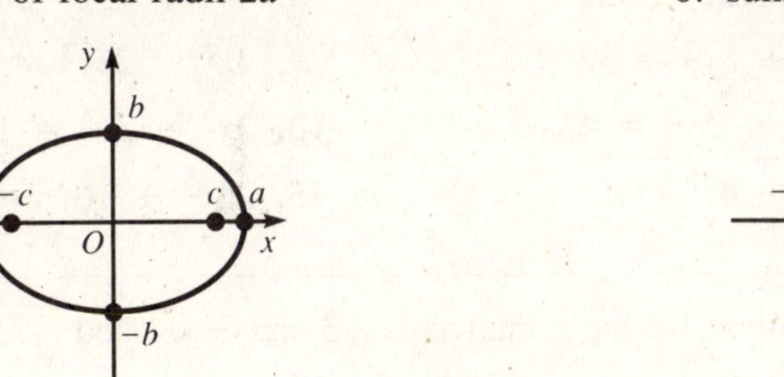

If an ellipse has an equation of the form

$$\frac{x^2}{b^2} + \frac{y^2}{a^2} = 1 \ (a > b > 0),$$

then the ellipse has:

1. center (0, 0)
2. x-intercepts $\pm b$
3. y-intercepts $\pm a$
4. vertical major axis
5. foci $(0, -c)$ and $(0, c)$, where $c^2 = a^2 - b^2$
6. sum of focal radii $2a$

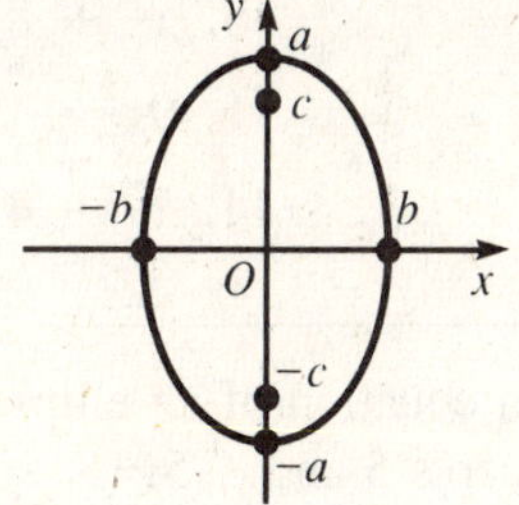

Example 1 Graph the ellipse $9x^2 + 16y^2 = 144$ and find its foci.

Solution

$$9x^2 + 16y^2 = 144$$

$$\frac{9x^2}{144} + \frac{16y^2}{144} = \frac{144}{144} \qquad \text{Divide both sides by 144.}$$

$$\frac{x^2}{16} + \frac{y^2}{9} = 1$$

Since the x^2-term has the larger denominator, $a = \sqrt{16} = 4$ and $b = \sqrt{9} = 3$. So the ellipse, which is centered at the origin, has x-intercepts ± 4 and y-intercepts ± 3.

(Solution continues on the next page.)

9–4 Ellipses *(continued)*

Solution *(continued)*

To see how the ellipse looks in the first quadrant, solve its equation for y and make a short table of first-quadrant points as shown at the right.

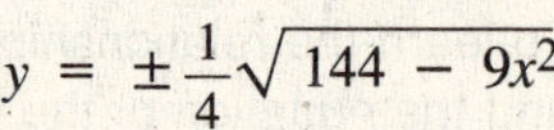

$$y = \pm\frac{1}{4}\sqrt{144 - 9x^2}$$

x	y
0	3
1	2.90
2	2.60
3	1.98
4	0

Due to the symmetry of the ellipse, the reflections of the first-quadrant points in the x-axis, the y-axis, and the origin are also points on the ellipse.

The graph of the ellipse is shown at the right.

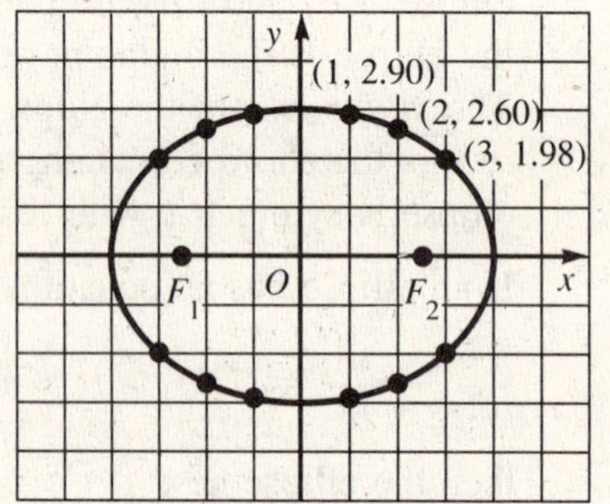

To find the foci, use the relationship:

$$c^2 = a^2 - b^2$$
$$c^2 = 16 - 9 = 7$$
$$c = \pm\sqrt{7}$$

Thus the foci of the ellipse are $F_1(-\sqrt{7}, 0)$ and $F_2(\sqrt{7}, 0)$.

Graph each ellipse and find its foci. You may wish to check your graphs on a computer or a graphing calculator.

1. $\frac{x^2}{4} + \frac{y^2}{9} = 1$
2. $\frac{x^2}{9} + \frac{y^2}{16} = 1$
3. $\frac{x^2}{16} + \frac{y^2}{4} = 1$
4. $4x^2 + y^2 = 16$
5. $x^2 + 9y^2 = 36$
6. $4x^2 + y^2 = 36$
7. $x^2 + 5y^2 = 25$
8. $x^2 + 8y^2 = 32$
9. $4x^2 + 25y^2 = 100$
10. $36x^2 + 4y^2 = 36$
11. $9x^2 + 5y^2 = 45$
12. $2x^2 + 3y^2 = 18$
13. $4x^2 + y^2 = 12$
14. $x^2 + 4y^2 = 1$
15. $25x^2 + 16y^2 = 25$

Example 2 Find an equation of an ellipse having x-intercepts $\sqrt{3}$ and $-\sqrt{3}$ and y-intercepts 5 and -5.

Solution Since $5^2 > (\sqrt{3})^2$, the y^2-term has the larger denominator, and so the major axis is vertical. The center is (0, 0), $a^2 = 5^2 = 25$, and $b^2 = (\sqrt{3})^2 = 3$.

$\therefore$ the equation is $\frac{x^2}{3} + \frac{y^2}{25} = 1$.

Find an equation of an ellipse having the given intercepts.

16. x-intercepts ± 4; y-intercepts ± 3
17. x-intercepts ± 6; y-intercepts ± 8
18. x-intercepts ± 3; y-intercepts $\pm\sqrt{2}$
19. x-intercepts $\pm\sqrt{5}$; y-intercepts $\pm 3\sqrt{2}$
20. x-intercepts ± 4; y-intercepts $\pm 2\sqrt{2}$
21. x-intercepts $\pm 3\sqrt{3}$; y-intercepts $\pm 4\sqrt{3}$

NAME ______________________ DATE ______________

9–5 Hyperbolas

Objective: To learn the relationship between the foci, intercepts, and asymptotes of a hyperbola and the equation of the hyperbola.

Vocabulary

Hyperbola The set of all points P in the plane such that the difference between the distances from two fixed points, called the *foci*, is a given constant. The distances from a point P to the foci are called *focal radii*. Hyperbolas are curves with two pieces, or *branches*, that get increasingly closer to two lines called *asymptotes*. The *center* of the hyperbola is the midpoint of the line segment joining its foci.

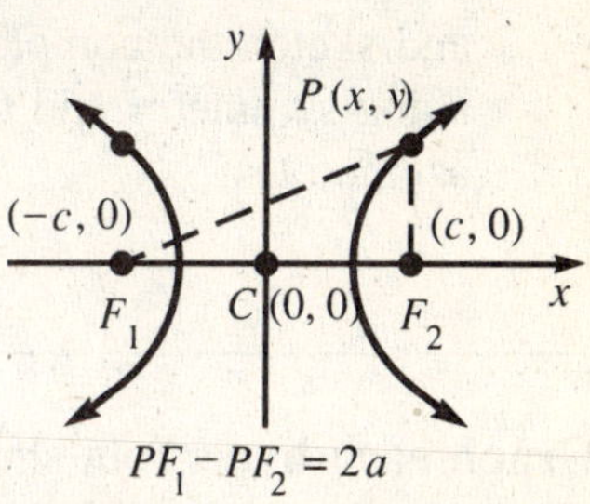

$PF_1 - PF_2 = 2a$

If a hyperbola has an equation of the form

$$\frac{x^2}{a^2} - \frac{y^2}{b^2} = 1,$$

then the hyperbola has:

1. center (0, 0)
2. foci *on the x-axis* at $(-c, 0)$ and $(c, 0)$, where $c^2 = a^2 + b^2$
3. difference of the focal radii $2a$
4. asymptotes $y = \frac{b}{a}x$ and $y = -\frac{b}{a}x$

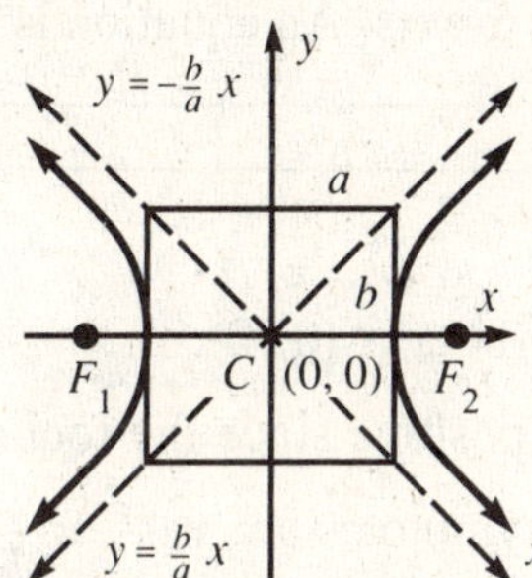

As you move away from the x-axis, the branches get closer to the asymptotes.

If a hyperbola has an equation of the form

$$\frac{y^2}{a^2} - \frac{x^2}{b^2} = 1,$$

then the hyperbola has:

1. center (0, 0)
2. foci *on the y-axis* at $(0, -c)$ and $(0, c)$, where $c^2 = a^2 + b^2$
3. difference of the focal radii $2a$
4. asymptotes $y = \frac{a}{b}x$ and $y = -\frac{a}{b}x$

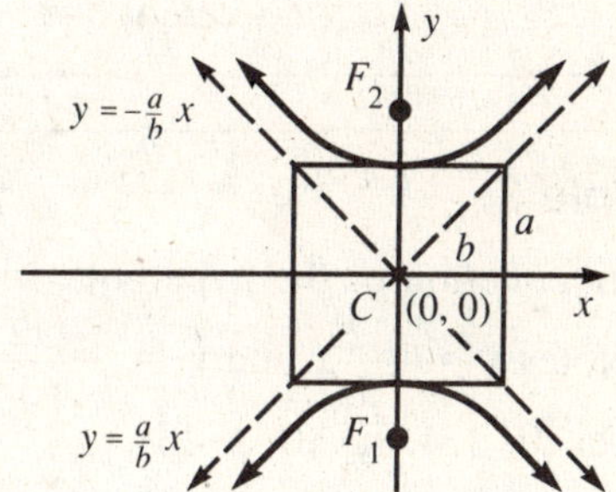

As you move away from the y-axis, the branches get closer to the asymptotes.

Example 1 Graph the hyperbola $9x^2 - 4y^2 = 36$, showing its asymptotes as dashed lines. Find its foci.

Solution

1. $9x^2 - 4y^2 = 36$

$$\frac{9x^2}{36} - \frac{4y^2}{36} = \frac{36}{36}$$

The x-intercepts are ± 2. There are no y-intercepts.

$$\frac{x^2}{4} - \frac{y^2}{9} = 1$$

2. The foci are on the x-axis. Since $a^2 = 4$ and $b^2 = 9$, $c^2 = a^2 + b^2 = 4 + 9 = 13$. Since $c = \pm\sqrt{13} = \pm 3.6$, the foci are $(-\sqrt{13}, 0)$ and $(\sqrt{13}, 0)$.

3. The asymptotes have equations $y = \pm\frac{\sqrt{9}}{\sqrt{4}}x$, or $y = \pm\frac{3}{2}x$.

(*Solution continues on the next page.*)

9–5 Hyperbolas *(continued)*

4. Make a short table of values and sketch the complete graph using symmetry and the asymptotes.

$y = \pm\frac{3}{2}\sqrt{x^2 - 4}$

x	y
± 2	0
± 3	± 3.35
± 4	± 5.20

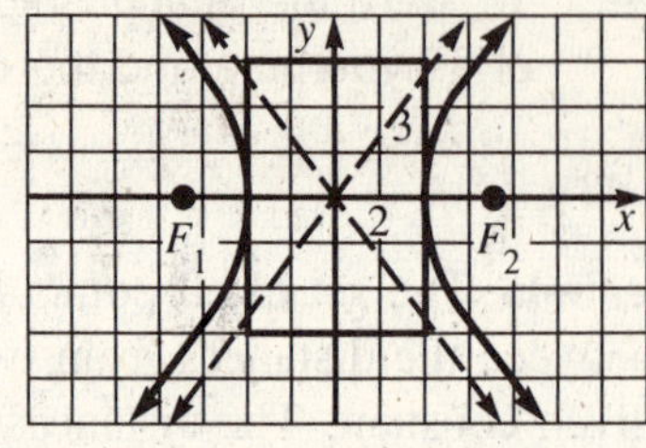

Graph each hyperbola showing its asymptotes as dashed lines and find the coordinates of the foci. You may wish to check your graphs on a computer or a graphing calculator.

1. $\frac{x^2}{9} - \frac{y^2}{16} = 1$ **2.** $\frac{y^2}{4} - \frac{x^2}{1} = 1$ **3.** $x^2 - 36y^2 = 36$ **4.** $x^2 - 4y^2 = -16$

5. $25y^2 - 4x^2 = 100$ **6.** $25x^2 - 36y^2 = 900$ **7.** $9x^2 - y^2 + 81 = 0$ **8.** $2x^2 - y^2 = 2$

Example 2

Find an equation of the hyperbola having foci at (4, 0) and (−4, 0) and difference of focal radii equal to 6.

Solution

1. Center of hyperbola is (0, 0).
2. Distance from each focus to center is 4, so $c = 4$.
3. Difference of focal radii is $2a$, so $2a = 6$, $a = 3$.
4. $b^2 = c^2 - a^2 = 4^2 - 3^2 = 16 - 9 = 7$.
5. Foci are on the x-axis. An equation is $\frac{x^2}{9} - \frac{y^2}{7} = 1$.

Example 3

Find an equation of the hyperbola with asymptotes $y = \frac{12}{5}x$ and $y = -\frac{12}{5}x$ and foci (0, 13) and (0, −13).

Solution

1. Center of hyperbola is (0, 0).
2. The y^2-term is positive since the foci are on the y-axis.
3. Equations of asymptotes are $y = \pm\frac{a}{b}x = \pm\frac{12}{5}x$. So $a^2 = 144$ and $b^2 = 25$.
4. An equation of the parabola is $\frac{y^2}{144} - \frac{x^2}{25} = 1$.

Find an equation of each hyperbola described.

9. Foci (0, −4) and (0, 4); difference of focal radii is 2.

10. Foci (−8, 0) and (8, 0); difference of focal radii is 12.

11. Asymptotes $y = \frac{5}{3}x$ and $-\frac{5}{3}x$; foci $(-\sqrt{34}, 0)$ and $(\sqrt{34}, 0)$.

12. Asymptotes $y = \frac{\sqrt{2}}{3}x$ and $y = -\frac{\sqrt{2}}{3}x$; foci $(0, -\sqrt{11})$ and $(0, \sqrt{11})$.

Mixed Review Exercises

Graph each equation.

1. $4x^2 + y^2 = 16$ **2.** $x^2 + y^2 - 2x + 4y - 11 = 0$ **3.** $4x^2 - 8x + y - 1 = 0$

NAME ______________________ DATE ______________

9–6 *More on Central Conics*

Objective: To find an equation of a conic section with center not at the origin and to identify a conic as a circle, ellipse, or hyperbola.

Vocabulary

Central conics Circles, ellipses, and hyperbolas. They are called central conics because they have centers.

Central conics with centers (h, k) Central conics that have their centers translated from $(0, 0)$ to (h, k) by sliding every point of their graphs h units horizontally and k units vertically.

If an **ellipse** has an equation of the form

$$\frac{(x-h)^2}{a^2} + \frac{(y-k)^2}{b^2} = 1,$$

then the ellipse has:

1. center (h, k)
2. horizontal major axis
3. foci $(h - c, k)$ and $(h + c, k)$, where $c^2 = a^2 - b^2$

If an **ellipse** has an equation of the form

$$\frac{(x-h)^2}{b^2} + \frac{(y-k)^2}{a^2} = 1,$$

then the ellipse has:

1. center (h, k)
2. vertical major axis
3. foci $(h, k - c)$ and $(h, k + c)$, where $c^2 = a^2 - b^2$

If a **hyperbola** has an equation of the form

$$\frac{(x-h)^2}{a^2} - \frac{(y-k)^2}{b^2} = 1,$$

then the hyperbola has:

1. center (h, k)
2. foci *on the same horizontal line* at $(h - c, k)$ and $(h + c, k)$, where $c^2 = a^2 + b^2$

If a **hyperbola** has an equation of the form

$$\frac{(y-k)^2}{a^2} - \frac{(x-h)^2}{b^2} = 1,$$

then the hyperbola has:

1. center (h, k)
2. foci *on the same vertical line* at $(h, k - c)$ and $(h, k + c)$, where $c^2 = a^2 + b^2$

Example 1 Find an equation of the ellipse having foci $(-1, 3)$ and $(5, 3)$ and sum of focal radii 10.

Solution

1. $2a = 10$, so $a = 5$. (sum of focal radii)
2. Foci are on the same horizontal line.
3. Center is $\left(\frac{-1+5}{2}, \frac{3+3}{2}\right) = (2, 3)$. (halfway between foci)
4. $c = 3$. (distance from center to each focus)
5. $b^2 = a^2 - c^2 = 25 - 9 = 16$.

$\therefore$ an equation of the ellipse is $\frac{(x-2)^2}{25} + \frac{(y-3)^2}{16} = 1$.

Find an equation of the ellipse having the given foci and sum of focal radii.

1. $(1, -3), (9, -3)$; 10

2. $(4, -4), (4, 2)$; 8

3. $(5, 3), (5, 7)$; 12

4. $(6, 7), (0, 7)$; 16

9–6 More on Central Conics *(continued)*

Example 2 Find an equation of the hyperbola having foci $(-4, 3)$ and $(-4, -5)$ and difference of focal radii 6.

Solution

1. $2a = 6$, so $a = 3$. (difference of focal radii)
2. Foci are on the same vertical line.
3. Center is $(-4, -1)$. (halfway between foci)
4. $c = 4$ (distance from center to each focus)
5. $b^2 = c^2 - a^2 = 16 - 9 = 7$.

$\therefore$ an equation of the hyperbola is $\dfrac{(y+1)^2}{9} - \dfrac{(x+4)^2}{7} = 1$.

Find an equation of the hyperbola having the given foci and difference of focal radii.

5. $(3, 9), (3, -1); 6$ **6.** $(0, -10), (0, 0); 2$ **7.** $(-1, -2), (5, -2); 4$ **8.** $(4, 8), (16, 8); 4$

Example 3 Identify the conic $9x^2 + y^2 + 18x - 6y + 9 = 0$. Find its center and foci. Then draw its graph.

Solution Complete the square using the x-terms and then the y-terms.

$$(9x^2 + 18x + \underline{?}) + (y^2 - 6y + \underline{?}) = -9 + \underline{?} + \underline{?}$$

$$9(x^2 + 2x + 1) + (y^2 - 6y + 9) = -9 + 9 + 9$$

$$9(x+1)^2 + (y-3)^2 = 9$$

$$\frac{(x+1)^2}{1} + \frac{(y-3)^2}{9} = 1 \quad \text{Divide both sides by 9.}$$

This equation matches the general form $\dfrac{(x-h)^2}{b^2} + \dfrac{(y-k)^2}{a^2} = 1$.

Therefore the conic is an ellipse with:

1. center $(-1, 3)$
2. vertical major axis
3. Since $a^2 = 9$ and $b^2 = 1$,
   $c^2 = a^2 - b^2 =$
   $9 - 1 = 8$ and $c = \sqrt{8} = 2\sqrt{2}$.
   Foci are $(-1, 3 - 2\sqrt{2})$ and $(-1, 3 + 2\sqrt{2})$,
   or approximately $(-1, 0.17)$ and $(-1, 5.82)$.

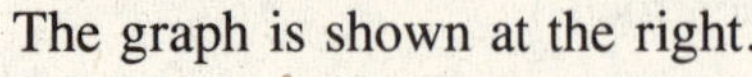

The graph is shown at the right.

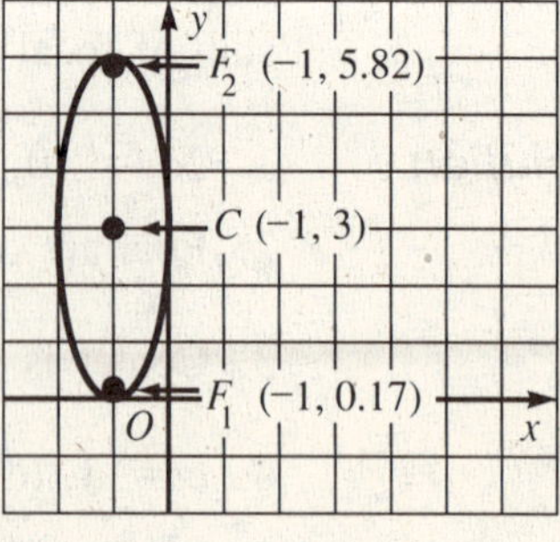

Identify each conic. Find its center and its foci (if any). Then draw its graph. You may wish to check your graphs on a computer or a graphing calculator.

9. $y^2 - x^2 - 2y + 4x - 4 = 0$ **10.** $x^2 + 25y^2 + 8x + 50y + 16 = 0$

11. $x^2 + y^2 - 4x + 4y - 41 = 0$ **12.** $16x^2 + 9y^2 - 32x - 36y - 92 = 0$

13. $x^2 - 4y^2 - 6x - 16y - 11 = 0$ **14.** $x^2 + y^2 + 12y + 27 = 0$

NAME ______________________________ DATE ____________

9–7 The Geometry of Quadratic Equations

Objective: To use graphs to determine the number of real solutions of a quadratic system and to estimate the solutions.

Vocabulary

Quadratic system A system containing only quadratic equations or a combination of linear and quadratic equations in the same two variables. A system of one quadratic and one linear equation may have 2, 1, or 0 real solutions. A system of two quadratic equations may have 4, 3, 2, 1, or 0 real solutions.

Example 1 By sketching a graph, find the *number* of real solutions each system has. Then estimate the real solutions (if any) to the nearest half unit.

a. $9x^2 + 4y^2 = 36$; $2x + y = 1$ **b.** $x^2 + 2y = 6$; $x^2 - 4y^2 = 16$ **c.** $x^2 + y^2 = 25$; $y^2 - 9x^2 = 9$

Solution **a.** Identify each equation in the system:

$9x^2 + 4y^2 = 36 \longrightarrow \frac{x^2}{4} + \frac{y^2}{9} = 1 \longrightarrow$ ellipse

$2x + y = 1 \longrightarrow y = -2x + 1 \longrightarrow$ line

The ellipse has x-intercepts $\pm\sqrt{4}$, or ± 2, and y-intercepts $\pm\sqrt{9}$, or ± 3. The graph of the system is shown at the right. Since there are two points of intersection, the system has 2 real solutions. To the nearest half unit, the solutions are $(1.5, -2)$ and $(-1, 3)$.

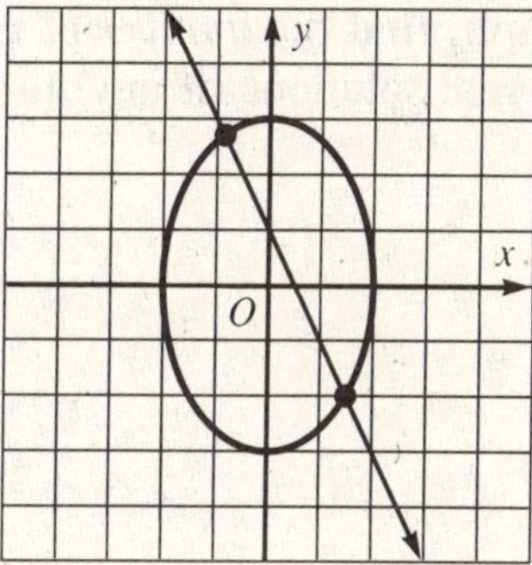

b. Identify each equation in the system:

$x^2 + 2y = 6 \longrightarrow y - 3 = -\frac{1}{2}x^2 \longrightarrow$ parabola

$x^2 - 4y^2 = 16 \longrightarrow \frac{x^2}{16} - \frac{y^2}{4} = 1 \longrightarrow$ hyperbola

The parabola has vertex $(0, 3)$ and opens downward. The hyperbola has x-intercepts $\pm\sqrt{16}$, or ± 4, and asymptotes $y = \pm\frac{\sqrt{4}}{\sqrt{16}}x$, or $y = \pm\frac{1}{2}x$.

The graph of the system is shown at the right. Since there are no points of intersection, the system has 0 real solutions.

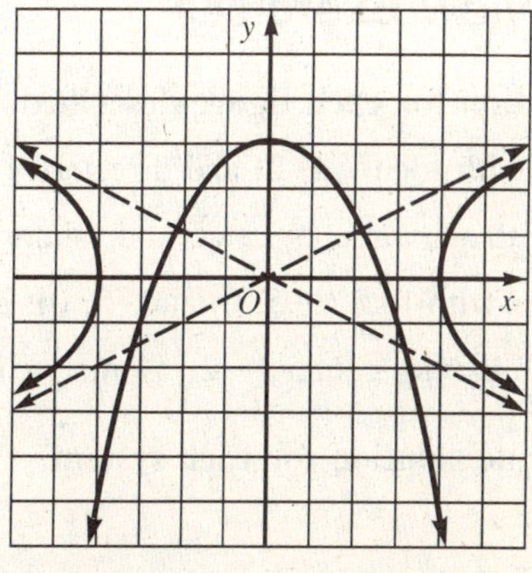

(Solution continues on the next page.)

9–7 The Geometry of Quadratic Systems *(continued)*

c. Identify each equation in the system:

$x^2 + y^2 = 25$ ⟶ circle

$y^2 - 9x^2 = 9$ ⟶ $\frac{y^2}{9} - \frac{x^2}{1} = 1$ ⟶ hyperbola

The circle has center (0, 0) and radius 5. The hyperbola has y-intercepts $\pm\sqrt{9}$, or ± 3, and asymptotes $y = \pm\frac{\sqrt{9}}{\sqrt{1}}x$, or $y = \pm 3x$. The graph of the system is shown at the right. Since there are four points of intersection, the system has 4 real solutions. To the nearest half unit, the solutions are (1.5, 5), (−1.5, 5), (−1.5, −5), and (1.5, −5).

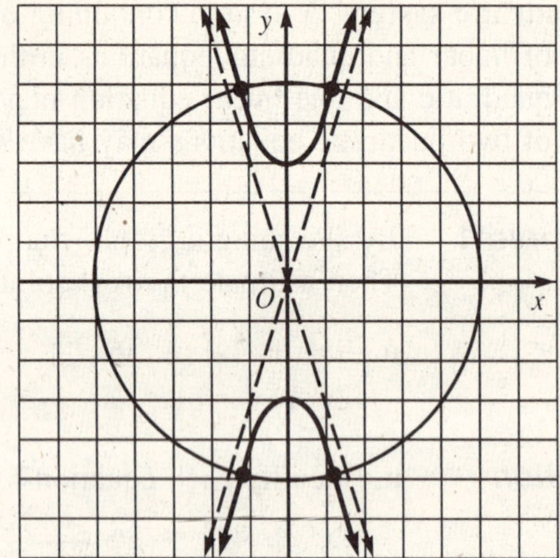

By sketching a graph, find the *number* of real solutions each system has. Then estimate the real solutions (if any) to the nearest half unit.

1. $x^2 + 5 = y$
$x + y = 4$

2. $x^2 + y^2 = 9$
$2x + y = 4$

3. $x^2 - y^2 = 4$
$x - y = 2$

4. $9x^2 + 4y^2 = 36$
$x^2 + y^2 = 4$

5. $x^2 - 4y^2 = 16$
$x^2 + y^2 = 25$

6. $4x^2 + y^2 = 16$
$y = x^2 - 2$

7. $y = x^2 - 4x + 9$
$y = 2x + 1$

8. $x^2 + y^2 = 16$
$x = y^2 + 5$

9. $y = x^2 - 2x$
$x = 2y^2$

10. $x^2 + y^2 = 9$
$x^2 + y^2 + 4y = 0$

Mixed Review Exercises

Find an equation for each figure described.

1. Parabola with focus (4, 2) and directrix $y = -4$.
2. Ellipse with x-intercepts $\pm\sqrt{5}$ and y-intercepts $\pm\sqrt{3}$.
3. Hyperbola with foci (−5, 0) and (5, 0), and difference of focal radii 6.
4. The circle having center (−2, 5) and radius 4.

Find the unique solution for each system.

5. $y = 2x - 1$
$4x - y = 3$

6. $2x - 5y = 9$
$3x + 2y = 4$

7. $5x + 2y = -16$
$3x - 4y = 6$

NAME ______________________ DATE ______________

9–8 Solving Quadratic Systems

Objective: To use algebraic methods to find exact solutions of quadratic systems.

Example 1 Solve this system: $x^2 - y^2 = 8$
$x - y = 2$

Solution **(Substitution Method)** Solve the linear equation for y: $y = x - 2$
Substitute $x - 2$ for y in the quadratic equation.

$$\begin{aligned} x^2 - y^2 &= 8 \\ x^2 - (x - 2)^2 &= 8 \\ x^2 - x^2 + 4x - 4 &= 8 \\ 4x &= 12 \\ x &= 3 \end{aligned}$$

Substitute 3 for x in $y = x - 2$: $y = 3 - 2 = 1$
Therefore (3, 1) is a solution.
Check the ordered pairs in both *given* equations. (The check is left for you.)
$\therefore$ the solution set is $\{(3, 1)\}$.
The graph of each equation (a hyperbola and a line) is shown at the right. As you can see, there is one point of intersection.

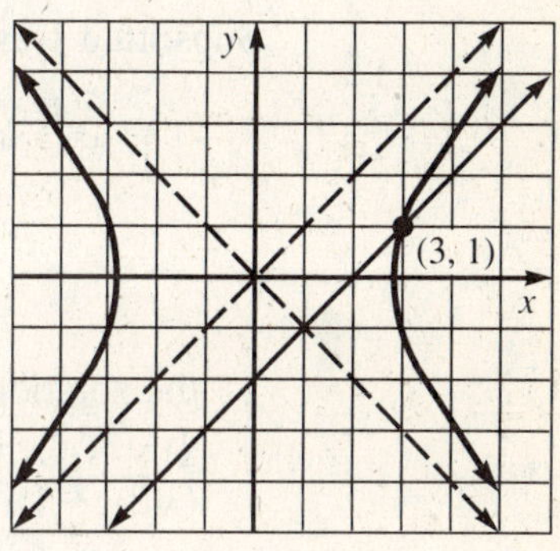

Example 2 Solve this system: $4x^2 + 9y^2 = 36$
$y^2 - 3 = x$

Solution **(Substitution Method)** Solve the second equation for y^2: $y^2 = x + 3$
Substitute $x + 3$ for y^2 in the first equation.

$$\begin{aligned} 4x^2 + 9y^2 &= 36 \\ 4x^2 + 9(x + 3) &= 36 \\ 4x^2 + 9x + 27 &= 36 \\ 4x^2 + 9x - 9 &= 0 \\ (4x - 3)(x + 3) &= 0 \quad \text{Factor the polynomial.} \end{aligned}$$

$$4x - 3 = 0 \quad \text{or} \quad x + 3 = 0$$
$$x = \frac{3}{4} \quad \text{or} \quad x = -3$$

Substitute the x-values in $y^2 = x + 3$.
If $x = \frac{3}{4}$, then $y^2 = \frac{15}{4}$ and $y = \pm\frac{\sqrt{15}}{2}$.
If $x = -3$, then $y^2 = 0$ and $y = 0$.
$\therefore$ the solution set is

$$\left\{(-3, 0), \left(\frac{3}{4}, \frac{\sqrt{15}}{2}\right), \left(\frac{3}{4}, -\frac{\sqrt{15}}{2}\right)\right\}.$$

The graph is shown at the right.

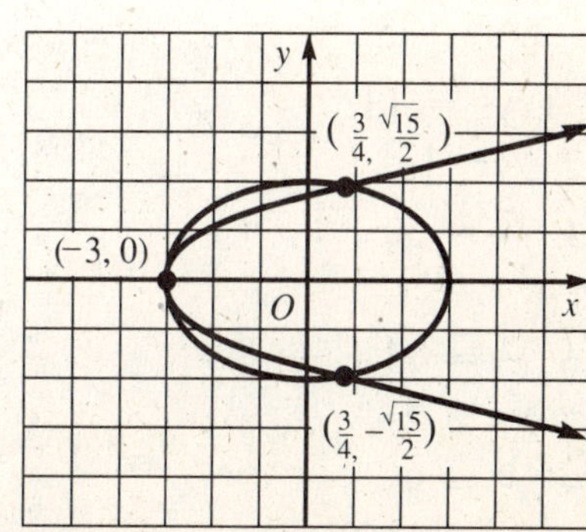

NAME ______________________ DATE ______________

9–8 Solving Quadratic Systems *(continued)*

Example 3 Solve this system: $x^2 + y^2 = 4$
$x^2 - 2y^2 = 1$

Solution **(Linear-Combination Method)** Multiply the second equation by -1 and add the two equations.

$$\begin{aligned} x^2 + y^2 &= 4 \\ -x^2 + 2y^2 &= -1 \\ 3y^2 &= 3 \quad \text{Solve the resulting equation.} \\ y^2 &= 1 \\ y &= \pm 1 \end{aligned}$$

Substitute 1 and -1 for y in $x^2 + y^2 = 4$ to find the corresponding values of x.

If $y = 1$:

$$\begin{aligned} x^2 + y^2 &= 4 \\ x^2 + 1^2 &= 4 \\ x^2 &= 3 \\ x &= \pm\sqrt{3} \end{aligned}$$

If $y = -1$:

$$\begin{aligned} x^2 + y^2 &= 4 \\ x^2 + (-1)^2 &= 4 \\ x^2 &= 3 \\ x &= \pm\sqrt{3} \end{aligned}$$

$\therefore$ the solution set is $\{(\sqrt{3}, 1), (-\sqrt{3}, 1), (\sqrt{3}, -1), (-\sqrt{3}, -1)\}$.

Notice that the linear-combination method (used in this example) is more appropriate for solving a quadratic-quadratic system, while the substitution method (used in the previous examples) is more appropriate for solving a linear-quadratic system.

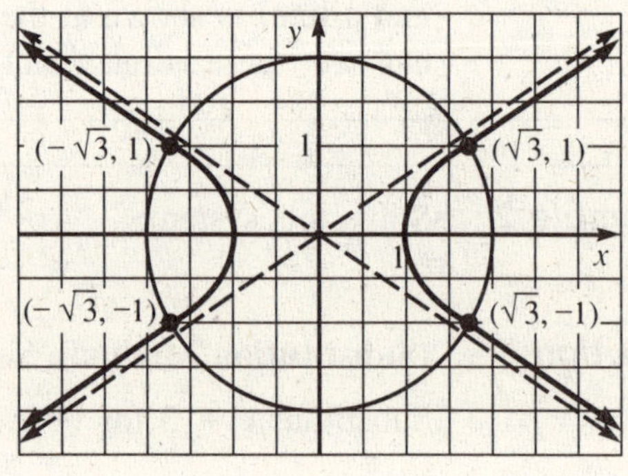

Find the real solutions, if any, of each system. You may wish to check your answers visually on a computer or a graphing calculator.

1. $y = x^2 - 3$
$y = 2x$

2. $x^2 - y^2 = 8$
$x - 3y = 0$

3. $x^2 + y^2 = 25$
$x - 2y = -5$

4. $y^2 = x + 5$
$x - y = 1$

5. $4x^2 + y^2 = 13$
$2x - y = 1$

6. $x^2 - y^2 = -3$
$3x - y = 1$

7. $x^2 + y^2 = 30$
$y = x^2$

8. $x^2 + 3y^2 = 7$
$x - y^2 = 1$

9. $x^2 + y^2 = 13$
$x^2 - y^2 = 5$

10. $4x^2 + y^2 = 16$
$x^2 + y^2 = 7$

11. $x^2 - y^2 = 9$
$x^2 + 9y^2 = 169$

12. $6x^2 + y^2 = 36$
$2x^2 - y^2 = -4$

NAME ______________________ DATE ______________

9–9 Systems of Linear Equations in Three Variables

Objective: To solve systems of linear equations in three variables.

Vocabulary

Linear equation in three variables Any equation of the form $Ax + By + Cz = D$ where A, B, C, and D are real numbers with A, B, and C not all zero.

Ordered triple A trio of numbers having a significant order. Example: The ordered triple $(x, y, z) = (2, 1, -4)$ is a solution of $3x - 2y + z = 0$ because $3(2) - 2(1) + (-4) = 0$.

Triangular system A system whose shape suggests a triangle. Example:

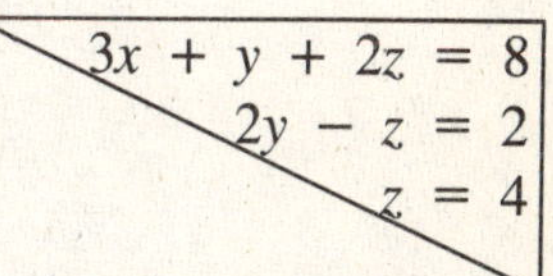

$$\begin{aligned} 3x + y + 2z &= 8 \\ 2y - z &= 2 \\ z &= 4 \end{aligned}$$

Example 1 Solve this system:

$$\begin{aligned} 3x + 2y + z &= 8 \\ y - z &= -5 \\ z &= 4 \end{aligned}$$

Solution **(Substitution Method)** Because the given system is triangular, you can use a sequence of substitutions to solve it. Since the third equation gives the value of z ($z = 4$), substitute 4 for z in the second equation and solve for y.

$$\begin{aligned} y - z &= -5 \\ y - \mathbf{4} &= -5 \\ y &= -1 \end{aligned}$$

Substitute 4 for z and -1 for y in the first equation and solve for x.

$$\begin{aligned} 3x + 2y + z &= 8 \\ 3x + 2(\mathbf{-1}) + \mathbf{4} &= 8 \\ 3x - 2 + 4 &= 8 \\ 3x &= 6 \\ x &= 2 \end{aligned}$$

Check the ordered triple $(2, -1, 4)$ in each equation.

$$\begin{aligned} 3x + 2y + z &\stackrel{?}{=} 8 \\ 3(\mathbf{2}) + 2(\mathbf{-1}) + \mathbf{4} &\stackrel{?}{=} 8 \\ 6 - 2 + 4 &\stackrel{?}{=} 8 \\ 8 &= 8 \checkmark \end{aligned} \qquad \begin{aligned} y - z &\stackrel{?}{=} -5 \\ \mathbf{-1} - \mathbf{4} &\stackrel{?}{=} -5 \\ -5 &= -5 \checkmark \end{aligned} \qquad \begin{aligned} z &= 4 \\ \mathbf{4} &= 4 \checkmark \end{aligned}$$

$\therefore$ the solution of the system is $(2, -1, 4)$.

Solve each system.

1. $\begin{aligned} 3x - 2y + z &= 13 \\ 3y - z &= -9 \\ z &= 3 \end{aligned}$

2. $\begin{aligned} 2x + 2y - z &= -5 \\ 3y - 5z &= 1 \\ z &= 1 \end{aligned}$

3. $\begin{aligned} x + y + z &= 5 \\ y + z &= -3 \\ z &= 4 \end{aligned}$

4. $\begin{aligned} 3x - 2y - z &= -1 \\ 3y - 2z &= 3 \\ 2z &= -6 \end{aligned}$

5. $\begin{aligned} 6x - 4y + z &= 1 \\ 4y - 3z &= 1 \\ 5z &= 25 \end{aligned}$

6. $\begin{aligned} 4x + 2y + z &= 0 \\ 5y + z &= 17 \\ 4z &= 8 \end{aligned}$

9–9 *Systems of Linear Equations in Three Variables* *(continued)*

Example 2 Solve this system:

$$\left.\begin{aligned} x + 3y + z &= 8 \\ 3x + 2y - z &= 14 \\ 2x - y + 3z &= 1 \end{aligned}\right\} \quad \textbf{(1)}$$

Solution **(Triangular-System Method)** The given system is not triangular, but it can be put in triangular form by eliminating one variable from one equation and two variables from another. In **(1)** add the first equation to the second and then multiply the first equation by -3 and add it to the third. (This eliminates z from the second and third equations.) The resulting system is:

$$\left.\begin{aligned} x + 3y + z &= 8 \\ 4x + 5y &= 22 \\ -x - 10y &= -23 \end{aligned}\right\} \quad \textbf{(2)}$$

In **(2)** multiply the second equation by 2 and add it to the third. (This eliminates y from the third equation.)

$$\left.\begin{aligned} x + 3y + z &= 8 \\ 4x + 5y &= 22 \\ 7x &= 21 \end{aligned}\right\} \quad \textbf{(3)}$$

Solve the system in **(3)** by the substitution method.

$$\begin{aligned} 7x &= 21 \\ x &= 3 \end{aligned} \qquad \begin{aligned} 4x + 5y &= 22 \\ 4(3) + 5y &= 22 \\ 12 + 5y &= 22 \\ 5y &= 10 \\ y &= 2 \end{aligned} \qquad \begin{aligned} x + 3y + z &= 8 \\ 3 + 3(2) + z &= 8 \\ 3 + 6 + z &= 8 \\ 9 + z &= 8 \\ z &= -1 \end{aligned}$$

The check is left for you.

$\therefore$ the solution is $(3, 2, -1)$.

Solve each system.

7. $\begin{aligned} x + 2y - z &= 5 \\ 2x - y + z &= 2 \\ 3x + y + 2z &= 5 \end{aligned}$

8. $\begin{aligned} 2x + y - 2z &= 4 \\ x - y + 2z &= 8 \\ 3x + 2y - z &= 10 \end{aligned}$

9. $\begin{aligned} x + y - z &= -1 \\ x - y + z &= -1 \\ x - y - z &= 1 \end{aligned}$

10. $\begin{aligned} 2x - 2y + z &= -3 \\ x + y - z &= -4 \\ 3x + 2y + 2z &= 2 \end{aligned}$

11. $\begin{aligned} 3x - y + 2z &= 8 \\ x + y + z &= 6 \\ x - y - z &= -2 \end{aligned}$

12. $\begin{aligned} x - 2y - z &= -2 \\ 2x + y + z &= 9 \\ x + 2y - z &= 6 \end{aligned}$

Mixed Review Exercises

Find the solution of each system.

1. $\begin{aligned} x^2 + y^2 &= 5 \\ x^2 - y^2 &= 3 \end{aligned}$

2. $\begin{aligned} 3x + 2y &= -7 \\ 4x - 3y &= -15 \end{aligned}$

3. $\begin{aligned} y &= x^2 + 2 \\ 2x + y &= 2 \end{aligned}$

Simplify.

4. $(4x^{-2}y^{-3})^{-2}$

5. $2\sqrt{24x^5}$

6. $(3x^2y^{-1})(-4x^2y^4)$

7. $(3y\sqrt{2})^2$

8. $\dfrac{12xy^{-2}}{8x^{-3}y^4}$

9. $\dfrac{20x^4y^2}{28x^3y^3}$

NAME ______________________________ DATE ______________

10 Exponential and Logarithmic Functions

10–1 Rational Exponents

Objective: To extend the meaning of exponents to include rational numbers.

Vocabulary

Exponential form A radical expression is in exponential form when it is expressed as a power or product of powers.

$b^{1/n}$ The exponential form of the nth root of b. The notation follows from the laws of exponents. Since $(b^{1/n})^n = b^{(1/n)\cdot n} = b^1 = b$, and $(\sqrt[n]{b})^n = b$, we can say that $b^{1/n} = \sqrt[n]{b}$. Examples: $9^{1/2} = \sqrt{9} = 3$ and $-8^{1/3} = -\sqrt[3]{8} = -2$

$b^{p/q}$ The qth root of b to the power p. In other words, if p and q are integers, with $q > 0$, and b is a positive real number, then $b^{p/q} = (\sqrt[q]{b})^p = \sqrt[q]{b^p}$.

CAUTION You will have to be familiar with the laws of exponents from Lesson 5–1 in order to do the exercises in this lesson.

Example 1 Simplify $64^{2/3}$.

Solution 1 $64^{2/3} = (\sqrt[3]{64})^2 = 4^2 = 16$

Solution 2 $64^{2/3} = \sqrt[3]{64^2} = \sqrt[3]{4096} = 16$

Example 2 Simplify.

a. $32^{-3/5}$ **b.** $4^{3.5}$ **c.** $(8^5)^{1/3}$ **d.** $(25^{1/2} + 36^{1/2})^2$

Solution

a. $32^{-3/5} = \dfrac{1}{32^{3/5}} = \dfrac{1}{(\sqrt[5]{32})^3} = \dfrac{1}{2^3} = \dfrac{1}{8}$

b. $4^{3.5} = 4^{7/2} = (\sqrt{4})^7 = 2^7 = 128$

c. $(8^5)^{1/3} = 8^{5/3} = (\sqrt[3]{8})^5 = 2^5 = 32$

d. $(25^{1/2} + 36^{1/2})^2 = (\sqrt{25} + \sqrt{36})^2 = (5 + 6)^2 = 11^2 = 121$

Simplify.

1. $16^{1/2}$ **2.** $81^{1/4}$ **3.** $16^{-1/2}$ **4.** $27^{-1/3}$ **5.** $81^{3/4}$ **6.** $36^{-3/2}$

7. $-64^{2/3}$ **8.** $4^{3.5}$ **9.** $\left(\dfrac{9}{16}\right)^{1.5}$ **10.** $25^{-0.5}$ **11.** $4^{-2.5}$ **12.** $(9^{-3})^{1/6}$

13. $(27^5)^{-2/15}$ **14.** $(27^{1/3} + 64^{1/3})^{-2}$ **15.** $(27^{2/3} - 32^{1/5})^2$

NAME ______________________ DATE ______________

10–1 Rational Exponents *(continued)*

Example 3 Write in exponential form: a. $\sqrt[6]{x^3y^{-2}}$ b. $\sqrt[5]{\dfrac{9a^5}{b^2}}$

Solution a. $\sqrt[6]{x^3y^{-2}} = (x^3y^{-2})^{1/6}$

$= x^{3/6}y^{-2/6}$

$= x^{1/2}y^{-1/3}$

b. $\sqrt[5]{\dfrac{9a^5}{b^2}} = \left(\dfrac{9a^5}{b^2}\right)^{1/5}$

$= \dfrac{9^{1/5}a^{5/5}}{b^{2/5}}$

$= 9^{1/5}ab^{-2/5}$

Write in exponential form.

16. $\sqrt{x^5y^4}$ 17. $\sqrt[5]{x^4y^{-6}}$ 18. $\sqrt[3]{m^{-3}n^{-6}}$ 19. $\sqrt[3]{27u^9v^{-6}}$ 20. $\sqrt[4]{\dfrac{3^4 \cdot a^{-6}}{b^2}}$

Example 4 Express $\sqrt{8} \cdot \sqrt[3]{4}$ in simplest radical form.

Solution

$\sqrt{8} \cdot \sqrt[3]{4} = \sqrt{2^3} \cdot \sqrt[3]{2^2}$ Express each radical in exponential form.

$= 2^{3/2} \cdot 2^{2/3}$

$= 2^{9/6 + 4/6}$ Use the rule $a^m \cdot a^n = a^{m+n}$.

$= 2^{13/6}$

$= 2^{12/6} \cdot 2^{1/6}$ Factor the power into two powers, one with an exponent that is the largest possible whole number.

$= 2^2 \cdot 2^{1/6}$

$= 4\sqrt[6]{2}$ Write in simplest radical form.

Express in simplest radical form.

21. $\sqrt[3]{25} \cdot \sqrt[3]{25}$ 22. $\sqrt{27} \cdot \sqrt[6]{27}$ 23. $\dfrac{\sqrt[3]{9}}{\sqrt[6]{3}}$ 24. $\dfrac{\sqrt[8]{16}}{\sqrt[6]{4}}$ 25. $\sqrt[3]{32} \cdot \sqrt{8}$

Mixed Review Exercises

Express in simplest form without negative exponents.

1. $\dfrac{x}{x-2} - \dfrac{8}{x^2-4}$ 2. $\sqrt{\dfrac{75x}{y^7}}$ 3. $(a^2 + 2a - 3) - (5 - a^2)$

4. $(-3x^3)(-3x)^3$ 5. $\left(\dfrac{x^2}{y}\right)^{-3} \cdot \left(\dfrac{y^2}{x^3}\right)^{-1}$ 6. $(3n + 2)(2n - 5)$

NAME ______________________ DATE ______________

10–2 Real Number Exponents

Objective: To extend the meaning of exponents to include irrational numbers and to define exponential functions.

Vocabulary

Exponential function If $b > 0$ and $b \neq 1$, the function defined by $y = b^x$ is called the exponential function with base b. Example: $y = 3^x$.

One-to-one function A function f is called one-to-one if for every p and q in the domain of f, $f(p) = f(q)$ if and only if $p = q$.

Exponential equation An equation in which a variable appears in an exponent.

Example 1 Simplify: **a.** $5^{\sqrt{2}} \cdot 5^{\sqrt{2}}$ **b.** $\dfrac{5^{\sqrt{2}+2}}{5^{\sqrt{2}-1}}$ **c.** $\sqrt[3]{2^{9\pi}}$ **d.** $\dfrac{16^{1.4}}{2^{1.6}}$

Solution

a. $5^{\sqrt{2}} \cdot 5^{\sqrt{2}} = 5^{\sqrt{2}+\sqrt{2}}$ Recall that $a^m \cdot a^n = a^{m+n}$.

$= 5^{2\sqrt{2}}$ Simplify the exponent.

$= (5^2)^{\sqrt{2}} = 25^{\sqrt{2}}$ Recall that $(a^m)^n = a^{mn}$.

b. $\dfrac{5^{\sqrt{2}+2}}{5^{\sqrt{2}-1}} = 5^{(\sqrt{2}+2)-(\sqrt{2}-1)}$ Recall that $\dfrac{a^m}{a^n} = a^{m-n}$.

$= 5^3 = 125$ Simplify the exponent.

c. $\sqrt[3]{2^{9\pi}} = 2^{9\pi/3}$ Write in exponential form.

$= 2^{3\pi}$ Simplify the exponent.

$= (2^3)^{\pi} = 8^{\pi}$

d. $\dfrac{16^{1.4}}{2^{1.6}} = \dfrac{(2^4)^{1.4}}{2^{1.6}}$ Write the numerator and denominator using the same base.

$= \dfrac{2^{5.6}}{2^{1.6}}$

$= 2^{5.6-1.6}$

$= 2^4 = 16$ Simplify the exponent.

Simplify.

1. $4^{\sqrt{3}} \cdot 4^{\sqrt{3}}$ **2.** $(4^{\sqrt{3}})^3$ **3.** $(4^{\sqrt{3}})^{\sqrt{3}}$ **4.** $\dfrac{4^{\sqrt{3}+1}}{4^{\sqrt{3}-1}}$

5. $6^{\sqrt{5}} \cdot 6^{\sqrt{3}}$ **6.** $(5^{\sqrt{2}})^2$ **7.** $(9^{\pi})^2$ **8.** $(5^{\sqrt{3}})^{-1/\sqrt{3}}$

9. $\sqrt{5^{4\pi}}$ **10.** $\sqrt[5]{3^{10\pi}}$ **11.** $9^{2.1} \cdot 3^{-3.2}$ **12.** $(\sqrt{7})^{\sqrt{3}}(\sqrt{7})^{-\sqrt{3}}$

13. $\dfrac{7^{\sqrt{5}-3}}{7^{\sqrt{5}-5}}$ **14.** $\dfrac{36^{1.2}}{6^{4.4}}$ **15.** $\sqrt[4]{\dfrac{25^{5+\pi}}{25^{5-\pi}}}$ **16.** $\dfrac{(1+\sqrt{2})^{2+\pi}}{(1+\sqrt{2})^{\pi}}$

NAME ______________________ DATE ______________

10–2 Real Number Exponents *(continued)*

Vocabulary

- **Steps for solving exponential equations**
  1. Express each side of the equation as a power of the same base.
  2. Set the exponents equal and then solve.
  3. Check the answer.

Example 2 Solve.

a. $27^x = \frac{1}{9}$ **b.** $7^{4-x} = 49^{x-1}$ **c.** $125^{x-3} = 5\sqrt{5}$

Solution

	Step 1	*Step 2*	*Step 3*
a.	$27^x = \frac{1}{9}$	$3x = -2$	$27^{-2/3} \stackrel{?}{=} \frac{1}{9}$
	$(3^3)^x = \frac{1}{3^2}$	$x = -\frac{2}{3}$	$(\sqrt[3]{27})^{-2} \stackrel{?}{=} \frac{1}{9}$
	$3^{3x} = 3^{-2}$		$3^{-2} \stackrel{?}{=} \frac{1}{9}$
	$\therefore$ the solution set is $\left\{-\frac{2}{3}\right\}$.		$\frac{1}{9} = \frac{1}{9}$ ✓
b.	$7^{4-x} = 49^{x-1}$	$4 - x = 2x - 2$	$7^{4-2} \stackrel{?}{=} 49^{2-1}$
	$7^{4-x} = (7^2)^{x-1}$	$-3x = -6$	$7^2 \stackrel{?}{=} 49^1$
	$7^{4-x} = 7^{2x-2}$	$x = 2$	$49 = 49$ ✓
	$\therefore$ the solution set is $\{2\}$.		
c.	$125^{x-3} = 5\sqrt{5}$	$3x - 9 = \frac{3}{2}$	$125^{(7/2)-3} \stackrel{?}{=} 5\sqrt{5}$
	$(5^3)^{x-3} = 5^1 \cdot 5^{1/2}$	$6x - 18 = 3$	$125^{1/2} \stackrel{?}{=} 5\sqrt{5}$
	$5^{3x-9} = 5^{3/2}$	$6x = 21$	$\sqrt{125} \stackrel{?}{=} 5\sqrt{5}$
		$x = \frac{21}{6} = \frac{7}{2}$	$5\sqrt{5} = 5\sqrt{5}$ ✓
	$\therefore$ the solution set is $\left\{\frac{7}{2}\right\}$.		

Solve. If an equation has no solution, say so.

17. $4^x = \frac{1}{16}$ **18.** $3^x = \sqrt{27}$ **19.** $9^{3+x} = 3$

20. $9^{1+x} = 27$ **21.** $64^{x-1} = 4$ **22.** $100^{2x} = 10^{3x+2}$

23. $27^{x-2} = 3\sqrt{3}$ **24.** $7^{2x-1} = 49^{x-3}$ **25.** $4^{-(1-x)} = 16^x$

26. $36^{x+1} = 6^{x-1}$ **27.** $25^{3-x} = 5^{2x+1}$ **28.** $8^{2x+3} = 64^{x-5}$

NAME ______________________ DATE ____________

10–3 Composition and Inverses of Functions

Objective: To find the composite of two given functions and to find the inverse of a given function.

Vocabulary

Composite of two functions The function whose value at x is $f(g(x))$ is called the composite of the functions f and g. The operation that combines f and g to produce their composite is called *composition*. Note: In general, $f(g(x)) \neq g(f(x))$.

Identity function The function that maps x to itself, that is, $I(x) = x$.

Inverse function The functions f and g are inverse functions if together they act as the identity function. In other words, f and g are inverse functions if

$f(g(x)) = x$ for all x in the domain of g

and $g(f(x)) = x$ for all x in the domain of f.

The graph of a function and its inverse are mirror images about the line $y = x$.

Horizontal-line test The inverse of a function is a function if and only if every horizontal line intersects the graph of the function in *at most* one point.

Symbols f^{-1} means the inverse of a function or "f inverse."

$f^{-1}(x)$ means the value of f inverse at x.

CAUTION The superscript $^{-1}$ used in inverse function notation is *not* an exponent.

Example 1 If $f(x) = 2x - 7$ and $g(x) = \sqrt{x} + 3$, find the following.

a. $f(g(16))$ **b.** $g(f(8))$ **c.** $f(g(x))$

Solution

a. First, find $g(16)$: $g(16) = \sqrt{16} + 3 = 4 + 3 = 7$

Substitute 7 for $g(16)$: $f(g(16)) = f(7) = 2(7) - 7 = 14 - 7$

$\therefore f(g(16)) = 7$

b. First, find $f(8)$: $f(8) = 2(8) - 7 = 16 - 7 = 9$

Substitute 9 for $f(8)$: $g(f(8)) = g(9) = \sqrt{9} + 3 = 3 + 3$

$\therefore g(f(8)) = 6$

c. First, find $g(x)$: $g(x) = \sqrt{x} + 3$

Substitute $\sqrt{x} + 3$ for $g(x)$: $f(g(x)) = f(\sqrt{x} + 3) = 2(\sqrt{x} + 3) - 7$

$\therefore f(g(x)) = 2\sqrt{x} - 1$

If $f(x) = 2x + 1$, $g(x) = \sqrt{x} - 1$, and $h(x) = \dfrac{x+1}{2}$, find a real-number value or an expression in x for each of the following. If no real value can be found, say so.

1. a. $f(g(4))$ **b.** $f(g(9))$ **c.** $f(g(0))$ **d.** $f(f(x))$

2. a. $g(f(4))$ **b.** $g(f(9))$ **c.** $g(f(0))$ **d.** $g(f(x))$

3. a. $f(h(5))$ **b.** $f(h(2))$ **c.** $h(f(\sqrt{3}))$ **d.** $f(h(x))$

4. a. $h(g(9))$ **b.** $h(g(-3))$ **c.** $g(h(15))$ **d.** $h(g(x))$

NAME ____________________ DATE ____________

10–3 Composition and Inverses of Functions *(continued)*

Example 2 Graph $f(x) = 4x - 5$ and use the horizontal-line test to determine whether f has an inverse function. If so, graph f^{-1} by reflecting f across the line $y = x$.

Solution The graph of the function $f(x) = 4x - 5$ is a line passing through the points $(0, -5)$ and $(1, -1)$. It passes the horizontal-line test, so f^{-1} is a function.

If a point (a, b) is on the graph of f, then (b, a) is on the graph of f^{-1}. Thus, the graph of f^{-1} passes through the points $(-5, 0)$ and $(-1, 1)$.

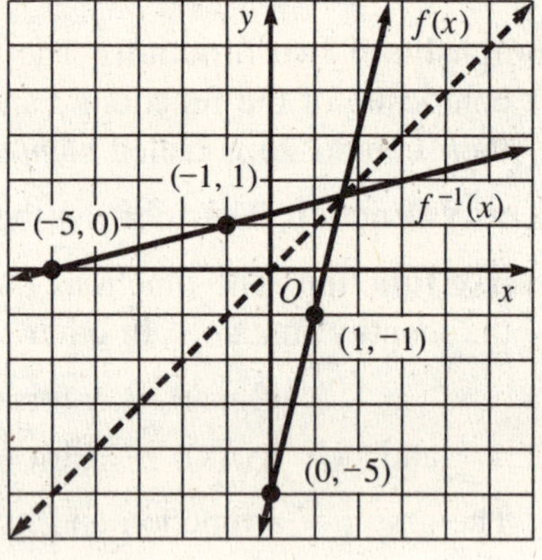

Example 3 Find f^{-1} for each of the following functions.

a. $f(x) = 2x - 6$ **b.** $f(x) = x^7$

Solution

a. Replace $f(x)$ by y: $y = 2x - 6$

Interchange x and y: $x = 2y - 6$

Solve for y: $x + 6 = 2y$

$y = \frac{1}{2}x + 3$

$\therefore f^{-1}(x) = \frac{1}{2}x + 3$

b. Replace $f(x)$ by y: $y = x^7$

Interchange x and y: $x = y^7$

Solve for y: $\sqrt[7]{x} = \sqrt[7]{y^7}$

$y = \sqrt[7]{x}$

$\therefore f^{-1}(x) = \sqrt[7]{x}$

Graph f and use the horizontal-line test to determine whether f has an inverse function. If so, graph f^{-1} by reflecting f across the line $y = x$, and find f^{-1}. If not, say so.

5. $f(x) = 3x + 2$ **6.** $f(x) = \frac{x + 1}{2}$ **7.** $f(x) = x^2$

8. $f(x) = -x^3$ **9.** $f(x) = \frac{4}{x}$ **10.** $f(x) = |x| + 1$

Mixed Review Exercises

Simplify.

1. $100^{-3/2}$ **2.** $\sqrt{12} + \sqrt{48}$ **3.** $(3 + 2i)^2$

4. $(2^{\sqrt{3}})^{2\sqrt{3}}$ **5.** $\sqrt{-18} \cdot \sqrt{-6}$ **6.** $\frac{\sqrt{3} - 2}{1 + \sqrt{3}}$

7. $\frac{6^2 - 1}{3 - 2^3}$ **8.** $\frac{3 + 2i}{1 - i}$ **9.** $16^{1.5} \cdot 2^{-2}$

NAME ______________________________ DATE ______________

10–4 Definition of Logarithms

Objective: To define logarithmic functions and to learn how they are related to exponential functions.

Vocabulary

Logarithmic function The inverse of an exponential function.
Example: The inverse function of $f(x) = 3^x$ is the logarithmic function $f(x) = \log_3 x$. It is read "the base 3 logarithm of x" or "log base 3 of x."

Logarithm ($\log_b N$) If b and N are positive numbers ($b \neq 1$), then $\log_b N = k$ if and only if $b^k = N$. $\log_b N = k$ is in *logarithmic* form and $b^k = N$ is in *exponential* form.

Properties of logarithms

1. Every positive number has a unique logarithm with base b.
2. $\log_b M = \log_b N$ if and only if $M = N$
3. $\log_b b^k = k$
4. $b^{\log_b N} = N$
5. $\log_b b = 1$
6. $\log_b 1 = 0$

Steps for simplifying logarithms

1. Set the logarithm that you want to simplify equal to k.
2. Write the equation in exponential form.
3. Rewrite each side of the equation using the same base.
4. Set the exponents equal.

Example 1

Exponential Form		*Logarithmic Form*
$3^4 = 81$	means	$\log_3 81 = 4$
$3^2 = 9$	means	$\log_3 9 = 2$
$3^0 = 1$	means	$\log_3 1 = 0$
$3^{-1} = \frac{1}{3}$	means	$\log_3 \frac{1}{3} = -1$
$3^k = N$	means	$\log_3 N = k$

Notice the pattern: $\log_3 81 = 4$ $\log_3 9 = 2$ $\log_3 1 = 0$

Write each equation in exponential form.

1. $\log_2 64 = 6$ **2.** $\log_4 \frac{1}{64} = -3$ **3.** $\log_{10} (0.01) = -2$

Write each equation in logarithmic form.

4. $2^5 = 32$ **5.** $5^{-1/2} = \frac{\sqrt{5}}{5}$ **6.** $10^{-1} = 0.1$

NAME ______________________ DATE ______________

10–4 Definition of Logarithms *(continued)*

Example 2 Simplify each logarithm.

a. $\log_4 64$ **b.** $\log_3 27\sqrt{3}$ **c.** $\log_2 \frac{1}{8}$

Solution

a. $\log_4 64 = k \longrightarrow 4^k = 64 \longrightarrow 4^k = 4^3 \longrightarrow k = 3$

$\therefore \log_4 64 = 3$

b. $\log_3 27\sqrt{3} = k \longrightarrow 3^k = 27\sqrt{3} \longrightarrow 3^k = 3^{7/2} \longrightarrow k = \frac{7}{2}$

$= 3^3 \cdot 3^{1/2}$

$= 3^{7/2}$

$\therefore \log_3 27\sqrt{3} = \frac{7}{2}$

c. $\log_2 \frac{1}{8} = k \longrightarrow 2^k = \frac{1}{8} \longrightarrow 2^k = 2^{-3} \longrightarrow k = -3$

$= \frac{1}{2^3}$

$= 2^{-3}$

$\therefore \log_2 \frac{1}{8} = -3$

Simplify each logarithm.

7. $\log_2 8$ **8.** $\log_8 64$ **9.** $\log_6 216$ **10.** $\log_7 7$

11. $\log_6 1$ **12.** $\log_8 2$ **13.** $\log_7 \frac{1}{49}$ **14.** $\log_9 \frac{1}{27}$

15. $\log_5 \sqrt{5}$ **16.** $\log_7 7\sqrt{7}$ **17.** $\log_9 \sqrt{3}$ **18.** $\log_8 \sqrt[3]{2}$

19. $\log_{1/2} 16$ **20.** $\log_{1/3} 9$ **21.** $\log_{1/3} 81$ **22.** $\log_3 \sqrt[3]{\frac{1}{9}}$

Example 3 Solve each equation.

a. $\log_5 x = 4$ **b.** $\log_x 32 = 5$

Solution

a. Write in exponential form.

$\log_5 x = 4$

$5^4 = x$

$625 = x$

$\therefore$ the solution set is $\{625\}$.

b. Write in exponential form.

$\log_x 32 = 5$

$x^5 = 32$

$x^5 = 2^5$

$x = 2$

$\therefore$ the solution set is $\{2\}$.

Solve for *x*.

23. $\log_6 x = 2$ **24.** $\log_5 x = 3$ **25.** $\log_{16} x = \frac{1}{2}$ **26.** $\log_8 x = 2.5$

27. $\log_9 x = \frac{3}{2}$ **28.** $\log_{1/4} x = -\frac{1}{2}$ **29.** $\log_x 64 = 3$ **30.** $\log_x 8 = -1$

NAME ______________________ DATE ______________

10–5 Laws of Logarithms

Objective: To learn and apply the basic properties of logarithms.

Vocabulary

Laws of Logarithms Let b be the base of a logarithmic function ($b > 0$, $b \neq 1$). Let M and N be positive numbers.

1. $\log_b MN = \log_b M + \log_b N$
2. $\log_b \frac{M}{N} = \log_b M - \log_b N$
3. $\log_b M^k = k \log_b M$

Example 1 Express $\log_5 M^4N^2$ in terms of $\log_5 M$ and $\log_5 N$.

Solution

$$\log_5 M^4N^2 = \log_5 M^4 + \log_5 N^2 \qquad \text{Use Law 1.}$$
$$= 4 \log_5 M + 2 \log_5 N \qquad \text{Use Law 3.}$$

Example 2 Express $\log_3 \sqrt[3]{\frac{M^2}{N}}$ in terms of $\log_3 M$ and $\log_3 N$.

Solution

$$\log_3 \sqrt[3]{\frac{M^2}{N}} = \log_3 \left(\frac{M^2}{N}\right)^{1/3} = \frac{1}{3} \log_3 \frac{M^2}{N} \qquad \text{Use Law 3.}$$
$$= \frac{1}{3} (\log_3 M^2 - \log_3 N) \qquad \text{Use Law 2.}$$
$$= \frac{1}{3} (2 \log_3 M - \log_3 N) \qquad \text{Use Law 3.}$$
$$= \frac{2}{3} \log_3 M - \frac{1}{3} \log_3 N \qquad \text{Multiply.}$$

Express each logarithm in terms of $\log_2 M$ and $\log_2 N$.

1. $\log_2 M^4N^3$ **2.** $\log_2 (MN)^3$ **3.** $\log_2 M \sqrt[3]{N}$ **4.** $\log_2 \sqrt{M^3N}$

5. $\log_2 \frac{M^5}{N^6}$ **6.** $\log_2 \left(\frac{N}{M}\right)^4$ **7.** $\log_2 \sqrt{\frac{M}{N}}$ **8.** $\log_2 \sqrt[3]{\frac{M}{N^4}}$

Example 3 Express as a single logarithm.

a. $\log_2 M + 2 \log_2 N$ **b.** $\frac{1}{2} \log_2 M - 3 \log_2 N$

Solution

a. $\log_2 M + 2 \log_2 N =$
$\log_2 M + \log_2 N^2 =$ (Law 3)
$\log_2 MN^2$ (Law 1)

b. $\frac{1}{2} \log_2 M - 3 \log_2 N =$
$\log_2 \sqrt{M} - \log_2 N^3 =$ (Law 3)
$\log_2 \frac{\sqrt{M}}{N^3}$ (Law 2)

NAME ______________________ DATE ______________

10–5 Laws of Logarithms *(continued)*

Express as a single logarithm.

9. $2 \log_2 M + 4 \log_2 N$ **10.** $\log_5 N - 3 \log_5 M$ **11.** $\log_{10} M + 5 \log_{10} N$

12. $7 \log_2 N - \log_2 M$ **13.** $\frac{1}{3} \log_3 M + 4 \log_3 N$ **14.** $\frac{3}{2} \log_4 M - \log_4 N$

Example 4 If $\log_{10} 4 = 0.60$ and $\log_{10} 3 = 0.48$, find:

a. $\log_{10} 40$ **b.** $\log_{10} 36$ **c.** $\log_{10} \frac{1}{\sqrt{3}}$

Solution

a. $\log_{10} 40 = \log_{10} (4 \times 10)$ $40 = 4 \times 10$

$= \log_{10} 4 + \log_{10} 10$ Use Law 1.

$= 0.60 + 1 = 1.60$ Substitute. ($\log_{10} 10 = 1$)

b. $\log_{10} 36 = \log_{10} (3^2 \times 4)$ $36 = 9 \times 4 = 3^2 \times 4$

$= \log_{10} 3^2 + \log_{10} 4$ Use Law 1.

$= 2 \log_{10} 3 + \log_{10} 4$ Use Law 3.

$= 2(0.48) + 0.60 = 1.56$ Substitute.

c. $\log_{10} \frac{1}{\sqrt{3}} = \log_{10} 1 - \log_{10} \sqrt{3}$ Use Law 2.

$= \log_{10} 1 - \log_{10} 3^{1/2}$ Write the radical in exponential form.

$= \log_{10} 1 - \frac{1}{2} \log_{10} 3$ Use Law 3.

$= 0 - \frac{1}{2}(0.48) = -0.24$ Substitute. ($\log_{10} 1 = 0$)

If $\log_{10} 4 = 0.60$ and $\log_{10} 3 = 0.48$ (accurate to two decimal places), find the following.

15. $\log_{10} 12$ **16.** $\log_{10} 16$ **17.** $\log_{10} 27$ **18.** $\log_{10} 48$

19. $\log_{10} \sqrt{3}$ **20.** $\log_{10} 2$ **21.** $\log_{10} \frac{1}{3}$ **22.** $\log_{10} \frac{2}{3}$

23. $\log_{10} \frac{30}{4}$ **24.** $\log_{10} 400$ **25.** $\log_{10} \frac{1}{4000}$ **26.** $\log_{10} \sqrt{\frac{3}{4}}$

Mixed Review Exercises

Solve.

1. $3^{x-1} = 9^{x+3}$ **2.** $\sqrt{x} + 6 = x$ **3.** $\log_x 27 = 3$ **4.** $2(x - 5) = 3x + 6$

5. $(x + 3)^2 = 7$ **6.** $1 - \frac{4}{x} = \frac{1}{x - 3}$ **7.** $2^x = \frac{\sqrt{2}}{8}$ **8.** $x^3 + 3x^2 - 4x - 12 = 0$

If $f(x) = 2x^2 + 1$ and $g(x) = \sqrt{x}$, find each of the following.

9. $f(3)$ **10.** $g(16)$ **11.** $f(g(4))$ **12.** $g(f(2))$ **13.** $f(g(a))$ **14.** $g(f(a))$

NAME ______________________________ DATE ______________

10–6 Applications of Logarithms

Objective: To use common logarithms to solve equations involving powers and to evaluate logarithms with any given base.

Vocabulary

Common logarithms Logarithms with base 10. The base 10 is usually not written. Example: log 3 means $\log_{10} 3$.

Characteristic Using scientific notation, the common logarithm of a number can be written as the sum of an integer and a nonnegative number less than 1. The characteristic is the integral part of a logarithm.

Mantissa The decimal, or fractional, part of a logarithm.

Antilogarithm If $\log y = a$, the number y is sometimes called the antilogarithm of a.

CAUTION If you will be using a calculator to find logarithms, skip over Examples 1 and 2.

Example 1 **Using a Table of Logarithms** The table below gives the common logarithms of some numbers between 1 and 10 rounded to four decimal places. The decimal point is omitted. To find an approximation for the logarithm of a number x between 0 and 10, find the first two digits of the number in the column marked N, and the third digit in the row to the right of N. The number in the spot where that row and column intersect is the logarithm of x.

To find log 3.27, find 32 in the column under N, and 7 in the row to the right of N. The value where row 32 intersects column 7 is 5145. Thus, log 3.27 = 0.5145.

N	0	1	2	3	4	5	6	7	8	9
30	4771	4786	4800	4814	4829	4843	4857	4871	4886	4900
31	4914	4928	4942	4955	4969	4983	4997	5011	5024	5038
32	5051	5065	5079	5092	5105	5119	5132	**5145**	5159	5172
33	5185	5198	5211	5224	5237	5250	5263	5276	5289	5302
34	5315	5328	5340	5353	5366	5378	5391	5403	5416	5428

Example 2 Find each logarithm using the table above: **a.** log 3400 **b.** log 0.0319

Solution Write each number in scientific notation and use the laws of logarithms.

a. $\log 3400 = \log (3.40 \times 10^3) = \log 3.40 + 3 \log 10 = 0.5315 + 3$

$\therefore \log 3400 = 3.5315$. The mantissa is 0.5315. The characteristic is 3.

b. $\log 0.0319 = \log (3.19 \times 10^{-2}) = \log 3.19 - 2 \log 10 = 0.5038 - 2$

$\therefore \log 0.0319 = -1.4962$. The mantissa is 0.5038. The characteristic is -2.

Use a calculator or the table above to find each logarithm.

1. log 32.5 **2.** log 33,200 **3.** log 0.348 **4.** log 0.00307

NAME ______________________ DATE ______________

10–6 Applications of Logarithms *(continued)*

Example 3 Find y to three significant digits if: **a.** $\log y = 0.4983$ **b.** $\log y = 3.4812$

Using Tables

a. Find the value closest or equal to 0.4983 among the entries in the table in Example 1. This is the entry in the row labeled 31 and in the column labeled 5.
$\therefore y = 3.15$

b. First find the number that has 0.4812 as its mantissa. The table entry with mantissa closest to 0.4812 (0.4814) is 3.03. Working backwards you have:

$$\log y = 3.4812 = 0.4812 + 3 = \log 3.03 + \log 10^3 = \log (3.03 \times 10^3)$$

$$\therefore y = 3.03 \times 10^3 = 3030$$

Using a Calculator

On some calculators you find the antilogarithm by using the inverse function key with the logarithm key. On others, you can use the 10^x key as shown.

a. If $\log y = 0.4983$, then $y = 10^{0.4983} = 3.15$.
b. If $\log y = 3.4812$, then $y = 10^{3.4812} = 3030$.

Use a calculator or the table in Example 1 to find x to three significant digits.

5. $\log x = 0.5378$ **6.** $\log x = 0.5009$ **7.** $\log x = 1.507$ **8.** $\log x = 2.51$

Example 4 Solve $(3.2)^x = 3$. First give the solution in calculation-ready form, and then to three significant digits.

Solution

$(3.2)^x = 3$	Take the logarithm of both sides.
$\log (3.2)^x = \log 3$	Use laws of logarithms to simplify.
$x \log 3.2 = \log 3$	Solve for x.
$x = \dfrac{\log 3}{\log 3.2}$	**(calculation-ready form)**
$x = \dfrac{0.4771}{0.5051}$	Find log 3 and log 3.2 with a calculator or the given table.
$= 0.945$	**(to three significant digits)**

Solve each equation. Give the solution (a) in calculation-ready form, and (b) to 3 significant digits.

9. $34^x = 30$ **10.** $3.1^x = 300$ **11.** $33^{-x} = 3$ **12.** $(3.09)^{2x} = 34$

Example 5 Solve $x^{3/5} = 32$. Give your answer to three significant digits.

Solution

1. Raise each side to the $\frac{5}{3}$ power. You find that $x = 32^{5/3}$.
2. To simplify $32^{5/3}$ do one of the following:

   (a) Use a calculator and the y^x key to obtain $x = 323$.

   (b) Use logarithms to obtain $\log x = \frac{5}{3} \log 32 = 2.5086$.
   Then find the antilogarithm: $x = 323$.

Solve.

13. $x^{3/5} = 31$

14. $x^{8/9} = 3$

15. $\sqrt[7]{x^3} = 32$

16. $\sqrt[3]{x^2} = 10$

NAME ______________________ DATE ______________

10–7 Problem Solving: Exponential Growth and Decay

Objective: To use exponential and logarithmic functions in growth and decay problems.

Vocabulary

Exponential growth A quantity exhibits exponential growth when its size after a period of time can be modeled by an exponential function whose base is greater than one.

Compound interest formula If an amount P (called the *principal*) is invested at an annual interest rate r compounded n times a year, then in t years the investment will grow to an amount A given by $A = P\left(1 + \frac{r}{n}\right)^{nt}$ where r is expressed as a decimal.

Doubling-time growth formula If a population of size N_0 doubles every d years (or hours, or days, or any other unit of time), then the number N in the population at time t is given by $N = N_0 \cdot 2^{t/d}$.

Example 1 **a.** One hundred dollars is invested at 7.8% interest compounded semiannually (twice a year). Determine how much the investment is worth after 3 years.

b. The value of a new \$300 VCR decreases 10% per year. Find its value after 2 years.

Solution **a.** Use the compound interest formula. Let $P = 100$, $r = 0.078$, and $n = 2$.

$$A = P\left(1 + \frac{r}{n}\right)^{nt} = 100\left(1 + \frac{0.078}{2}\right)^{2t} = 100(1 + 0.039)^{2t}$$

Since $t = 3$, $A = 100(1 + 0.039)^{2(3)} = 100(1.039)^6 \approx 100(1.26) = 126.$

$\therefore$ the investment is worth \$126 after 3 years.

Note: To evaluate $(1.039)^6$ without a calculator, write $x = (1.039)^6$, take the logarithm of both sides, and solve for x.

b. Modify the compound interest formula. Replace P with the value of a new VCR: $V_0 = 300$. Let $r = -0.10$ (since the value *decreases* over time), and $n = 1$.

$$A = V_0\left(1 + \frac{r}{n}\right)^{nt} = 300(1 - 0.10)^t, \text{ or } 300(0.9)^t$$

Since $t = 2$, value $= 300(0.9)^2 = 300(0.81) = 243.$

$\therefore$ the value of the VCR after 2 years is \$243.

Solve. Give answers to the nearest dollar. A calculator may be helpful.

1. One thousand dollars is invested at 8% interest compounded annually. Determine how much the investment is worth after:
a. 1 year **b.** 4 years **c.** 7 years

2. Redo Problem 1, assuming that the interest is compounded quarterly (four times a year).

3. The value of a new \$500 television decreases 10% per year. Find its value after:
a. 1 year **b.** 5 years **c.** 10 years

4. The value of a new \$400 bicycle decreases 20% per year. Find its value after:
a. 1 year **b.** 2 years **c.** 6 years

NAME ______________________________ DATE ______________

10–7 Problem Solving: Exponential Growth and Decay *(continued)*

Vocabulary

Exponential decay A quantity exhibits exponential decay when its size after a period of time can be modeled by an exponential function whose base is between 0 and 1.

Half-life The amount of time it takes for exactly half of a radioactive substance to decay.

Half-life decay formula If an amount N_0 has a half-life h, then the amount remaining at time t is $N = N_0\left(\frac{1}{2}\right)^{t/h}$.

Example 2 A certain bacteria population doubles in size every 16 hours. By how much will it grow in 4 days?

Solution Use the doubling-time growth formula. Let $t = 96$ hours (4 days) and let the doubling time $d = 16$ hours.

$$N = N_0 \cdot 2^{t/d} = N_0 \cdot 2^{96/16} = N_0 \cdot 2^6 = 64N_0$$

$\therefore$ the population grows by a factor of 64 in 4 days.

Example 3 The half-life of carbon 14 (C-14) is approximately 5730 years. Determine how much of 15.0 mg of this substance will remain after 2500 years.

Solution Use the half-life decay formula. Let $N_0 = 15$, $h = 5730$, and $t = 2500$.

$$N = N_0\left(\frac{1}{2}\right)^{t/h} = 15\left(\frac{1}{2}\right)^{2500/5730}$$

If you have a calculator, use the y^x key to obtain N. Otherwise, take the logarithm of both sides and use Table 3 on pages 812 and 813 of your textbook to obtain $\log N = 1.0448$. Then find the antilogarithm. In either case, you will find that 11.1 mg of this substance will remain after 2500 years (to three significant digits).

Solve. Give answers to three significant digits. A calculator may be helpful.

5. The population of a certain country doubles in size every 60 years. The population is now N_0. Find its size, in terms of N_0, in:
   a. 120 years **b.** 180 years **c.** y years
6. The half-life of carbon 14 is approximately 5730 years. Determine how much of 1000 kg of this substance will remain after 8000 years.
7. A radioactive substance has a half-life of approximately 18 hours. About how much of a 10 g sample will remain after 4 days?

Mixed Review Exercises

If the domain of each function is $D = \{1, 2, 4\}$, find the range.

1. $h(x) = \sqrt{x} - 1$ **2.** $F(x) = \log_2 x$ **3.** $G(x) = |x - 5|$ **4.** $H(x) = (\sqrt{3})^x$

NAME ______________________ DATE ______________

10–8 The Natural Logarithm Function

Objective: To define and use the natural logarithm function.

Vocabulary

Natural logarithm function The logarithm function with base *e*, where *e* is an irrational number approximately equal to 2.71828. The natural logarithm is used extensively in advanced science and mathematics.

Symbol $\ln x$ (natural logarithm of x, or $\log_e x$)

CAUTION You will have to be familiar with the properties of logarithms found in Lesson 10–4, and the laws of logarithms found in Lesson 10–5 in order to do the exercises in this lesson. These properties and laws apply to natural logarithms just as they do to logarithms with other bases.

Example 1	*Exponential Form*		*Logarithmic Form*
	$e^{2.64} = 14$	means	$\ln 14 = 2.64$
	$e^{3.09} = 22$	means	$\ln 22 = 3.09$
	$e^{-1.1} = \frac{1}{3}$	means	$\ln \frac{1}{3} = -1.1$
	$\sqrt[5]{e} = 1.22$	means	$\ln 1.22 = \frac{1}{5}$

Write each equation in exponential form.

1. $\ln 10 = 2.30$ **2.** $\ln 50 = 3.91$ **3.** $\ln \frac{1}{4} = -1.39$ **4.** $\ln \frac{1}{e^3} = -3$

Write each equation in logarithmic form.

5. $e^4 = 54.6$ **6.** $e^9 = 8103$ **7.** $e^{1/4} = 1.28$ **8.** $\sqrt{e} = 1.65$

Example 2 Simplify: **a.** $\ln e^5$ **b.** $\ln \frac{1}{e^3}$ **c.** $e^{\ln 3}$

Solution

a. $\ln e^5 = 5 \ln e$
$= 5 \cdot 1$
$= 5$

b. $\ln \frac{1}{e^3} = \ln e^{-3}$
$= -3 \ln e$
$= -3 \cdot 1$
$= -3$

c. $e^{\ln 3} = 3$
A property of logarithms:
$b^{\log_b N} = N$

Simplify. If the expression is undefined, say so.

9. $\ln e^4$ **10.** $\ln e^7$ **11.** $\ln \frac{1}{e^5}$ **12.** $\ln \sqrt{e}$

13. $\ln 1$ **14.** $\ln (-1)$ **15.** $e^{\ln 1.2}$ **16.** $e^{\ln \sqrt{3}}$

NAME ________________________ DATE ____________

10–8 The Natural Logarithm Function *(continued)*

Example 3 Write as a single logarithm: **a.** $\ln 2 + \ln 5$ **b.** $2 \ln 3 - \ln 4 + 1$

Solution **a.** $\ln 2 + \ln 5 = \ln (2 \cdot 5)$
$= \ln 10$

b. $2 \ln 3 - \ln 4 + 1 = \ln 3^2 - \ln 4 + \ln e$
$= \ln \left(\frac{3^2}{4} \cdot e\right) = \ln \frac{9e}{4}$

Write as a single logarithm.

17. $\ln 3 + \ln 7$ **18.** $\ln 12 - \ln 3$ **19.** $\ln 11 + \frac{1}{2} \ln 4$ **20.** $3 \ln 3 - \ln 2 + 2$

Example 4 Solve: **a.** $\ln \frac{1}{x} = 4$ **b.** $\ln (x + 3) = 1$

Solution Write in exponential form. Then solve for x.

a.
$$\ln \frac{1}{x} = 4$$
$$e^4 = \frac{1}{x}$$
$$x = \frac{1}{e^4}$$
$$x = e^{-4}$$

b.
$$\ln (x + 3) = 1$$
$$e^1 = x + 3$$
$$e - 3 = x$$
$$x = e - 3$$

Solve for x. Leave answers in terms of e.

21. $\ln x = 5$ **22.** $\ln \frac{1}{x} = 3$ **23.** $\ln (x - 2) = 2$ **24.** $\ln \sqrt{x + 5} = 1$

Example 5 Solve: **a.** $e^x = 7$ **b.** $e^{x + 3} = 8$

Solution Take the natural logarithm of both sides of the equation. Then solve for x.

a.
$$e^x = 7$$
$$\ln e^x = \ln 7$$
$$x \cdot \ln e = \ln 7$$
$$x \cdot 1 = \ln 7$$
$$x = \ln 7$$

b.
$$e^{2x + 8} = 9$$
$$\ln e^{2x + 8} = \ln 9$$
$$(2x + 8) \cdot \ln e = \ln 9$$
$$(2x + 8) \cdot 1 = \ln 9$$
$$2x + 8 = \ln 9$$
$$2x = -8 + \ln 9$$
$$x = \frac{-8 + \ln 9}{2}, \text{ or } -4 + \ln 3$$

Solve for x. Leave answers in terms of natural logarithms.

25. $e^x = 5$ **26.** $e^{-x} = 2$ **27.** $e^{2x} = 36$ **28.** $e^{x-4} = 3$

29. $e^{3x-6} = 27$ **30.** $\sqrt{e^x} = 4$ **31.** $e^{-3x} = \frac{1}{8}$ **32.** $(e^x)^4 = 81$

NAME ______________________ DATE ______________

11 Sequences and Series

11–1 Types of Sequences

Objective: To determine whether a sequence is arithmetic, geometric, or neither and to supply missing terms of a sequence.

Vocabulary

Sequence A function whose domain consists of consecutive positive integers. Each corresponding value is a *term* of a sequence.

Finite sequence A sequence that has a limited number of terms.

Infinite sequence A sequence that has an unlimited number of terms.

Arithmetic sequence (or **arithmetic progression**) A sequence in which the difference between any two successive terms is constant. This constant difference is called the *common difference* and is usually denoted by d.

Geometric sequence (or **geometric progression**) A sequence in which the ratio of every pair of successive terms is constant. This constant ratio is called the *common ratio* and is usually denoted by r.

Symbols

d (common difference) r (common ratio) t_n (nth term of a sequence)

Example 1 For each arithmetic sequence, find the common difference and the next two terms of the sequence.

a. 5, 8, 11, 14, 17, . . . **b.** 23, 16, 9, 2, −5, . . .

Solution **a.** Find the common difference by subtracting any term from the term that follows it:

$$d = 8 - 5 = 3$$

$\therefore$ the next two terms are $17 + 3 = 20$ and $20 + 3 = 23$.

b. The common difference is $d = 16 - 23 = -7$.

$\therefore$ the next two terms are $-5 + (-7) = -12$ and $-12 + (-7) = -19$.

Example 2 For each geometric sequence, find the common ratio and the next two terms of the sequence.

a. 3, 6, 12, 24, . . . **b.** $90, -30, 10, -\frac{10}{3}, \ldots$

Solution **a.** Find the common ratio by dividing any term by the term before it:

$$r = \frac{6}{3} = 2$$

$\therefore$ the next two terms are $24 \cdot 2 = 48$ and $48 \cdot 2 = 96$.

b. The common ratio is $r = \frac{-30}{90} = -\frac{1}{3}$.

$\therefore$ the next two terms are $\left(-\frac{10}{3}\right)\left(-\frac{1}{3}\right) = \frac{10}{9}$ and $\left(\frac{10}{9}\right)\left(-\frac{1}{3}\right) = -\frac{10}{27}$.

NAME ______________________ DATE ______________

11-1 Types of Sequences *(continued)*

Example 3 Using the given formula for the nth term, find t_1, t_2, t_3, and t_4. Then tell whether the sequence is arithmetic, geometric, or neither.

a. $t_n = 3 + 2n$ **b.** $t_n = \sqrt{n}$ **c.** $t_n = -2 \cdot 3^n$

Solution Substitute 1, 2, 3, and 4 for n in turn.

a.

n	$t_n = 3 + 2n$
1	$t_1 = 3 + 2(1) = 5$
2	$t_2 = 3 + 2(2) = 7$
3	$t_3 = 3 + 2(3) = 9$
4	$t_4 = 3 + 2(4) = 11$

The sequence 5, 7, 9, 11, . . . is arithmetic since the common difference is 2.

b.

n	$t_n = \sqrt{n}$
1	$t_1 = \sqrt{1} = 1$
2	$t_2 = \sqrt{2}$
3	$t_3 = \sqrt{3}$
4	$t_4 = \sqrt{4} = 2$

The sequence 1, $\sqrt{2}$, $\sqrt{3}$, 2, . . . is neither arithmetic nor geometric.

c.

n	$t_n = -2 \cdot 3^n$
1	$t_1 = -2 \cdot 3^1 = -6$
2	$t_2 = -2 \cdot 3^2 = -18$
3	$t_3 = -2 \cdot 3^3 = -54$
4	$t_4 = -2 \cdot 3^4 = -162$

The sequence -6, -18, -54, -162, . . . is geometric since the common ratio is 3.

Tell whether each sequence is arithmetic, geometric, or neither. Then supply the missing terms of the sequence.

1. 13, 16, 19, 22, _?_, _?_ **2.** 1, 4, 16, 64, _?_, _?_ **3.** 48, 24, 12, 6, _?_, _?_

4. -3, 0, 3, 6, _?_, _?_ **5.** 31, 27, 23, _?_, 15, _?_ **6.** -1, 3, -9, _?_, -81, _?_

7. $\frac{1}{3}, \frac{2}{5}, \frac{3}{7}, \frac{4}{9}$, _?_, _?_ **8.** -2, _?_, 12, 19, _?_, 33 **9.** 2, $2^{3/2}$, _?_, _?_, 2^3

Find the first four terms of the sequence with the given formula. Then tell whether the sequence is arithmetic, geometric, or neither.

10. $t_n = 2 + 5n$ **11.** $t_n = 2^n$ **12.** $t_n = 3^{1-n}$ **13.** $t_n = n^3$

14. $t_n = 3n - 1$ **15.** $t_n = \frac{1}{n}$ **16.** $t_n = (-3)^n$ **17.** $t_n = \frac{n+1}{2n}$

Mixed Review Exercises

Graph each equation.

1. $y = x^2 - 4x - 1$ **2.** $y = 3^{-x}$ **3.** $x^2 + 9y^2 = 36$ **4.** $(x - 1)^2 - y^2 = 16$

Solve each system.

5. $x + 4y = 1$, $3x + y = 14$ **6.** $y = x^2$, $x + y = 2$ **7.** $y = 4x^2 - 2$, $x^2 + y^2 = 4$ **8.** $x^2 + y^2 = 21$, $x^2 - y^2 = 11$

NAME ______________________ DATE ______________

11–2 Arithmetic Sequences

Objective: To find a formula for the *n*th term of an arithmetic sequence and to find specified terms of arithmetic sequences.

Vocabulary

***n*th (or general) term of an arithmetic sequence** In an arithmetic sequence with first term t_1 and common difference d, the nth (or *general*) term is given by

$$t_n = t_1 + (n - 1)d.$$

Arithmetic means The term(s) between two given terms of an arithmetic sequence. Example: For the arithmetic sequence 5, 9, 13, 17, . . . , the numbers 9 and 13 are two arithmetic means between 5 and 17.

The arithmetic mean A single arithmetic mean between two numbers. The arithmetic mean, or *average,* of two numbers a and b is the number $\frac{a + b}{2}$.

Example: The arithmetic mean of 5 and 17 is $\frac{5 + 17}{2}$, or 11.

Example 1 Find a formula for the nth term of the arithmetic sequence 11, 17, 23, 29, ...

Solution The first term, t_1, is 11. The common difference, d, is $17 - 11$, or 6. Substitute 11 for t_1 and 6 for d in the formula $t_n = t_1 + (n - 1)d$.

$$\begin{aligned} t_n &= 11 + (n - 1)6 \\ &= 11 + 6n - 6 \\ &= 5 + 6n \end{aligned}$$

$\therefore t_n = 5 + 6n$

Find a formula for the *n*th term of each arithmetic sequence.

1. 7, 10, 13, 16, . . .
2. 5, 10, 15, 20, . . .
3. 4, −1, −6, −11, . . .
4. −1, 4, 9, 14, . . .
5. −2, −6, −10, −14, . . .
6. 21, 29, 37, 45, . . .

Example 2 Find t_{15} for the arithmetic sequence 11, 17, 23, 29, ...

Solution Use the formula $t_n = 5 + 6n$ from the solution of Example 1.

$$t_{15} = 5 + 6(15) = 95$$

Example 3 Find t_{25} for the arithmetic sequence in which $t_2 = -5$ and $t_6 = 7$.

Solution Substitute -5 for t_2 and 7 for t_6 in the formula $t_n = t_1 + (n - 1)d$ to obtain a system of equations in t_1 and d.

$$t_2 = t_1 + (2 - 1)d \longrightarrow -5 = t_1 + d$$
$$t_6 = t_1 + (6 - 1)d \longrightarrow 7 = t_1 + 5d$$

Solve the first equation for t_1: $t_1 = -d - 5$.

(Solution continues on the next page.)

11–2 Arithmetic Sequences *(continued)*

Substitute $-d - 5$ for t_1 in the second equation and solve for d.

$$7 = -d - 5 + 5d$$
$$12 = 4d$$
$$d = 3$$

Then $t_1 = -3 - 5 = -8$. Now using $t_1 = -8$ and $d = 3$ in the formula $t_n = t_1 + (n - 1)d$, find t_{25}.

$$t_{25} = -8 + (25 - 1)(3)$$
$$= -8 + 24(3)$$
$$= 64$$

$\therefore t_{25} = 64$

Find the specified term of each arithmetic sequence.

7. $4, 7, 10, 13, \ldots; t_{19}$

8. $2, 10, 18, 26, \ldots; t_{56}$

9. $90, 87, 84, 81, \ldots; t_{20}$

10. $1, 1.25, 1.5, 1.75, \ldots; t_{33}$

11. $-3, -12, -21, \ldots; t_{50}$

12. $19, 8, -3, \ldots; t_{41}$

13. $t_1 = 2, t_4 = 8; t_{14}$

14. $t_2 = -5, t_4 = -11; t_{10}$

15. $t_6 = 22, t_{10} = 38; t_1$

16. $t_{10} = 70, t_{15} = 60; t_5$

Example 4 **a.** Find the arithmetic mean of -3 and 8.

b. Insert three arithmetic means between 10 and 26.

Solution **a.** The arithmetic mean is the average of -3 and 8.

$$\frac{-3 + 8}{2} = \frac{5}{2} = 2.5$$

b. Set up the sequence: 10, _?_, _?_, _?_, 26

In this sequence, 10 is the *first* term and 26 is the *fifth* term. So to find d, substitute 10 for t_1 and 26 for t_5 in the formula $t_n = t_1 + (n - 1)d$.

$$26 = 10 + (5 - 1)d$$
$$16 = 4d$$
$$d = 4$$

The three arithmetic means are obtained by adding 4 to successive terms:

10, **14, 18, 22,** 26

Find the arithmetic mean of each pair of numbers.

17. $-2, 9$

18. $3.2, 6.4$

19. $\frac{2}{3}, \frac{3}{5}$

20. $\sqrt{3}, 2\sqrt{3}$

Insert (a) two and (b) three arithmetic means between each pair of numbers.

21. $-1, 5$

22. $30, 50$

23. $14, 39$

24. $0, 40$

NAME ______________________ DATE ______________

11–3 Geometric Sequences

Objective: To find a formula for the *n*th term of a geometric sequence and to find specified terms of geometric sequences.

Vocabulary

***n*th** (or **general**) term of a geometric sequence In a geometric sequence with first term t_1 and common ratio r, the *n*th (or *general*) term is given by $t_n = t_1 \cdot r^{n-1}$.

Geometric means The terms between two given terms of a geometric sequence. Example: For the geometric sequence 2, −6, 18, −54, . . . , the numbers −6 and 18 are two geometric means between 2 and −54.

The geometric mean The geometric mean of a and b is commonly taken to be $\sqrt{ab}$ if a and b are positive and $-\sqrt{ab}$ if a and b are negative. Example: The geometric mean of 2 and 8 is $\sqrt{2 \cdot 8}$, or 4, and of −2 and −8 is $-\sqrt{(-2)(-8)}$, or −4.

Example 1 Find a formula for the *n*th term of the sequence 2, −8, 32, −128, ...

Solution The first term, t_1, is 2. The common ratio, r, is $\frac{-8}{2}$, or −4. Substitute 2 for t_1 and −4 for r in the formula $t_n = t_1 \cdot r^{n-1}$.

$$t_n = 2(-4)^{n-1}$$

Find a formula for the *n*th term of each geometric sequence.

1. 4, 12, 36, 108, . . .
2. 400, 100, 25, 6.25, . . .
3. 1, $\sqrt{3}$, 3, $3\sqrt{3}$, . . .
4. 25, 20, 16, 12.8, . . .
5. 80, −60, 45, −33.75, . . .
6. −3, 0.3, −0.03, 0.003, . . .

Example 2 Find t_7 for the geometric sequence 2, −8, 32, −128, ...

Solution Use the formula $t_n = 2(-4)^{n-1}$ from the solution of Example 1.

$$t_7 = 2(-4)^{7-1} = 2(-4)^6 = 2(4096) = 8192$$

Example 3 Find t_9 for the geometric sequence in which $t_2 = 15$ and $t_5 = 405$.

Solution Substitute 15 for t_2 and 405 for t_5 in the formula $t_n = t_1 \cdot r^{n-1}$ to obtain a system of equations in t_1 and r.

$$t_2 = t_1 \cdot r^{2-1} \longrightarrow 15 = t_1 \cdot r$$
$$t_5 = t_1 \cdot r^{5-1} \longrightarrow 405 = t_1 \cdot r^4$$

Solve the first equation for t_1: $t_1 = \frac{15}{r}$.

Substitute $\frac{15}{r}$ for t_1 in the second equation and solve for r.

(*Solution continues on the next page.*)

11–3 Geometric Sequences *(continued)*

$$405 = \frac{15}{r} \cdot r^4$$
$$405 = 15r^3$$
$$27 = r^3, \text{ so } r = \sqrt[3]{27} = 3$$

Then $t_1 = \frac{15}{3} = 5$. Now use $t_1 = 5$ and $r = 3$ in the formula $t_n = t_1 \cdot r^{n-1}$ to find t_9.

$$t_9 = 5(3^{9-1}) = 5(3^8) = 5(6561) = 32{,}805$$

Find the specified term of each geometric sequence.

7. 7, 14, 28, 56, . . . ; t_{11} **8.** 100, 60, 36, $\frac{108}{5}$, . . . ; t_8 **9.** $-\frac{1}{4}, \frac{1}{2}, -1, 2, \ldots; t_{12}$

10. $t_2 = 0.1$, $t_5 = 0.0001$; t_9 **11.** $t_3 = 12$, $t_6 = -96$; t_8 **12.** $t_1 = 3$, $t_3 = 75$; t_5

Example 4 **a.** Find the geometric mean of 16 and 25.
b. Insert three geometric means between 32 and 162.

Solution **a.** The geometric mean of 16 and 25 is $\sqrt{16 \cdot 25} = \sqrt{400} = 20$.

b. Set up the sequence: 32, __?__, __?__, __?__, 162

In this sequence, 32 is the *first* term and 162 is the *fifth* term. To find r, substitute 32 for t_1 and 162 for t_5 in the formula $t_n = t_1 \cdot r^{n-1}$.

$$162 = 32r^{5-1}$$
$$\frac{81}{16} = r^4, \text{ so } r = \pm\sqrt[4]{\frac{81}{16}} = \pm\frac{3}{2}$$

For $r = \frac{3}{2}$: 32, **48, 72, 108,** 162. For $r = -\frac{3}{2}$: 32, **−48, 72, −108,** 162.

Find the geometric mean of each pair of numbers.

13. 3, 27 **14.** $\frac{1}{2}, \frac{1}{8}$ **15.** $\sqrt{5}, \sqrt{45}$ **16.** $\frac{2}{3}, \frac{27}{2}$

Insert the given number of geometric means between the pairs of numbers.

17. Two; 1, 729 **18.** Four; −7, 224 **19.** Three; −5, −405 **20.** Three; $\frac{1}{36}, \frac{9}{4}$

Mixed Review Exercises

Solve each inequality and graph the solution set.

1. $|3x + 1| > 5$ **2.** $x^2 + 2x \le 3$ **3.** $-1 \le \frac{x}{3} - 2 \le 4$ **4.** $6 - 2x \ge -4$

Find an equation for each figure described.

5. The line containing (−1, 4) and (3, −2)

6. The parabola with focus (0, 2) and directrix $y = -2$

NAME ______________________________ DATE ______________

11–4 Series and Sigma Notation

Objective: To identify series and to use sigma notation.

Vocabulary

Series The indicated sum of the terms of a sequence.

Arithmetic series A series whose related sequence is arithmetic.
Example: 5 + 8 + 11 + 14 is a *finite* arithmetic series (with four terms).

Geometric series A series whose related sequence is geometric.
Example: 3 + 6 + 12 + 24 + · · · is an *infinite* geometric series.

Sigma notation The Greek letter Σ (*sigma*), called the *summation sign*, can be used to write a series in abbreviated form. Examples: The finite arithmetic series 5 + 8 + 11 + 14 can be written $\sum_{n=1}^{4} (3n + 2)$, which is read "the sum of $3n + 2$ for values of n from 1 to 4." The infinite geometric series 3 + 6 + 12 + 24 + · · · can be written $\sum_{k=1}^{\infty} 3(2^{k-1})$, which is read "the sum of $3(2^{k-1})$ for values of k from 1 to infinity." In the first example, $3n + 2$ is called a *summand*; the letter n is called the *index*; and 1 and 4 are called the *upper* and *lower limits of summation*, respectively. Any letter can be used as the index in a summation.

Symbol

Σ (sigma; used as a summation sign)   ∞ (infinity; used to denote an infinite series)

Example 1 Write the series $\sum_{j=1}^{3} (-1)^j(3j + 1)$ in expanded form.

Solution Replace j with 1, 2, and 3 in turn.

$$\sum_{j=1}^{3} (-1)^j(3j + 1) = (-1)^1(3 \cdot \mathbf{1} + 1) + (-1)^2(3 \cdot \mathbf{2} + 1) + (-1)^3(3 \cdot \mathbf{3} + 1)$$
$$= (-1)(4) + (1)(7) + (-1)(10) = -4 + 7 - 10$$

Write each series in expanded form.

1. $\sum_{n=1}^{5} (2n - 3)$ **2.** $\sum_{j=0}^{4} 5^{-j}$ **3.** $\sum_{k=3}^{9} |6 - k|$ **4.** $\sum_{n=5}^{10} \frac{(-1)^n}{n + 2}$

Example 2 Use sigma notation to write each series.
a. 4 + 7 + 10 + · · · + 61   **b.** 3 + 6 + 12 + · · · + 1536

Solution **a.** Since the series is arithmetic with common difference 3, the nth term is:

$$t_n = t_1 + (n - 1)d = 4 + (n - 1)3 = 3n + 1$$

(*Solution continues on the next page.*)

NAME ______________________ DATE ______________

Now find n such that the last term is 61:

$$\begin{aligned} t_n &= 3n + 1 \\ 61 &= 3n + 1 \\ 60 &= 3n \\ n &= 20 \end{aligned}$$

$\therefore$ the series is $\sum_{n=1}^{20} (3n + 1)$.

b. Since the series is geometric with common ratio 2, the nth term is $t_n = 3(2^{n-1})$.

Now find n such that the last term is 1536:

$$\begin{aligned} t_n &= 3(2^{n-1}) \\ 1536 &= 3(2^{n-1}) \\ 512 &= 2^{n-1} \\ 2^9 &= 2^{n-1} \\ 9 &= n - 1 \quad \text{Equate exponents.} \\ n &= 10 \end{aligned}$$

$\therefore$ the series is $\sum_{n=1}^{10} 3(2^{n-1})$.

Write each series using sigma notation.

5. $4 + 8 + 12 + \cdots + 100$

6. $25 + 30 + 35 + \cdots + 205$

7. $2 \cdot 5 + 2 \cdot 5^2 + 2 \cdot 5^3 + \cdots + 2 \cdot 5^{15}$

8. $1 + 0.1 + 0.01 + \cdots + 0.0000001$

9. $8 + 5 + 2 + \cdots + (-40)$

10. $\frac{2}{3} + \frac{4}{9} + \frac{8}{27} + \cdots + \frac{256}{6561}$

Example 3 Use sigma notation to write the series $\frac{2}{3} + \frac{4}{5} + \frac{6}{7} + \frac{8}{9} + \cdots$.

Solution Since the infinite series is neither arithmetic nor geometric, you need to look for patterns.

1. Since the numerators are consecutive even integers, a general expression for the numerators is $2n$.
2. Since each denominator is 1 more than the corresponding numerator, a general expression for the denominators is $2n + 1$.

So the expression for the summand is $\frac{2n}{2n + 1}$.

$\therefore$ the series is $\sum_{n=1}^{\infty} \frac{2n}{2n + 1}$.

Write each series in sigma notation.

11. $\sqrt{11} + \sqrt{22} + \sqrt{33} + \cdots + \sqrt{143}$

12. $1^2 + 3^4 + 5^6 + 7^8 + \cdots + 99^{100}$

13. $\frac{1}{3} + \frac{2}{5} + \frac{3}{7} + \frac{4}{9} + \cdots$

14. $\frac{3}{2} + \frac{4}{3} + \frac{5}{4} + \frac{6}{5} + \cdots$

NAME ______________________________ DATE ______________

11–5 Sums of Arithmetic and Geometric Series

Objective: To find sums of finite arithmetic and geometric series.

Vocabulary

Theorem 1: Sum of a finite arithmetic series The sum of the first n terms of an arithmetic series is $S_n = \frac{n(t_1 + t_n)}{2}$.

Theorem 2: Sum of a finite geometric series The sum of the first n terms of a geometric series with common ratio r $(r \neq 1)$ is $S_n = \frac{t_1(1 - r^n)}{1 - r}$. (*Note*: If $r = 1$, the sum of the first n terms is $S_n = nt_1$.)

Symbol

S_n (the sum of the first n terms of a series)

Example 1 Find the sum of the first 50 terms of the arithmetic series $4 + 8 + 12 + 16 + \cdots$.

Solution First find the 50th term: $t_{50} = t_1 + (50 - 1)d$
$= 4 + (49)4 = 200$

Then use the formula $S_n = \frac{n(t_1 + t_n)}{2}$: $S_{50} = \frac{50(4 + 200)}{2} = 5100$

Example 2 Evaluate $\sum_{k=1}^{25} (3k - 1)$.

Solution Evaluating $\sum_{k=1}^{25} (3k - 1)$ means finding S_{25} for this arithmetic series:

$$2 + 5 + 8 + \cdots + 74$$

$\therefore S_{25} = \frac{25(2 + 74)}{2} = 950$

Find the sum of each arithmetic series.

1. $n = 20, t_1 = 3, t_{20} = 98$
2. $n = 50, t_1 = -8, t_{50} = 286$
3. $n = 30, t_1 = 7, t_{30} = 94$
4. $n = 100, t_1 = 196, t_{100} = -2$
5. $\sum_{n=1}^{24} (6n + 1)$
6. $\sum_{n=1}^{30} (4n - 3)$
7. $\sum_{n=1}^{24} (1 - 2n)$
8. $\sum_{n=1}^{50} (6 - n)$
9. The first 75 terms of the series $-3 + 1 + 5 + 9 + \cdots$
10. The first 100 terms of the series $200 + 195 + 190 + 185 + \cdots$
11. $3 + 6 + 9 + \cdots + 132$
12. $8 + 14 + 20 + \cdots + 128$

11–5 Sums of Arithmetic and Geometric Series *(continued)*

Example 3 Find the sum of the first 10 terms of the geometric series $\frac{3}{4} + \frac{3}{2} + 3 + 6 \cdots$.

Solution First find the common ratio: $r = \frac{3}{2} \div \frac{3}{4} = \frac{3}{2} \cdot \frac{4}{3} = 2$

Since $r \neq 1$, use the formula $S_n = \frac{t_1(1 - r^n)}{1 - r}$: $S_{10} = \frac{\frac{3}{4}(1 - 2^{10})}{1 - 2}$

$$= \frac{\frac{3}{4}(1 - 1024)}{-1}$$

$$= \frac{3069}{4} = 767.25$$

Example 4 Evaluate $\sum_{n=1}^{9} 5(-2)^{n-1}$.

Solution You are finding S_9 for the series $5 - 10 + 20 - \cdots$, whose common ratio is -2.

Since $r \neq 1$, use the formula $S_n = \frac{t_1(1 - r^n)}{1 - r}$: $S_9 = \frac{5(1 - (-2)^9)}{1 - (-2)}$

$$= \frac{5(513)}{3} = 855$$

Find the sum of each geometric series.

13. $n = 7, r = 3, t_1 = 1$

14. $n = 7, r = -4, t_1 = 5$

15. $n = 8, r = \frac{1}{2}, t_1 = 10$

16. $n = 10, r = -2, t_1 = 2$

17. $\sum_{n=1}^{8} 3^n$

18. $\sum_{k=1}^{7} \left(-\frac{2}{3}\right)^k$

19. Find S_{10} if the series $8 + 4 + \cdots$ is (a) arithmetic and (b) geometric.

Mixed Review Exercises

Evaluate if $x = -3$ and $y = 9$.

1. y^{-x}

2. $3x^2 + y$

3. $x^2 - y^{-1}$

4. $\sqrt[3]{xy}$

5. $x^{-3}y$

6. $\sqrt{x^2 + y^2}$

7. $|xy|$

8. $\log_3 y^2$

9. $\frac{y - x}{x}$

10. Find the slope of the line $2x + 5y = 5$.

11. Find the radius of the circle $x^2 + y^2 - 2x + 4y - 4 = 0$.

12. Find the y-intercepts of the hyperbola $9y^2 - 4x^2 = 36$.

NAME ______________________________ DATE ______________

11–6 Infinite Geometric Series

Objective: To find sums of infinite geometric series having ratios with absolute value less than one.

Vocabulary

Theorem: Sum of an infinite geometric series An infinite geometric series with common ratio r has a sum S if $|r| < 1$. This sum is $S = \frac{t_1}{1 - r}$.

Example 1 Find the sum of each infinite geometric series. If the series has no sum, say so.

a. $8 - 6 + \frac{9}{2} - \frac{27}{8} + \cdots$ **b.** $2 - \frac{8}{3} + \frac{32}{9} - \frac{128}{27} + \cdots$

Solution **a.** First find the common ratio r: $r = -\frac{6}{8} = -\frac{3}{4}$

Since $\left| -\frac{3}{4} \right| < 1$, the series has a sum.

Use the formula $S = \frac{t_1}{1 - r}$:

$$S = \frac{8}{1 - \left(-\frac{3}{4}\right)} = \frac{8}{1 + \frac{3}{4}} = \frac{8}{\frac{7}{4}} = 8 \cdot \frac{4}{7} = \frac{32}{7}$$

$\therefore$ the sum is $\frac{32}{7}$.

b. First find the common ratio r: $r = -\frac{8}{3} \div 2 = -\frac{8}{3} \cdot \frac{1}{2} = -\frac{8}{6} = -\frac{4}{3}$

Since $\left| -\frac{4}{3} \right| \not< 1$, the series has no sum.

For each geometric series, find the sum. If the series has no sum, say so.

1. $36 + 24 + 16 + \frac{32}{3} + \cdots$

2. $12 - 9 + \frac{27}{4} - \frac{81}{16} + \cdots$

3. $4 + 6 + 9 + \frac{27}{2} + \cdots$

4. $\frac{3}{4} + \frac{3}{5} + \frac{12}{25} + \frac{48}{125} + \cdots$

5. $18 - 12 + 8 - \frac{16}{3} + \cdots$

6. $8 - 12 + 18 - 27 + \cdots$

7. $10 - 9 + 8.1 - 7.29 + \cdots$

8. $5 + 3 + 1.8 + 1.08 + \cdots$

9. $3^{1/2} - 3^{-1/2} + 3^{-3/2} - 3^{-5/2} + \cdots$

10. $4 - 2\sqrt{2} + 2 - \sqrt{2} + \cdots$

11. $\sum_{n=1}^{\infty} 5\left(\frac{1}{3}\right)^n$

12. $\sum_{j=0}^{\infty} 7\left(\frac{9}{10}\right)^j$

NAME ______________________ DATE ______________

11-6 Infinite Geometric Series *(continued)*

Example 2 Given the geometric series $3 + 1 + \frac{1}{3} + \frac{1}{9} + \frac{1}{27} + \cdots$:

a. Find S_1, S_2, S_3, S_4, and S_5.

b. Use the sums from part (a) to approximate S.

c. Use the formula to find S.

d. Compare your approximation of S with the value obtained from the formula.

Solution **a.** $S_1 = 3$

$S_2 = 3 + 1 = 4$

$S_3 = 3 + 1 + \frac{1}{3} = 4\frac{1}{3}$

$S_4 = 3 + 1 + \frac{1}{3} + \frac{1}{9} = 4\frac{4}{9}$

$S_5 = 3 + 1 + \frac{1}{3} + \frac{1}{9} + \frac{1}{27} = 4\frac{13}{27}$

b. The sums 3, 4, $4\frac{1}{3}$, $4\frac{4}{9}$, $4\frac{13}{27}$ seem to approach $4\frac{1}{2}$. Therefore, $S \approx 4\frac{1}{2}$.

c. The common ratio is $r = \frac{1}{3}$.

$$\therefore S = \frac{3}{1 - \frac{1}{3}} = \frac{3}{\frac{2}{3}} = 3 \cdot \frac{3}{2} = \frac{9}{2}$$

d. The actual and approximate values of S are equal.

For each geometric series find S_1, S_2, S_3, S_4, and S_5. Use these sums to approximate S. Then use the formula to find S. Compare your approximation with the value obtained from the formula.

13. $4 + 2 + 1 + \frac{1}{2} + \frac{1}{4} + \cdots$

14. $3 + \frac{3}{10} + \frac{3}{100} + \frac{3}{1000} + \frac{3}{10,000} + \cdots$

15. $\frac{1}{2} - \frac{1}{4} + \frac{1}{8} - \frac{1}{16} + \frac{1}{32} - \cdots$

16. $25 + 5 + 1 + \frac{1}{5} + \frac{1}{25} + \cdots$

Example 3 Write 0.271271271... as a common fraction.

Solution The infinite repeating decimal can be written as the infinite series:

$$0.271 + 0.000271 + 0.000000271 + \cdots$$

Since the common ratio is 0.001 and $|0.001| < 1$, the series has a sum.

$$S = \frac{t_1}{1 - r} = \frac{0.271}{1 - 0.001} = \frac{0.271}{0.999} = \frac{271}{999}$$

Write each repeating decimal as a common fraction.

17. 0.6666... **18.** 0.5555... **19.** 0.423423...

20. 0.727272... **21.** 0.59999... **22.** 5.20520520...

NAME ______________________ DATE ____________

11–7 Powers of Binomials

Objective: To expand powers of binomials.

Vocabulary

Binomial expansion The sum of terms that results from multiplying out a power of a binomial. Example: The expansion of $(a + b)^2$ is $a^2 + 2ab + b^2$.

Pascal's triangle A triangular array of numbers representing the coefficients of the expansion of $(a + b)^n$, where n is a nonnegative integer.

Triangle	Row
1	0
1 1	1
1 2 1	2
1 3 3 1	3
1 4 6 4 1	4
1 5 10 (10) (5) 1	5
1 6 15 20 (15) 6 1	6

The triangle has 1's at the beginning and end of each row. Each of the other numbers is the sum of the two numbers above it. Example: 15 = 10 + 5

Example 1 Expand $(a + b)^5$. Use row 5 of Pascal's triangle.

Solution

The coefficients are:	1	5	10	10	5	1
The powers of a are:	a^5	a^4	a^3	a^2	a^1	a^0
The powers of b are:	b^0	b^1	b^2	b^3	b^4	b^5

Combining this information, you get:

$$(a + b)^5 = a^5 + 5a^4b + 10a^3b^2 + 10a^2b^3 + 5ab^4 + b^5$$

Example 2 Expand and simplify $(2x - y^2)^7$.

Solution First find row 7 of Pascal's triangle. (Remember: Except for the 1's at the ends of a given row, each number in the row is the sum of the two numbers above it in the triangle.)

Triangle	Row
1 6 15 20 15 6 1	6
1 7 21 35 35 21 7 1	7

Now expand $(a + b)^7$:

$$(a + b)^7 = a^7 + 7a^6b + 21a^5b^2 + 35a^4b^3 + 35a^3b^4 + 21a^2b^5 + 7ab^6 + b^7$$

Replace a by $2x$ and b by $-y^2$ and simplify:

$$(2x - y^2)^7 = (2x)^7 + 7(2x)^6(-y^2) + 21(2x)^5(-y^2)^2 + 35(2x)^4(-y^2)^3 + 35(2x)^3(-y^2)^4 + 21(2x)^2(-y^2)^5 + 7(2x)(-y^2)^6 + (-y^2)^7$$

$$= 128x^7 - 448x^6y^2 + 672x^5y^4 - 560x^4y^6 + 280x^3y^8 - 84x^2y^{10} + 14xy^{12} - y^{14}$$

NAME ______________________ DATE ______________

11-7 Powers of Binomials *(continued)*

Expand and simplify each expression.

1. $(x + y)^4$ **2.** $(x + y)^5$ **3.** $(x - y)^3$

4. $(x + 2)^5$ **5.** $(y - 3)^4$ **6.** $(2a + 1)^3$

7. $(3a + 2)^5$ **8.** $(x^2 - 1)^7$ **9.** $(\sqrt{x} + \sqrt{y})^4$

10. $(x^2 - y^3)^8$ **11.** $(x + \sqrt{2})^6$ **12.** $(x - x^{-1})^5$

Example 3 The first three terms in the expansion of $(a + b)^{15}$ are $a^{15} + 15a^{14}b + 105a^{13}b^2$. Write the last three terms.

Solution For any row of Pascal's triangle, notice that the pattern of the coefficients is symmetric. That is, the first and last coefficients are equal, the second and next-to-last coefficients are equal, and so on. So if the coefficients of the first three terms in the expansion of $(a + b)^{15}$ are 1, 15, and 105, then the coefficients of the last three terms must be 105, 15, and 1. Also, the powers of a are decreasing to 0 and the powers of b are increasing to 15.
$\therefore$ the last three terms are $105a^2b^{13} + 15ab^{14} + b^{15}$.

In Exercises 13–17, use the symmetry of the coefficients.

13. The first three terms of $(x + y)^{18}$ are $x^{18} + 18x^{17}y + 153x^{16}y^2$. Write the last three terms.

14. The first three terms of $(x - y)^{24}$ are $x^{24} - 24x^{23}y + 276x^{22}y^2$. Write the last three terms.

15. The first three terms of $(x + y)^{12}$ are $x^{12} + 12x^{11}y + 66x^{10}y^2$. Write the last three terms.

16. The ninth term of $(x + 1)^{18}$ is $43{,}758x^{10}$. Write the eleventh term. (*Hint*: The tenth term is the middle term.)

17. The tenth term of $(x - y)^{19}$ is $-92{,}378x^{10}y^9$. Write the eleventh term.

Mixed Review Exercises

Tell whether each sequence is arithmetic, geometric, or neither. Then find a formula for the *n*th term of the sequence.

1. $\frac{3}{2}, \frac{6}{5}, \frac{9}{8}, \frac{12}{11}, \ldots$ **2.** 24, 12, 6, 3, . . . **3.** $-6, -1, 4, 9, \ldots$

4. $\frac{2}{3}, 2, 6, 18, \ldots$ **5.** 1, 4, 9, 16, . . . **6.** 100, 92, 84, 76, . . .

Find the sum of each series.

7. $8 + 11 + 14 + \cdots + 59$ **8.** $4 + \frac{8}{3} + \frac{16}{9} + \cdots$

9. $\sum_{n=1}^{20} (2n + 5)$ **10.** $\sum_{n=1}^{8} 3(2^{-n})$

NAME ________________________ DATE ____________

11–8 The General Binomial Expansion

Objective: To use the binomial theorem to find a particular term of a binomial expansion.

Vocabulary

Factorial notation The symbol $r!$, read "r factorial," is defined as:

$$r! = \begin{cases} r(r-1)(r-2)\cdot \ldots \cdot 3\cdot 2\cdot 1 \text{ if } r \text{ is a positive integer} \\ 1 \text{ if } r = 0 \end{cases}$$

Example: $4! = 4\cdot 3\cdot 2\cdot 1 = 24$

The binomial theorem If n is a positive integer, then

$$(a+b)^n = a^n + \frac{n}{1}a^{n-1}b + \frac{n(n-1)}{1\cdot 2}a^{n-2}b^2 + \frac{n(n-1)(n-2)}{1\cdot 2\cdot 3}a^{n-3}b^3 + \cdots + b^n$$

$$= a^n + \frac{n!}{(n-1)!\,1!}a^{n-1}b + \frac{n!}{(n-2)!\,2!}a^{n-2}b^2 + \frac{n!}{(n-3)!\,3!}a^{n-3}b^3 + \cdots + b^n$$

$$= \sum_{k=0}^{n} \frac{n!}{(n-k)!\,k!}a^{n-k}b^k.$$

The $(k+1)$st term of $(a+b)^n$ is $\frac{n!}{(n-k)!\,k!}a^{n-k}b^k$.

Symbol

! (factorial)

Example 1 Evaluate: **a.** $\frac{12!}{10!}$ **b.** $\frac{10!}{7!\,3!}$ **c.** $\frac{(n+2)!}{(n-1)!}$

Solution

a. $\frac{12!}{10!} = \frac{12\cdot 11\cdot \cancel{10!}}{\cancel{10!}} = 132$

b. $\frac{10!}{7!\,3!} = \frac{10\cdot 9\cdot 8\cdot 7!}{7!\,3!} = \frac{10\cdot \overset{3}{\cancel{9}}\cdot \overset{4}{\cancel{8}}}{\cancel{3}\cdot \cancel{2}\cdot 1} = 120$

c. $\frac{(n+2)!}{(n-1)!} = \frac{(n+2)(n+1)(n)(n-1)!}{(n-1)!}$
$= (n+2)(n+1)n$
$= (n^2+3n+2)n$
$= n^3+3n^2+2n$

Evaluate.

1. $5!$
2. $8!$
3. $11!$
4. $0!$
5. $\frac{6!}{4!}$
6. $\frac{50!}{49!}$
7. $\frac{25!}{23!}$
8. $\frac{7!}{2!\,5!}$
9. $\frac{12!}{9!\,3!}$
10. $\frac{10!}{5!\,5!}$
11. $\frac{n!}{(n-1)!}$
12. $\frac{(n+2)!}{n!}$
13. $\frac{(n+1)!}{(n-1)!}$
14. $\frac{n!}{2!\,(n-2)!}$
15. $\frac{(n+3)!}{n!\,3!}$
16. $\frac{(n-1)!}{1!\,(n-2)!}$

NAME ______________________ DATE ______________

11-8 The General Binomial Expansion *(continued)*

Example 2 Find the first four terms in the expansion $(x - y)^{15}$. Simplify each coefficient.

Solution Use the binomial theorem, substituting x for a and $-y$ for b. The first four terms are:

$$x^{15} + \frac{15!}{(15-1)!\,1!}x^{14}(-y) + \frac{15!}{(15-2)!\,2!}x^{13}(-y)^2 + \frac{15!}{(15-3)!\,3!}x^{12}(-y)^3 =$$

$$x^{15} - \frac{15!}{14!\,1!}x^{14}y + \frac{15!}{13!\,2!}x^{13}y^2 - \frac{15!}{12!\,3!}x^{12}y^3 =$$

$$x^{15} - \frac{15 \cdot \cancel{14!}}{\cancel{14!}\,1!}x^{14}y + \frac{15 \cdot 14 \cdot \cancel{13!}}{\cancel{13!}\,2!}x^{13}y^2 - \frac{15 \cdot 14 \cdot 13 \cdot \cancel{12!}}{\cancel{12!}\,3!}x^{12}y^3 =$$

$$x^{15} - \frac{15}{1}x^{14}y + \frac{15 \cdot \overset{7}{\cancel{14}}}{\cancel{2} \cdot 1}x^{13}y^2 - \frac{\overset{5}{\cancel{15}} \cdot \overset{7}{\cancel{14}} \cdot 13}{\cancel{3} \cdot \cancel{2} \cdot 1}x^{12}y^3 =$$

$$x^{15} - 15x^{14}y + 105x^{13}y^2 - 455x^{12}y^3$$

Write the first four terms in the expansion of each of the following.

17. a. $(a + b)^{11}$ **b.** $(a - b)^{11}$

18. a. $(x + y)^{17}$ **b.** $(x - y)^{17}$

19. a. $(c + d)^{14}$ **b.** $(c - d^2)^{14}$

20. a. $(r + s)^{20}$ **b.** $(r - 3s)^{20}$

Example 3 Find and simplify the twelfth term in the expansion of $(x - 2y)^{15}$.

Solution By the binomial theorem, the $(k + 1)$st term of $(a + b)^n$ is $\frac{n!}{(n-k)!\,k!}a^{n-k}b^k$. So the *twelfth* term of $(a + b)^{15}$ can be obtained by letting $k = 11$ and $n = 15$:

$$\frac{15!}{(15-11)!\,11!}a^{15-11}b^{11} = \frac{15!}{4!\,11!}a^4b^{11}$$

$$= \frac{15 \cdot 14 \cdot 13 \cdot 12 \cdot \cancel{11!}}{4!\,\cancel{11!}}a^4b^{11}$$

$$= \frac{15 \cdot \overset{7}{\cancel{14}} \cdot 13 \cdot \cancel{12}}{\cancel{4} \cdot \cancel{3} \cdot \cancel{2} \cdot 1}a^4b^{11}$$

$$= 1365a^4b^{11}$$

To obtain the twelfth term of $(x - 2y)^{15}$, substitute x for a and $-2y$ for b.

$$1365x^4(-2y)^{11} = 1365x^4(-2048y^{11})$$
$$= -2{,}795{,}520x^4y^{11}$$

Find and simplify the specified term in each expansion.

21. The sixth term of $(x + 1)^{10}$

22. The twelfth term of $(2x + y)^{15}$

23. The seventh term of $(x - y)^9$

24. The fourth term of $(x - 3y)^{12}$

NAME ______________________________ DATE ______________________

12 Triangle Trigonometry

12–1 *Angles and Degree Measure*

Objective: To use degrees to measure angles.

Vocabulary

Degree A measure of rotation. One complete revolution measures 360 degrees.

Angle A figure formed by two rays that have the same endpoint.

Directed angle An angle generated by the rotation of a ray (the *initial side*) onto another ray (the *terminal side*).

Positive angle An angle generated by a counterclockwise rotation.

Negative angle An angle generated by a clockwise rotation.

Standard position An angle is in standard position when its initial side coincides with the positive x-axis.

Quadrantal angle An angle whose terminal side lies on a coordinate axis.

Coterminal angles Two angles whose terminal sides coincide when the angles are in standard position.

Symbols $1°$ (1 degree $= \frac{1}{360}$ revolution) $1'$ (1 minute $= \frac{1}{60}°$) $1''$ (1 second $= \frac{1}{3600}°$)

Example 1 **a.** Sketch $-315°$ in standard position. Indicate its rotation by a curved arrow and classify the angle by the quadrant containing its terminal side.

b. Sketch $\frac{3}{4}$ of a counterclockwise revolution and find its measure.

Solution **a.**

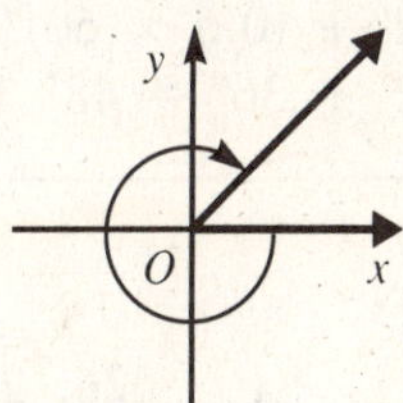

Quadrant I

b.

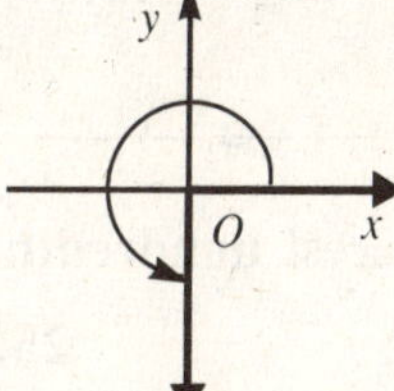

$\frac{3}{4} \times 360° = 270°$

Sketch each angle in standard position. Indicate its rotation by a curved arrow. Classify each angle by its quadrant. If the angle is a quadrantal angle, say so.

1. $60°$	**2.** $-60°$	**3.** $120°$	**4.** $-120°$	**5.** $200°$
6. $-200°$	**7.** $330°$	**8.** $-390°$	**9.** $450°$	**10.** $-820°$

Sketch in standard position the angle described and find its measure.

11. $\frac{1}{6}$ of a counterclockwise revolution

12. $\frac{2}{3}$ of a clockwise revolution

13. $\frac{3}{4}$ of a counterclockwise revolution

14. $1\frac{1}{8}$ of a clockwise revolution

NAME ______________________ DATE ______________

12–1 *Angles and Degree Measure* *(continued)*

Example 2 **a.** Write a formula for the measure of all angles coterminal with a 60° angle.

b. Use the formula to find two positive angles and two negative angles that are coterminal with a 60° angle.

Solution **a.** $60° + n \cdot 360°$, where n is an integer.

b. To find the measure of the angles, let $n = 1, 2, -1, -2$.

$60° + 1(360°) = 420°$ $\qquad$ $60° + 2(360°) = 780°$

$60° + (-1)(360°) = -300°$ $\qquad$ $60° + (-2)(360°) = -660°$

For Exercises 15–22: (a) Write a formula for the measures of all angles coterminal with the given angle. (b) Use the formula to find two angles (one positive and one negative) that are coterminal with the given angle.

15. 40° **16.** 120° **17.** −60° **18.** −235°

19. 450° **20.** 210° **21.** 900° **22.** −720°

Example 3 **a.** Express 28° 12′ 38″ in decimal degrees.

b. Express 48.21° in degrees, minutes, and seconds to the nearest second.

Solution Use these facts: $1' = \left(\frac{1}{60}\right)^{\circ}$ and $1'' = \left(\frac{1}{60}\right)' = \left(\frac{1}{3600}\right)^{\circ}$

a. $28°\ 12'\ 38'' = 28° + \left(\frac{12}{60}\right)^{\circ} + \left(\frac{38}{3600}\right)^{\circ} = 28° + 0.2° + 0.011° = 28.211°$

b. $48.21° = 48° + (0.21 \times 60)' = 48° + 12.6' = 48° + 12' + (0.6 \times 60)''$
$= 48° + 12' + 36'' = 48°\ 12'\ 36''$

Express in decimal degrees to the nearest hundredth of a degree.

23. 16° 40′ **24.** 49° 27′ **25.** 83° 52′ **26.** 28° 45′

Express in degrees, minutes, and seconds to the nearest second.

27. 26.52° **28.** 39.45° **29.** 85.39° **30.** 55.73°

Mixed Review Exercises

Find the third term in the expansion of each binomial.

1. $(x - 3y)^4$ **2.** $(x + \sqrt{2y})^6$ **3.** $(x^3 + y^2)^{10}$

Simplify.

4. $\sqrt{-5} \cdot \sqrt{-20}$ **5.** $\log_2 28 - \log_2 7$ **6.** $(9^{3/4} - 9^{5/4})^2$

7. $\ln \sqrt{e}$ **8.** $(5 + 2i)(3 - 4i)$ **9.** $\sqrt{27} - \sqrt{12} + \sqrt{48}$

NAME ______________________ DATE ______________

12–2 Trigonometric Functions of Acute Angles

Objective: To define trigonometric functions of acute angles.

Vocabulary

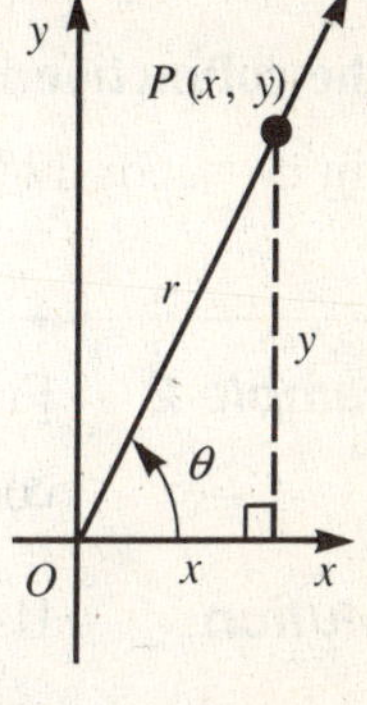

Trigonometric functions For an acute angle θ in standard position, with a point $P(x, y)$, other than the origin, on the terminal side and r = the distance OP:

Sine (sin)	**Cosine (cos)**	**Tangent (tan)**
$\sin\theta = \frac{y}{r}$	$\cos\theta = \frac{x}{r}$	$\tan\theta = \frac{y}{x}$

Cotangent (cot) The cotangent of an angle θ is the reciprocal of $\tan\theta$.

Secant (sec) The secant of an angle θ is the reciprocal of $\cos\theta$.

Cosecant (csc) The cosecant of an angle θ is the reciprocal of $\sin\theta$.

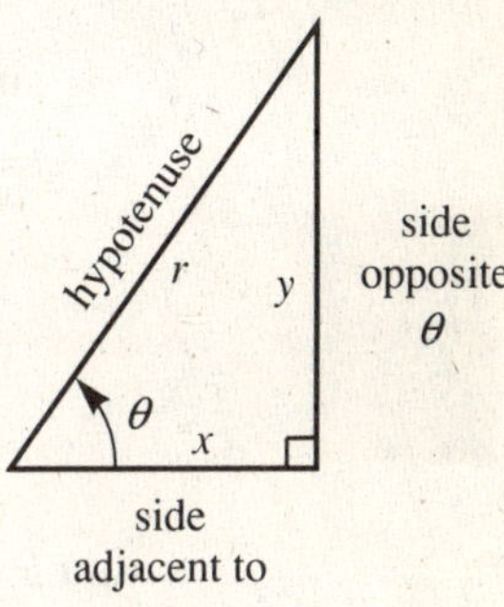

Formulas Given a right triangle with sides of lengths x, y, and r, and angle θ opposite side y, then:

$$\sin\theta = \frac{y}{r} = \frac{\text{length of side opposite } \theta}{\text{length of the hypotenuse}} \qquad \csc\theta = \frac{r}{y} \text{ or } \frac{1}{\sin\theta}$$

$$\cos\theta = \frac{x}{r} = \frac{\text{length of side adjacent to } \theta}{\text{length of the hypotenuse}} \qquad \sec\theta = \frac{r}{x} \text{ or } \frac{1}{\cos\theta}$$

$$\tan\theta = \frac{y}{x} = \frac{\text{length of side opposite } \theta}{\text{length of side adjacent to } \theta} \qquad \cot\theta = \frac{x}{y} \text{ or } \frac{1}{\tan\theta}$$

Cofunction identities in right triangle *ABC* Given complementary angles A and B (angles whose sum is 90°), the following identities hold:

$\sin A = \cos B$	$\tan A = \cot B$	$\sec A = \csc B$
$\cos A = \sin B$	$\cot A = \tan B$	$\csc A = \sec B$

Example: If $\sin A = \cos 60°$, then A = the complement of $60° = 30°$.

Pythagorean identity $\sin^2\theta + \cos^2\theta = 1$, where $\sin^2\theta$ means $(\sin\theta)^2$, and $\cos^2\theta$ means $(\cos\theta)^2$.

Example 1 Find the values of the six trigonometric functions of an angle in standard position whose terminal side passes through $P(8, 6)$.

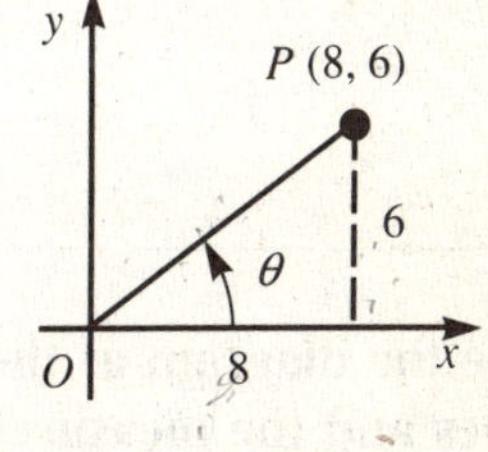

Solution Sketch θ in standard position.
By the Pythagorean theorem $r^2 = x^2 + y^2$:

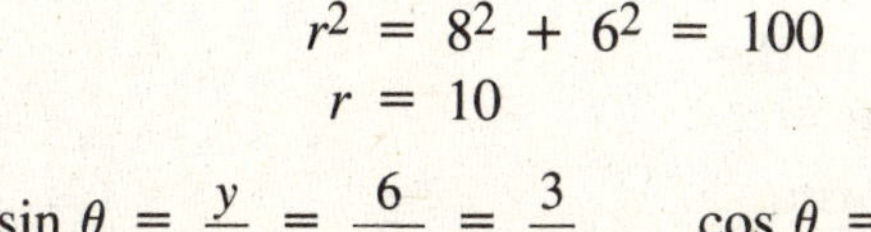

$$r^2 = 8^2 + 6^2 = 100$$
$$r = 10$$

$$\sin\theta = \frac{y}{r} = \frac{6}{10} = \frac{3}{5} \qquad \cos\theta = \frac{x}{r} = \frac{8}{10} = \frac{4}{5} \qquad \tan\theta = \frac{y}{x} = \frac{6}{8} = \frac{3}{4}$$

$$\csc\theta = \frac{r}{y} = \frac{10}{6} = \frac{5}{3} \qquad \sec\theta = \frac{r}{x} = \frac{10}{8} = \frac{5}{4} \qquad \cot\theta = \frac{x}{y} = \frac{8}{6} = \frac{4}{3}$$

NAME ______________________ DATE ______________

12–2 Trigonometric Functions of Acute Angles *(continued)*

Find the values of the six trigonometric functions of an angle θ in standard position whose terminal side passes through point P.

1. $P(4, 3)$ **2.** $P(3, 3)$ **3.** $P(12, 5)$ **4.** $P(1, 4)$

Use the cofunction identities to find the measure of the acute angle ϕ.

5. $\sin \phi = \cos 12°$ **6.** $\cos \phi = \sin 65°$ **7.** $\tan \phi = \cot 45°$ **8.** $\csc \phi = \sec 73°$

Example 2 Find $\sin \theta$ and $\tan \theta$ if θ is an acute angle and $\cos \theta = \frac{1}{2}$.

Solution Use the Pythagorean identity.

$$\sin^2 \theta + \cos^2 \theta = 1$$

$$\sin^2 \theta + \frac{1}{4} = 1$$

$$\sin^2 \theta = \frac{3}{4}$$

$$\sin \theta = \sqrt{\frac{3}{4}} = \frac{\sqrt{3}}{2}$$

$$\tan \theta = \frac{\sin \theta}{\cos \theta} = \frac{\sqrt{3}}{2} \div \frac{1}{2} = \sqrt{3}$$

Complete the table. In each case, θ is an acute angle.

	$\sin \theta$	$\cos \theta$	$\tan \theta$
9.	$\frac{1}{3}$	?	?
10.	?	$\frac{5}{6}$	?
11.	$\frac{3}{4}$	?	?
12.	?	$\frac{\sqrt{7}}{3}$	?
13.	$\frac{\sqrt{85}}{11}$	?	?

Example 3 Use the diagram to find the lengths of side $\overline{BC}$ and side $\overline{AB}$.

Solution Use the table on page 558 of your textbook to find tan 45° and cos 45°.

$$\tan 45° = \frac{BC}{AC} \qquad \cos 45° = \frac{AC}{AB}$$

$$1 = \frac{a}{8} \qquad \frac{\sqrt{2}}{2} = \frac{8}{c}$$

$$a = 8 \qquad \sqrt{2}c = 16$$

$$\therefore \text{ side } BC = 8 \qquad c = \frac{16}{\sqrt{2}} = 8\sqrt{2}$$

$$\therefore \text{ side } AB = 8\sqrt{2}$$

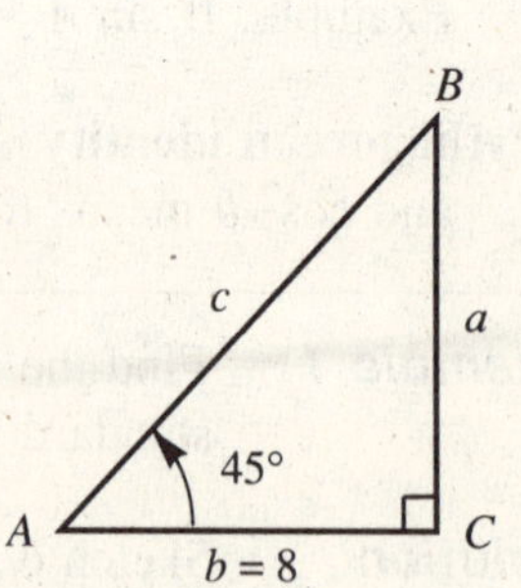

Use the diagram at the right to find the lengths of the sides and the measures of the angles that are not given. Leave your answers in simplest radical form. You may wish to use the table on page 558 of the text.

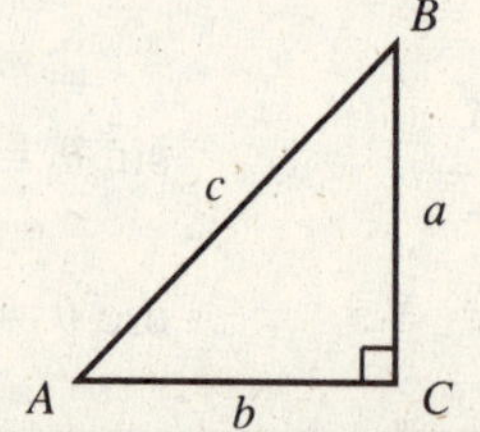

14. $a = 8, \angle A = 30°$ **15.** $b = 5, \angle A = 60°$ **16.** $c = 12, \angle A = 45°$

17. $c = 50, \angle B = 30°$ **18.** $a = 4, c = 8$ **19.** $a = 2, b = 2$

NAME ______________________ DATE ______________

12–3 Trigonometric Functions of General Angles

Objective: To define trigonometric functions of general angles.

Vocabulary

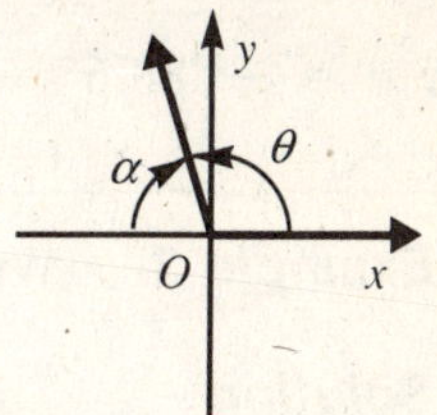

Reference angle If θ is not a quadrantal angle, there is a unique acute angle α, corresponding to θ, such that either $\theta + \alpha$ or $\theta - \alpha$ is an integral multiple of 180°. α is called the reference angle of θ.
Example: If θ measures 115°, its reference angle, α, measures 65°.

Formulas For any angle θ and any point $P(x, y)$ on the terminal side of θ:

$\sin \theta = \frac{y}{r}$ $\qquad$ $\cos \theta = \frac{x}{r}$ $\qquad$ $\tan \theta = \frac{y}{x}$, if $x \neq 0$

$\csc \theta = \frac{r}{y}$, if $y \neq 0$ $\qquad$ $\sec \theta = \frac{r}{x}$, if $x \neq 0$ $\qquad$ $\cot \theta = \frac{x}{y}$, if $y \neq 0$

Example 1 Find the values of the six trigonometric functions of an angle θ in standard position whose terminal side passes through $(-5, -12)$.

Solution First make a sketch.
Here $x = -5$ and $y = -12$.
Since r is the distance between (0, 0) and $(-5, -12)$, you can find r using the distance formula.

$$r = \sqrt{(x_2 - x_1)^2 + (y_2 - y_1)^2}$$

$$r = \sqrt{(-5 - 0)^2 + (-12 - 0)^2}$$

$$r = \sqrt{25 + 144} = \sqrt{169} = 13$$

$\sin \theta = -\frac{12}{13}$ $\qquad$ $\cos \theta = -\frac{5}{13}$ $\qquad$ $\tan \theta = \frac{-12}{-5} = \frac{12}{5}$

$\csc \theta = -\frac{13}{12}$ $\qquad$ $\sec \theta = -\frac{13}{5}$ $\qquad$ $\cot \theta = \frac{-5}{-12} = \frac{5}{12}$

Find the values of the six trigonometric functions of an angle θ in standard position whose terminal side passes through point P. If any value is undefined, say so.

1. $P(4, -3)$ $\qquad$ **2.** $P(-12, 5)$ $\qquad$ **3.** $P(-2, -2)$ $\qquad$ **4.** $P(15, -8)$ $\qquad$ **5.** $P(0, 1)$

Example 2 Find the measure of the reference angle α for each given angle θ.

a. $\theta = 155°$ $\qquad$ **b.** $\theta = 310°30'$ $\qquad$ **c.** $\theta = -125°$

Solution

a. Since θ is in quadrant II: $\alpha = 180° - \theta = 180° - 155° = 25°$

b. Since θ is in quadrant IV:
$\alpha = 360° - \theta = 360° - 310°30' = 359°60' - 310°30' = 49°30'$

c. First find the positive angle that is coterminal with $-125°$:
$-125° + 360° = 235°$
Since 235° is in quadrant III: $\alpha = \theta - 180° = 235° - 180° = 55°$

12–3 Trigonometric Functions of General Angles *(continued)*

Find the measure of the reference angle α of the given angle θ.

6. $\theta = 250°$ **7.** $\theta = 172°$ **8.** $\theta = -200°$ **9.** $\theta = -142°$

10. $\theta = 730°$ **11.** $\theta = 480°$ **12.** $\theta = 98.2°$ **13.** $\theta = -166.2°$

14. $\theta = -288.6°$ **15.** $\theta = 149°20'$ **16.** $\theta = 243°15'$ **17.** $\theta = 300°50'$

Example 3 Write sin 230° as a function of an acute angle.

Solution The reference angle of 230° is $230° - 180° = 50°$.
Since 230° is in the third quadrant, its sine is negative (see note below).
$\therefore \sin 230° = -\sin 50°$.

Note: The signs of the trigonometric functions vary with the quadrants. All values are positive in the first quadrant. Sine and cosecant are positive in the second quadrant. Tangent and cotangent are positive in the third quadrant. Cosine and secant are positive in the fourth quadrant.

Write each of the following as a function of an acute angle.

18. sin 312° **19.** cos 215° **20.** tan (−163°) **21.** sin (−29°)

22. sec 276.8° **23.** csc (−198.3°) **24.** cot 312°15′ **25.** cos 259°12′

Example 4 Find the exact values of the six trigonometric functions of 225°. Use the table of exact trigonometric values found in Lesson 12-2 of your textbook.

Solution The reference angle of 225° is 45°.
Since 225° is in the third quadrant, only the tangent and cotangent are positive.

$\sin 225° = -\sin 45° = -\frac{\sqrt{2}}{2}$ $\csc 225° = -\csc 45° = -\sqrt{2}$

$\cos 225° = -\cos 45° = -\frac{\sqrt{2}}{2}$ $\sec 225° = -\sec 45° = -\sqrt{2}$

$\tan 225° = \tan 45° = 1$ $\cot 225° = \cot 45° = 1$

Find the exact values of the six trigonometric functions of each angle.

26. 120° **27.** −135° **28.** 210° **29.** −225° **30.** 420°

Mixed Review Exercises

Find the zeros of each function. If the function has no zeros, say so.

1. $f(x) = 2x + 6$ **2.** $g(x) = \log (x - 2)$ **3.** $h(x) = 6x^2 + x - 2$

4. $F(x) = \frac{x^3 - 9x}{x^2 + 2}$ **5.** $G(x) = x^2 + 3x + 8$ **6.** $H(x) = 2^x$

NAME ______________________________ DATE ______________

12–4 Values of Trigonometric Functions

Objective: To use a calculator or trigonometric tables to find values of trigonometric functions.

Example 1 Find each function value to four significant digits: **a.** cos 29°40′ **b.** csc 57.7°

Solution 1 **Using a Calculator**

a. Some calculators operate only with decimal degrees, so you may need to change 29°40′ to decimal degrees. Divide 40′ by 60 (60′ = 1°).

$$\cos 29°40' = \cos\left(29 + \frac{40}{60}\right)^{\circ} = \cos 29.67°$$

$\cos 29.67° = 0.8688908$

$\therefore$ to four significant digits $\cos 29.67° = 0.8689$.

b. If your calculator does not have keys labeled sec, csc, and cot, you may need to use reciprocal functions and the reciprocal key on your calculator.

$$\csc 57.7° = \frac{1}{\sin 57.7°} = \frac{1}{0.8452618} = 1.1830654$$

$\therefore$ to four significant digits $\csc 57.7° = 1.183$.

Solution 2 **Using Tables**

a. In Table 5 of your textbook look *down* the column under "cos θ" until you find the entry opposite 29°40′ at the left. You should find "0.8689."

$\therefore \cos 29°40' = 0.8689$.

b. In Table 4 of the text look *up* the column over "csc θ" until you find the entry opposite 57.7° at the right. You should find "1.183."

$\therefore \csc 57.7° = 1.183$.

Example 2 Find cos 161.3° and sin 161.3°.

Solution 1 **Using a Calculator**

You may obtain the values directly. $\cos 161.3° = -0.9472$; $\sin 161.3° = 0.3206$

Solution 2 **Using Tables**

The reference angle for 161.3° is $180° - 161.3° = 18.7°$. Since 161.3° is a second-quadrant angle, its cosine is negative and its sine is positive.

$\therefore \cos 161.3° = -\cos 18.7° = -0.9472$; $\sin 161.3° = \sin 18.7° = 0.3206$

Find each function value to four significant digits.

1. tan 51.3°	**2.** cos 54.7°	**3.** sin 38.6°	**4.** cot 12.8°
5. sec 13.3°	**6.** csc 41.5°	**7.** tan 43°10′	**8.** cos 57°40′
9. sin 81°20′	**10.** cot 63°50′	**11.** sec 41°40′	**12.** csc 39°30′
13. tan 112.3°	**14.** cos 201.3°	**15.** sin 142°10′	**16.** cot 255°20′

NAME ______________________ DATE ______________

12–4 Values of Trigonometric Functions *(continued)*

Example 3 Find sin 65°44′.

Solution 1 **Using a Calculator**
You may need to change 65°44′ to decimal degrees by dividing 44 by 60.
$\sin 65°44' = \sin 65.73° = 0.9116186$

Solution 2 **Using Tables**
Notice that $\sin 65°40' < \sin 65°44' < \sin 65°50'$.
Using a vertical arrangement for the linear interpolation, you can write:

		θ	sin θ		
10′		65° 50′	0.9124		0.0012
	4′	65° 44′	?	d	
		65° 40′	0.9112		

Note: You must read upward to find sin 65°44′ in the table.

$\frac{d}{0.0012} = \frac{4}{10}$; $d = \frac{4}{10}(0.0012) = 0.00048$ or 0.0005

$\sin 65°44' = 0.9112 + 0.0005 = 0.9117$

Find each function value to four significant digits.

17. sin 31° 27′ **18.** cos 42° 53′ **19.** tan 15.28° **20.** sec 51.34°

Example 4 Find the measure of the acute angle θ to the nearest tenth of a degree, and to the nearest minute, when $\cos \theta = 0.6253$.

Solution 1 **Using a Calculator**
To find an acute angle when given one of its function values, you use the inverse function keys ($\sin^{-1}$, $\cos^{-1}$, $\tan^{-1}$, or inv sin, inv cos, inv tan).
If $\cos \theta = 0.6253$, then

$$\theta = \text{inv cos } 0.6253 = 51.29579°$$

∴ to the nearest tenth of a degree, $\theta = 51.3°$.

To convert 0.29579° to minutes, multiply by 60 to obtain 17.7474.
∴ to the nearest minute, $\theta = 51°18'$.

Solution 2 **Using Tables**
Reverse the process described in Example 1. Look in the cosine columns of Table 4 until you find the entry nearest 0.6253. This entry is 0.6252, and it is opposite 51.3° on the right. Convert to minutes as you did in Solution 1.

∴ to the nearest tenth of a degree, $\theta = 51.3°$. To the nearest minute, $\theta = 51°18'$.

Find the measure of the acute angle θ (a) to the nearest tenth of a degree, and (b) to the nearest minute.

21. $\sin \theta = 0.2763$ **22.** $\tan \theta = 1.112$

23. $\cos \theta = 0.6724$ **24.** $\csc \theta = 1.762$

25. $\sec \theta = 3.253$ **26.** $\cot \theta = 2.458$

NAME ______________________ DATE ____________

12–5 Solving Right Triangles

Objective: To find the sides and angles of a right triangle.

Vocabulary

Solving a triangle Finding the values for all the sides and angles of the triangle.

Angle of elevation The angle θ by which an observer's line of sight must be raised from the horizontal to see an object.

Angle of depression The angle ϕ by which an observer's line of sight must be lowered from the horizontal to see an object.

Example 1 Solve the right triangle ABC if $\angle A = 47.3°$, $b = 48.5$, and $\angle C = 90°$. Give the lengths to three significant digits.

Solution 1 Make a diagram.

Since $\angle A + \angle B = 90°$,

$$\angle B = 90° - \angle A = 90° - 47.3° = 42.7°.$$

$$\tan A = \frac{a}{b} \qquad \sec A = \frac{c}{b}$$

$$\tan 47.3° = \frac{a}{48.5} \qquad \sec 47.3° = \frac{c}{48.5}$$

$$a = 48.5(\tan 47.3°) = 48.5(1.084) = 52.6 \qquad c = 48.5(\sec 47.3°) = 48.5(1.475) = 71.5$$

$\therefore \angle B = 42.7°$, $a = 52.6$, and $c = 71.5$

Solution 2 Again $\angle B = 90° - 47.3° = 42.7°$.

$$\cot A = \frac{b}{a} \qquad \cos A = \frac{b}{c}$$

$$\cot 47.3° = \frac{48.5}{a} \qquad \cos 47.3° = \frac{48.5}{c}$$

$$a = \frac{48.5}{\cot 47.3°} = \frac{48.5}{0.9228} = 52.6 \qquad c = \frac{48.5}{\cos 47.3°} = \frac{48.5}{0.6782} = 71.5$$

$\therefore \angle B = 42.7°$, $a = 52.6$, and $c = 71.5$

CAUTION Before you can solve a triangle given only the lengths of two sides, you must find the measure of the included angle by using the inverse of the trigonometric function relating the two sides.

NAME ______________________ DATE ____________

12–5 Solving Right Triangles *(continued)*

Solve each right triangle *ABC*. Give lengths to three significant digits and angle measures to the nearest tenth of a degree or nearest ten minutes. You may wish to use a calculator.

1. $\angle A = 26.5°$, $c = 56$
2. $\angle B = 16.8°$, $c = 10.8$
3. $\angle B = 59.4°$, $a = 3.46$
4. $\angle A = 78.6°$, $b = 358$
5. $\angle B = 56.9°$, $b = 83.5$
6. $a = 156$, $c = 248$
7. $b = 49.6$, $c = 91.3$
8. $a = 4.60$, $b = 7.90$
9. $\angle B = 46° \, 20'$, $c = 125$
10. $\angle A = 25° \, 30'$, $a = 21.3$
11. $\angle B = 32° \, 50'$, $b = 46.2$
12. $\angle A = 80° \, 40'$, $b = 0.725$

Example 2 What is the angle of elevation of the sun when a tree 20 ft tall casts a shadow 29.3 ft long?

Solution Make a sketch of the problem.
$\angle B$ is the angle of elevation of the sun.

$$\tan B = \frac{20}{29.3}$$

$$\tan B = 0.6826$$

$$\angle B = 34.3°$$

$\therefore$ the angle of elevation of the sun is 34.3°.

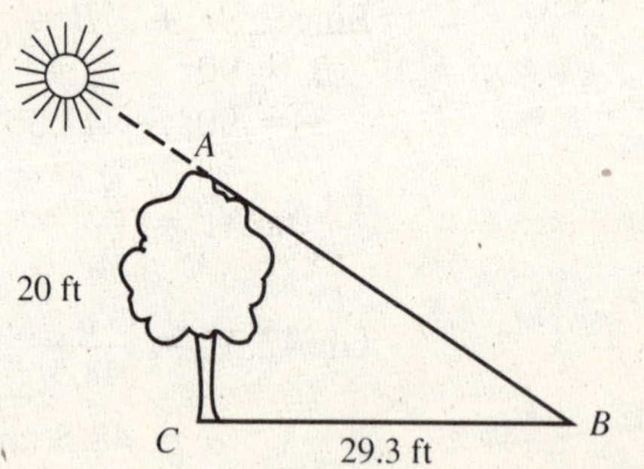

13. What is the angle of elevation of the sun when a flagpole 10 m tall casts a shadow 12.6 m long?
14. A window washer 50 ft above the ground sees a parked car 153 ft away. What is the angle of depression from the man to the car?
15. A ladder 8 ft long is leaning against a wall. It makes an angle of 65° with the ground. How far up the wall is the top of the ladder?

Mixed Review Exercises

Give the exact values of the six trigonometric functions of each angle. If any function is not defined for the angle, say so.

1. 270°
2. 225°
3. 390°
4. −210°

Write in simplest form without negative exponents.

5. $\dfrac{x^2 - 1}{x - x^{-1}}$
6. $\sqrt{72x^6}$
7. $\dfrac{1}{x + 3} + \dfrac{6}{x^2 - 9}$
8. $(-4x^3y^2)^2(3xy^2)^3$
9. $(x^{3/4})^{-4}$
10. $(2x + 3)(4x^2 - 6x + 9)$

NAME ______________________ DATE ______________

12–6 The Law of Cosines

Objective: To use the law of cosines to find sides and angles of triangles.

Vocabulary

The Law of Cosines In any triangle ABC,

$$c^2 = a^2 + b^2 - 2ab \cos C$$
$$b^2 = a^2 + c^2 - 2ac \cos B$$
$$a^2 = b^2 + c^2 - 2bc \cos A$$

Example 1 Use the law of cosines to give an equation involving the side x of the given triangle.

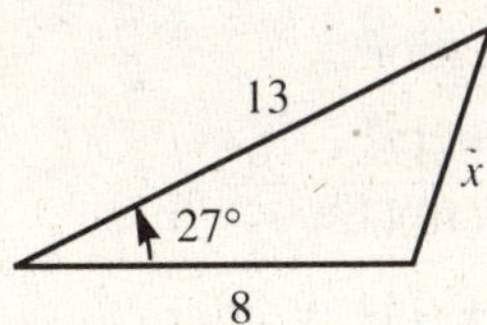

Solution $x^2 = 13^2 + 8^2 - 2(13)(8) \cos 27°$

Use the law of cosines to give an equation involving the side or angle labeled x.

1.

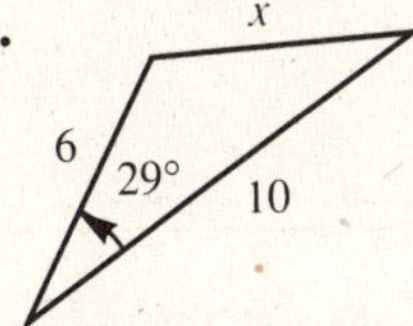

2.

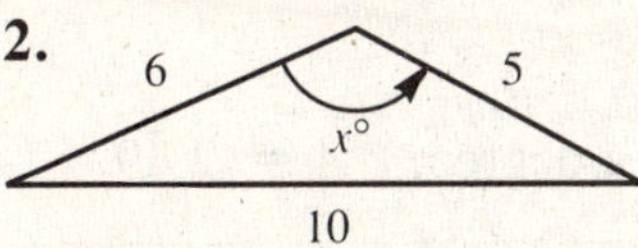

3.

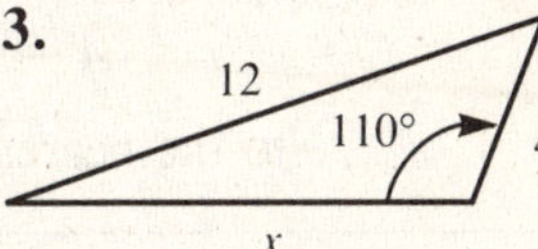

4.

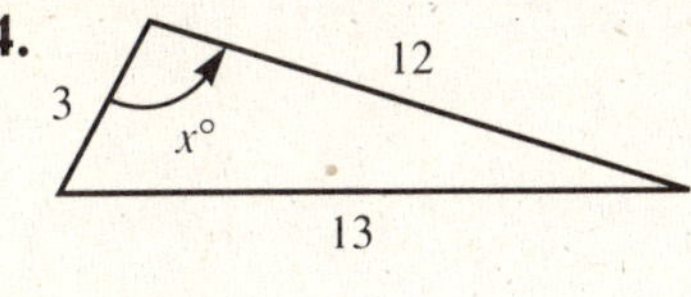

5.

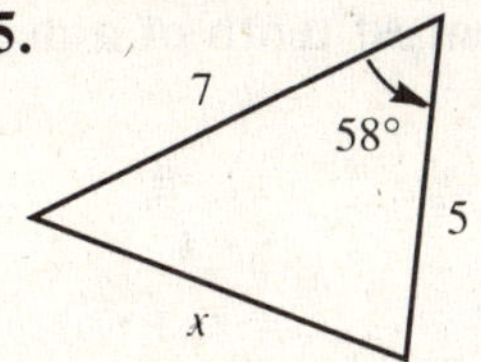

6.

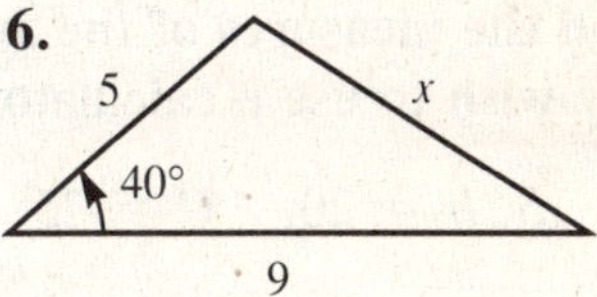

Example 2 In ΔABC, $a = 12$, $c = 14$, and $\angle B = 43°$. Find b to three significant digits.

Solution

$$b^2 = a^2 + c^2 - 2ac \cos B$$
$$b^2 = 12^2 + 14^2 - 2(12)(14) \cos 43°$$
$$= 144 + 196 - 336(0.7314)$$
$$= 340 - 245.7504$$
$$= 94.2496$$
$$b = \sqrt{94.2496} = 9.7082233$$

$\therefore$ to three significant digits, $b = 9.71$.

C, 12, 43°, B, 14, A

NAME ______________________________ DATE ______________

12–6 The Law of Cosines *(continued)*

Example 3 The lengths of the three sides of triangle ABC are 6, 10, and 12. Find the smallest angle of the triangle to the nearest tenth of a degree.

Solution Make a sketch of triangle ABC. Let $a = 6$, $b = 10$, and $c = 12$. The smallest angle of the triangle is opposite the shortest side. So, angle A is the angle you must find.

Use the third form of the law of cosines.

$$a^2 = b^2 + c^2 - 2bc \cos A$$

$$\cos A = \frac{b^2 + c^2 - a^2}{2bc}$$

Substitute into the equation.

$$\cos A = \frac{10^2 + 12^2 - 6^2}{2(10)(12)}$$

$$= \frac{100 + 144 - 36}{240}$$

$$= \frac{208}{240}$$

$$= 0.8667$$

$$\angle A = 29.92°$$

$\therefore$ to the nearest tenth of a degree, $\angle A = 29.9°$.

Find the indicated part of ΔABC. Find lengths to three significant digits and the measures of the angles to the nearest tenth of a degree. You may wish to use a calculator.

7. $a = 15$, $b = 20$, $\angle C = 75°$, $c = \underline{?}$

8. $a = 22$, $c = 31$, $\angle B = 64.3°$, $b = \underline{?}$

9. $b = 54$, $c = 42$, $\angle A = 39.5°$, $a = \underline{?}$

10. $a = 36$, $b = 39$, $\angle C = 140°$, $c = \underline{?}$

11. $a = 12$, $c = 16$, $\angle B = 160°$, $b = \underline{?}$

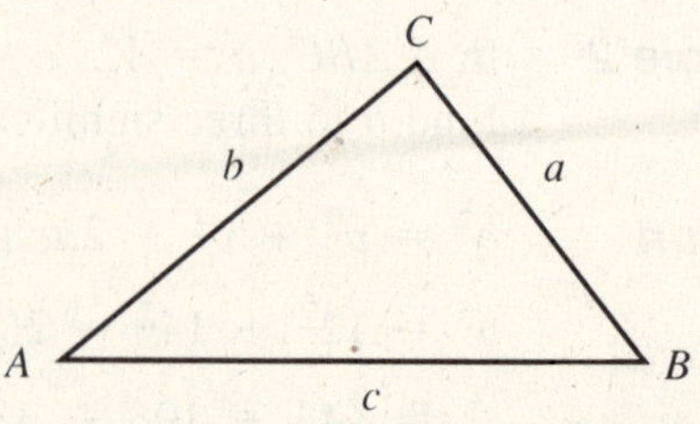

12. $a = 40$, $b = 34$, $c = 52$, $\angle C = \underline{?}$

13. $a = 10$, $b = 14$, $c = 16$, $\angle A = \underline{?}$

14. $a = 5.3$, $b = 4.7$, $c = 7.2$, $\angle B = \underline{?}$

15. $a = 12$, $b = 16$, $c = 19$, smallest angle $= \underline{?}$

16. $a = 2.4$, $b = 3.6$, $c = 4.8$, largest angle $= \underline{?}$

NAME ______________________ DATE ______________

12–7 The Law of Sines

Objective: To use the law of sines to find sides and angles of triangles.

Vocabulary

The Law of Sines In any triangle ABC, $\frac{\sin A}{a} = \frac{\sin B}{b} = \frac{\sin C}{c}$.

CAUTION When using the law of sines to find angles in triangles, be sure that the sum of the angles does not exceed 180°.

Example 1 In ΔABC, $\angle A = 50°$ and $a = 12$.

a. Find $\angle C$ if $c = 14$. **b.** Find $\angle C$ if $c = 9$.

Solution Use the formula $\frac{\sin A}{a} = \frac{\sin C}{c}$.

a. $\frac{\sin 50°}{12} = \frac{\sin C}{14}$

$$\sin C = \frac{14 \sin 50°}{12}$$

$$= \frac{14(0.7660)}{12}$$

$$= 0.8937$$

$$\angle C = 63.3° \text{ or } \angle C = 116.7°$$

Since $50° + 63.3° = 113.3°$, and $50° + 116.7° = 166.7°$, and both 113.3° and 166.7° are less than 180°, there are two solutions.

$\therefore \angle C = 63.3°$ or $\angle C = 116.7°$

b. $\frac{\sin 50°}{12} = \frac{\sin C}{9}$

$$\sin C = \frac{9 \sin 50°}{12}$$

$$= \frac{9(0.7660)}{12}$$

$$= 0.5745$$

$$\angle C = 35.1° \text{ or } \angle C = 144.9°$$

144.9° is not a solution because $\angle A + \angle C = 50° + 144.9° = 194.9°$, which is greater than 180°.

$\therefore \angle C = 35.1°$

Find the indicated part of ΔABC to the nearest tenth of a degree. If there are two solutions, give both. You may wish to use a calculator.

1. $a = 6, b = 5, \angle A = 50°, \angle B =$ _?_

2. $a = 10, b = 13, \angle B = 75°, \angle A =$ _?_

3. $a = 12, c = 14, \angle C = 76°, \angle A =$ _?_

4. $b = 17, c = 10, \angle C = 32°, \angle B =$ _?_

12–7 The Law of Sines *(continued)*

Example 2 In $\triangle ABC$, $\angle A = 28°$ and $\angle B = 65°$.

a. Find a if $b = 12$. **b.** Find c if $a = 6$.

Solution **a.** $\frac{\sin A}{a} = \frac{\sin B}{b}$

$$\frac{\sin 28°}{a} = \frac{\sin 65°}{12}$$

$$a = \frac{12 \sin 28°}{\sin 65°}$$

$$= \frac{12(0.4695)}{0.9063} = 6.216$$

$\therefore a = 6.22$ to three significant digits.

b. First find $\angle C$. $\angle A + \angle B + \angle C = 180°$ so $\angle C = 180° - 28° - 65° = 87°$.

$$\frac{\sin A}{a} = \frac{\sin C}{c}$$

$$\frac{\sin 28°}{6} = \frac{\sin 87°}{c}$$

$$c = \frac{6 \sin 87°}{\sin 28°}$$

$$= \frac{6(0.9986)}{0.4695} = 12.76$$

$\therefore c = 12.8$ to three significant digits.

Find the indicated part of $\triangle ABC$ to three significant digits. You may wish to use a calculator.

5. $a = 21, \angle A = 28°, \angle B = 62°, b = \underline{?}$
6. $c = 28, \angle A = 52°, \angle C = 72°, a = \underline{?}$
7. $b = 2.6, \angle A = 106°, \angle C = 50°, a = \underline{?}$
8. $c = 27, \angle A = 41°, \angle C = 115°, b = \underline{?}$
9. $a = 3.75, \angle B = 50°, \angle C = 110°, c = \underline{?}$
10. $b = 11.6, \angle B = 85°, \angle C = 45°, a = \underline{?}$

Mixed Review Exercises

Give the following values to four significant digits.

1. cos 31.5°
2. tan (−75° 10′)
3. sin 243.6°
4. csc 321° 50′
5. cot 127.8°
6. sec 431° 10′

NAME ______________________ DATE ______________

12–8 Solving General Triangles

Objective: To solve any given triangle.

Vocabulary

Solving triangles The problem of solving triangles can be divided into four cases:

SSS: Given three sides

SAS: Given two sides and the included angle

SSA: Given two sides and the angle opposite one of them

ASA and AAS: Given two angles and one side

CAUTION The SSA case is sometimes called the *ambiguous* case because no triangle, one triangle, or two triangles are possible. Suppose you are given a, b, and acute $\angle A$ of $\Delta\, ABC$ as shown at the right. Notice that the altitude from C to the side opposite $\angle C$ is $b \sin A$.

1. If $a < b \sin A$, then $\overline{BC}$ will not connect with the side opposite $\angle C$ to form a triangle.
2. If $a = b \sin A$, then $\Delta\, ABC$ will be a right triangle.
3. If $b \sin A < a < b$, then $\overline{BC}$ will connect with the side opposite $\angle C$ in two places to form two different triangles.
4. If $a > b$, then $\overline{BC}$ will connect with the side opposite $\angle C$ in one place to form one triangle.

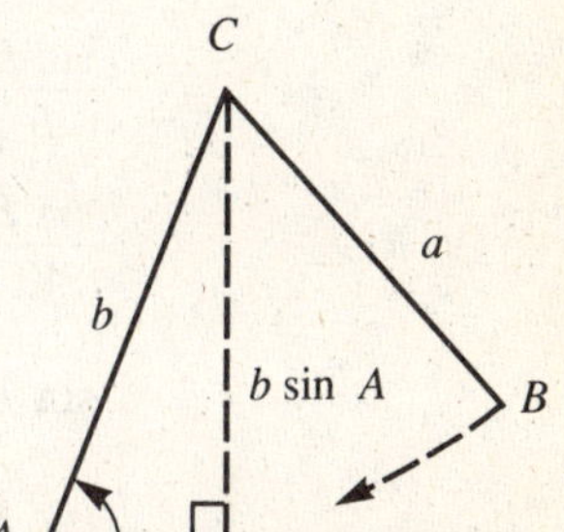

Example 1 (SSS case) Solve $\Delta\, ABC$ if $a = 5$, $b = 11$, and $c = 8$.

Solution Use the law of cosines to find two of the angles, say $\angle A$ and $\angle B$.

$$\cos A = \frac{b^2 + c^2 - a^2}{2bc} = \frac{11^2 + 8^2 - 5^2}{2(11)(8)} = \frac{160}{176} = 0.9091, \text{ so } \angle A = 24.6°$$

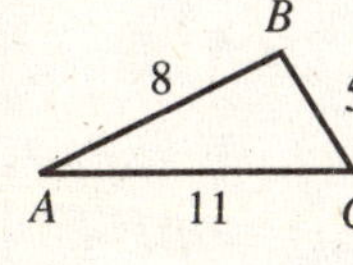

$$\cos B = \frac{a^2 + c^2 - b^2}{2ac} = \frac{5^2 + 8^2 - 11^2}{2(5)(8)} = -\frac{32}{80} = -0.4, \text{ so } \angle B = 113.6°$$

(*Note*: Although you can also use the law of sines to find $\angle A$ and $\angle B$, you must be sure to recognize that $\angle B$ is an obtuse angle.)

$\angle C = 180° - \angle A - \angle B = 180° - 24.6° - 113.6° = 41.8°$

$\therefore \angle A = 24.6°, \angle B = 113.6°$, and $\angle C = 41.8°$

Example 2 (SAS case) Solve $\Delta\, ABC$ if $a = 10$, $c = 12$, and $\angle B = 40°$.

Solution Use the law of cosines to find the third side, b, and one of the angles, say $\angle A$.

$b^2 = a^2 + c^2 - 2ac \cos B = 10^2 + 12^2 - 2(10)(12) \cos 40°$

$= 60.16$, so $b = \sqrt{60.16} = 7.76$

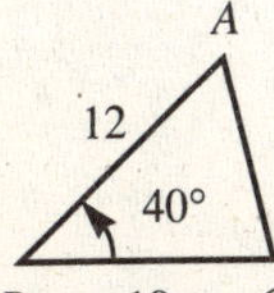

$$\cos A = \frac{b^2 + c^2 - a^2}{2bc} = \frac{7.76^2 + 12^2 - 10^2}{2(7.76)(12)} = \frac{104.22}{186.24} = 0.5596; \angle A = 56.0°$$

$\angle C = 180° - 40° - 56.0° = 84.0°$

$\therefore b = 7.76, \angle A = 56.0°$, and $\angle C = 84.0°$

NAME ______________________ DATE ____________

Example 3 **(SSA case)** Solve ΔABC if $a = 35$, $b = 40$, and $\angle A = 55°$.

Solution This is the case that may have no solution, one solution, or two solutions. Since the altitude from C to the side opposite $\angle C$ is

$$b \sin A = 40 \sin 55° = 32.8,$$

you have $32.8 < 35 < 40$ (that is, $b \sin A < a < b$), so two triangles are possible, as shown at the right.

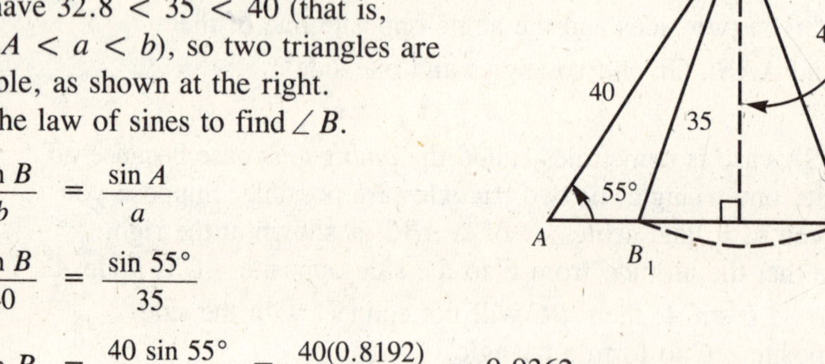

Use the law of sines to find $\angle B$.

$$\frac{\sin B}{b} = \frac{\sin A}{a}$$

$$\frac{\sin B}{40} = \frac{\sin 55°}{35}$$

$$\sin B = \frac{40 \sin 55°}{35} = \frac{40(0.8192)}{35} = 0.9362$$

$$\angle B = 69.4° \text{ or } 110.6°$$

If $\angle B = 69.4°$:

$$\angle C = 180° - 55° - 69.4° = 55.6°$$

$$c = \frac{a \sin C}{\sin A} = \frac{35 \sin 55.6°}{\sin 55°} = 35.3$$

If $\angle B = 110.6°$:

$$\angle C = 180° - 55° - 110.6° = 14.4°$$

$$c = \frac{a \sin C}{\sin A} = \frac{35 \sin 14.4°}{\sin 55°} = 10.6$$

$\therefore \angle B = 69.4°, \angle C = 55.6°$, and $c = 35.3$;
or $\angle B = 110.6°, \angle C = 14.4°$, and $c = 10.6$

Example 4 **(AAS case)** Solve ΔABC if $a = 30$, $\angle A = 62°$, and $\angle B = 75°$.

Solution $\angle C = 180° - \angle A - \angle B = 180° - 62° - 75° = 43°$

Use the law of sines to find b and c.

$$b = \frac{a \sin B}{\sin A} = \frac{30 \sin 75°}{\sin 62°} = \frac{30(0.9659)}{0.8829} = 32.8$$

$$c = \frac{a \sin C}{\sin A} = \frac{30 \sin 43°}{\sin 62°} = \frac{30(0.6820)}{0.8829} = 23.2$$

$\therefore b = 32.8$, $c = 23.2$, and $\angle C = 43°$

Solve triangle ABC. If there are two solutions, find both. If there are no solutions, say so. Give lengths to three significant digits and angle measures to the nearest tenth of a degree. You may wish to use a calculator.

1. $a = 5, b = 7, \angle C = 80°$
2. $a = 2, \angle B = 40°, \angle C = 100°$
3. $a = 12, \angle A = 20°, \angle B = 60°$
4. $a = 5, b = 6, c = 7$
5. $a = 14, b = 9, \angle A = 62°$
6. $a = 8, b = 6, \angle B = 43°$
7. $b = 32, c = 30, \angle C = 78°$
8. $a = 50, c = 100, \angle A = 30°$

NAME ______________________ DATE ______________

12–9 Areas of Triangles

Objective: To apply triangle area formulas.

Vocabulary

Triangle area formulas The area K of ΔABC is given by the following formulas:

$$K = \frac{1}{2}bc \sin A \qquad K = \frac{1}{2}ac \sin B \qquad K = \frac{1}{2}ab \sin C$$

$$K = \frac{1}{2}a^2 \frac{\sin B \sin C}{\sin A} \qquad K = \frac{1}{2}b^2 \frac{\sin A \sin C}{\sin B} \qquad K = \frac{1}{2}c^2 \frac{\sin A \sin B}{\sin C}$$

Hero's formula: $K = \sqrt{s(s - a)(s - b)(s - c)}$, where $s = \frac{1}{2}(a + b + c)$

Example 1 **(SAS case)** Find the area of ΔABC if $a = 24$, $b = 20$, and $\angle C = 76°$.

Solution Use the formula involving a, b, and $\angle C$.

$$K = \frac{1}{2}ab \sin C = \frac{1}{2}(24)(20) \sin 76°$$

$$= 240 \sin 76° = 240(0.9703) = 232.9$$

$\therefore$ to three significant digits, the area is 233 square units.

Example 2 **(ASA case)** Find the area of ΔABC if $a = 17.3$, $\angle B = 48°$, and $\angle C = 100°$.

Solution You can find $\angle A$ given $\angle B$ and $\angle C$: $\angle A = 180° - 48° - 100° = 32°$

Use the formula involving a, $\angle A$, $\angle B$, and $\angle C$.

$$K = \frac{1}{2}a^2 \frac{\sin B \sin C}{\sin A} = \frac{1}{2}(17.3)^2 \frac{\sin 48° \sin 100°}{\sin 32°}$$

$$= 0.5(299.29) \frac{(0.7431)(0.9848)}{0.5299} = 206.7$$

$\therefore$ to three significant digits, the area is 207 square units.

Example 3 **(SSS case)** Find the area of ΔABC if $a = 30$, $b = 40$, and $c = 60$.

Solution Use Hero's formula: $K = \sqrt{s(s - a)(s - b)(s - c)}$

Find s: $s = \frac{1}{2}(a + b + c)$

$= 0.5(30 + 40 + 60) = 65$

Then: $s - a = 65 - 30 = 35$

$s - b = 65 - 40 = 25$

$s - c = 65 - 60 = 5$

Now substitute: $K = \sqrt{s(s - a)(s - b)(s - c)}$

$$= \sqrt{65(35)(25)(5)} = 25\sqrt{455} = 533.3$$

$\therefore$ to three significant digits, the area is 533 square units.

NAME ______________________ DATE ____________

12–9 *Areas of Triangles* *(continued)*

Example 4 (SSA case) Find the area of ΔABC if $a = 26$, $b = 34$, and $\angle A = 25°$.

Solution This is the ambiguous case discussed in Lesson 12–8. Since $b \sin A = 14$, it is less than both a and b. Thus two triangles are formed. Using the law of sines you will find that $\angle B$ is either 33.5° or 146.5°. Therefore:

$\angle C = 121.5°$ or $\angle C = 8.5°$

$K = \frac{1}{2}ab \sin C$	$K = \frac{1}{2}ab \sin C$
$= \frac{1}{2}(26)(34) \sin 121.5°$	$= \frac{1}{2}(26)(34) \sin 8.5°$
$= 442(0.8526)$	$= 442(0.1478)$
$= 376.8$	$= 65.32$

$\therefore$ to three significant digits, the area is 377 square units or 65.3 square units.

Find the area of ΔABC using the given data. Give area measure to three significant digits. You may wish to use a calculator.

1. $\angle A = 30°, \angle B = 65°, b = 25$
2. $a = 20, b = 15, \angle C = 48°$
3. $a = 12, c = 16, \angle B = 59°$
4. $\angle A = 80°, b = 16, c = 35$
5. $a = 14, b = 19, c = 23$
6. $a = 6, b = 7, c = 9$
7. $\angle B = 25°, \angle C = 110°, c = 14$
8. $b = 23, c = 14, \angle A = 145°$
9. $a = 12, b = 18, \angle C = 45°$
10. $a = 11, b = 12, \angle A = 60°$

Solve.

11. Find the area of a parallelogram that has a 65° angle and sides with lengths 8 and 15. (Hint: Divide the parallelogram into two equal triangles.)
12. Find the area of a rhombus that has perimeter 36 and an angle of 45°.

Mixed Review Exercises

Find the indicated part of ΔABC to three significant digits or to the nearest tenth of a degree.

1. $a = 10, b = 8, \angle C = 65°, c =$ ___?___
2. $\angle A = 110°, \angle B = 25°, b = 18, a =$ ___?___
3. $a = 8, b = 6, c = 10, \angle C =$ ___?___
4. $a = 20, \angle B = 80°, \angle C = 52°, b =$ ___?___

Find the five other trigonometric functions of θ.

5. $\cos \theta = -\frac{1}{2}, 90° < \theta < 180°$
6. $\sin \theta = -\frac{3}{4}, 180° < \theta < 270°$
7. $\tan \theta = -1, 90° < \theta < 180°$
8. $\cos \theta = \frac{5}{13}, 270° < \theta < 360°$

NAME ______________________ DATE __________

13 Trigonometric Graphs; Identities

13–1 Radian Measure

Objective: To use radians to measure angles.

Vocabulary

Radian A unit for measuring angles. If a circle of radius r is centered at the vertex of angle θ, then the radian measure of θ is defined as the ratio of s, the length of the arc intercepted by θ, to r: $\theta = \frac{s}{r}$.

Conversion between radian and degree measure Since $180° = \pi$ radians,

$$1° = \frac{\pi}{180} \text{ radians} \approx 0.0175 \text{ radians} \quad \text{and} \quad 1 \text{ radian} = \frac{180°}{\pi} \approx 57.3°$$

Formulas for arc length and area of a sector If a circle of radius r is centered at the vertex of an angle θ measured in radians, then the length of the arc intercepted by θ is given by $s = r\theta$ and the area of the related sector is given by

$$A = \frac{1}{2}r^2\theta = \frac{1}{2}rs.$$

Example 1 Express in radians: **a.** 75° **b.** −225°

Solution **a.** $75° = 75 \cdot \frac{\pi}{180}$ radians $= \frac{5\pi}{12}$ radians

b. $-225° = -225 \cdot \frac{\pi}{180}$ radians $= -\frac{5\pi}{4}$ radians

---

Express each degree measure in radians. Leave your answer in terms of π.

1. 35° **2.** 252° **3.** 90° **4.** −30°

5. −60° **6.** 420° **7.** −135° **8.** 315°

9. 330° **10.** −150° **11.** 200° **12.** −306°

Example 2 Express in degrees: **a.** $\frac{7\pi}{6}$ **b.** $-\frac{3\pi}{2}$

Solution **a.** $\frac{7\pi}{6}$ radians $= \frac{7\pi}{6} \cdot \frac{180°}{\pi} = 210°$

b. $-\frac{3\pi}{2}$ radians $= -\frac{3\pi}{2} \cdot \frac{180°}{\pi} = -270°$

Express each radian measure in degrees.

13. $\frac{11\pi}{12}$ **14.** $\frac{19\pi}{15}$ **15.** $\frac{\pi}{4}$ **16.** $-\frac{13\pi}{6}$

17. $-\frac{4\pi}{3}$ **18.** $-\frac{2\pi}{3}$ **19.** 4π **20.** $\frac{17\pi}{12}$

NAME ______________________ DATE ____________

13–1 Radian Measure *(continued)*

Express each radian measure in degrees. Leave your answer in terms of π.

21. 5 **22.** 2 **23.** -3 **24.** $\frac{1}{2}$

> ***Example 3*** **a.** Express 48° in radians to the nearest hundredth of a radian.
> **b.** Express 2.3 radians in degrees to the nearest tenth of a degree.
>
> ***Solution***
> **a.** $48° = 48 \cdot \frac{\pi}{180}$ radians
> $\approx \frac{48(3.1416)}{180}$ radians
> ≈ 0.84 radians
>
> **b.** $2.3 \text{ radians} = 2.3 \cdot \frac{180°}{\pi}$
> $\approx \frac{2.3(180°)}{3.1416}$
> $\approx 131.8°$

Express each degree measure in radians. Give answers to the nearest hundredth of a radian.

25. 15° **26.** $-70°$ **27.** 83° **28.** 280°

29. 54° **30.** 312° **31.** $-162°$ **32.** 221°

Express each radian measure in degrees. Give answers to the nearest tenth of a degree.

33. 2.6 **34.** 4.5 **35.** 0.6 **36.** -1.3

37. 4 **38.** -6 **39.** 3.1 **40.** 7

> ***Example 4*** A central angle of a circle of radius 2 cm measures 1.3 radians. Find (a) the length of the intercepted arc and (b) the area of the related sector.
>
> ***Solution***
> **a.** Use the formula $s = r\theta$.
> $s = 2(1.3)$
> $= 2.6$ cm
>
> **b.** Use the formula $A = \frac{1}{2}r^2\theta$.
> $A = \frac{1}{2}(2^2)(1.3)$
> $= 2.6 \text{ cm}^2$

A central angle of a circle of radius r measures θ radians. For the given values of r and θ, find (a) the length of the intercepted arc and (b) the area of the related sector.

41. $r=4,\ \theta=2$ **42.** $r=5,\ \theta=0.5$ **43.** $r=6,\ \theta=3$ **44.** $r=10,\ \theta=2.5$

Mixed Review Exercises

Solve each triangle. Give lengths to three significant digits and angle measures to the nearest tenth of a degree.

1. $a=10,\ b=6,\ \angle C = 53°$ **2.** $\angle A=80°,\ \angle B=35°,\ b=10$

3. $b=12,\ c=8,\ \angle B=110°$ **4.** $a=4,\ b=5,\ c=7$

5–8. Find the area of each triangle in Exercises 1–4. Give answers to three significant digits.

NAME ______________________ DATE ______________

13–2 Circular Functions

Objective: To define the circular functions.

Vocabulary

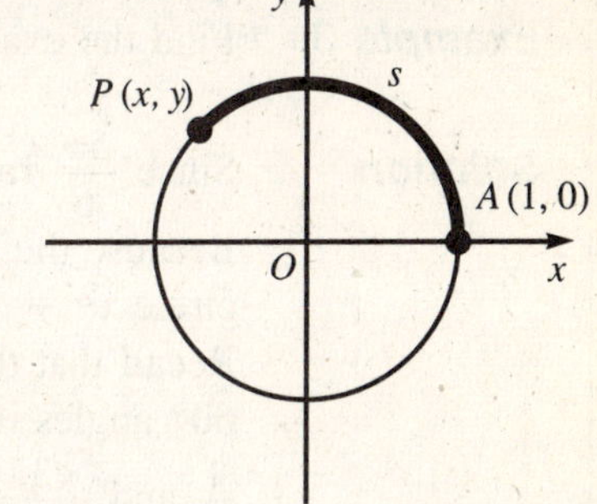

Circular functions Let O be the unit circle $x^2 + y^2 = 1$, and let A be the point $(1, 0)$. Given any real number s, start at A and measure $|s|$ units around O in a counterclockwise direction if $s \geq 0$ and in a clockwise direction if $s < 0$, arriving at a point $P(x, y)$. Then the circular functions of s are defined as follows:

$\sin s = y$ $\qquad$ $\cos s = x$

$\tan s = \dfrac{\sin s}{\cos s}$ if $\cos s \neq 0$ $\qquad$ $\cot s = \dfrac{\cos s}{\sin s}$ if $\sin s \neq 0$

$\sec s = \dfrac{1}{\cos s}$ if $\cos s \neq 0$ $\qquad$ $\csc s = \dfrac{1}{\sin s}$ if $\sin s \neq 0$

Note: A circular function of a number s is equal to the corresponding trigonometric function of any angle having radian measure s. Therefore, the circular functions can be referred to as trigonometric functions; the context should make it clear whether functions of angles or functions of numbers are intended.

Example 1 For some number s, the point $\left(-\frac{1}{3}, \frac{2\sqrt{2}}{3}\right)$ is s units from $A(1, 0)$ along a unit circle. Find the exact values of the six circular functions of s.

Solution Use the definitions of the circular functions given above.

$\sin s = \dfrac{2\sqrt{2}}{3}$ $\qquad$ $\cos s = -\dfrac{1}{3}$ $\qquad$ $\tan s = \dfrac{\frac{2\sqrt{2}}{3}}{-\frac{1}{3}} = -2\sqrt{2}$

$\csc s = \dfrac{1}{\frac{2\sqrt{2}}{3}} = \dfrac{3}{2\sqrt{2}} = \dfrac{3\sqrt{2}}{4}$ $\qquad$ $\sec s = \dfrac{1}{-\frac{1}{3}} = -3$ $\qquad$ $\cot s = \dfrac{-\frac{1}{3}}{\frac{2\sqrt{2}}{3}} = -\dfrac{1}{2\sqrt{2}} = -\dfrac{\sqrt{2}}{4}$

For some number s, the point P is s units from $A(1, 0)$ along the unit circle $x^2 + y^2 = 1$. Find the exact values of the six circular functions of s.

1. $P\left(-\frac{3}{5}, \frac{4}{5}\right)$ $\qquad$ **2.** $P\left(\frac{5}{13}, -\frac{12}{13}\right)$ $\qquad$ **3.** $P\left(-\frac{1}{2}, -\frac{\sqrt{3}}{2}\right)$ $\qquad$ **4.** $P\left(\frac{\sqrt{5}}{3}, \frac{2}{3}\right)$

Example 2 Find sin 1.35 to four significant digits.

Solution 1 **Using a Calculator** Set the calculator in radian mode.
$\sin 1.35 = 0.9757$ to four significant digits.

Solution 2 **Using Tables** Use Table 6 to find the sine of an angle having radian measure 1.35.
$\sin 1.35 = 0.9757$

13–2 Circular Functions (continued)

Find the sine, cosine, and tangent of each number to four significant digits.

5. 1.17 **6.** 1.32 **7.** 0.41 **8.** 0.8

Example 3 Find the exact values of the six trigonometric functions of $\frac{5\pi}{6}$.

Solution Since $\frac{5\pi}{6}$ radians = 150°, use a 30° – 60° – 90° triangle. Because the hypotenuse of the triangle is a radius of the circle $x^2 + y^2 = 1$, the length of the hypotenuse is 1 unit. Recall that the lengths of the legs opposite the 30° and 60° angles in a 30° – 60° – 90° triangle are, respectively, $\frac{1}{2}$ and $\frac{\sqrt{3}}{2}$ times the length of the hypotenuse.

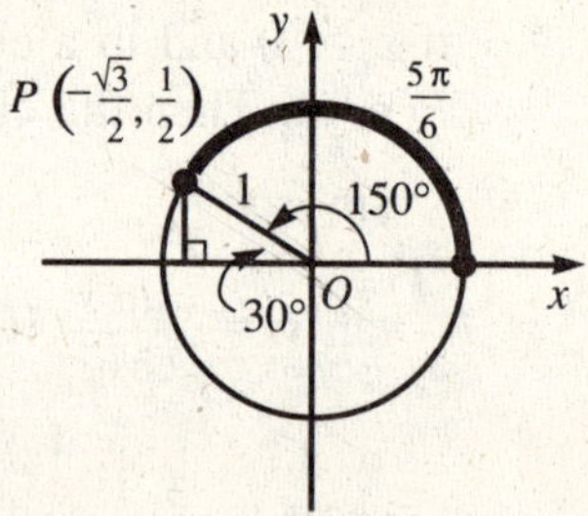

$\sin \frac{5\pi}{6} = \frac{1}{2}$ $\cos \frac{5\pi}{6} = -\frac{\sqrt{3}}{2}$ $\tan \frac{5\pi}{6} = -\frac{\sqrt{3}}{3}$

$\csc \frac{5\pi}{6} = 2$ $\sec \frac{5\pi}{6} = -\frac{2\sqrt{3}}{3}$ $\cot \frac{5\pi}{6} = -\sqrt{3}$

Find the exact values of the six trigonometric functions of the given number. If a function is undefined for the number, say so.

9. $\frac{2\pi}{3}$ **10.** $\frac{\pi}{2}$ **11.** $-\frac{\pi}{6}$ **12.** $\frac{7\pi}{6}$

13. 2π **14.** $-\frac{\pi}{4}$ **15.** $\frac{5\pi}{4}$ **16.** $\frac{5\pi}{3}$

Example 4 Find x to two decimal places if $\csc x = 1.077$ and $0 \le x \le \frac{\pi}{2}$.

Solution 1 **Using a Calculator**

Since $\csc x = 1.077$, $\sin x = \frac{1}{1.077} = 0.928505106$. Set the calculator in radian mode and use the inverse function key $\sin^{-1}$ (or inv sin).

$$x = \sin^{-1} 0.928505106 = 1.190366469$$

$\therefore x = 1.19$ to two decimal places.

Solution 2 **Using Tables**

Use Table 6 and read across 1.077 in the csc θ column. You should find that csc 1.1903 = 1.077.

$\therefore x = 1.19$ to two decimal places.

Find the number x to two decimal places if $0 \le x \le \frac{\pi}{2}$ and x has the given function value.

17. $\sin x = 0.6518$ **18.** $\tan x = 1.185$ **19.** $\csc x = 1.196$ **20.** $\cos x = 0.3809$

NAME ______________________ DATE ______________

13–3 Periodicity and Symmetry

Objective: To use periodicity and symmetry in graphing functions.

Vocabulary

Periodic function A function f is periodic if, for some positive constant p,

$$f(x + p) = f(x)$$

for every x in the domain of f. The smallest such p is the *period* of f.

Odd function A function f is odd if $f(-x) = -f(x)$ for every x in the domain of f. When a function f is odd, its graph is symmetric with respect to the origin. That is, if the point (x, y) is on the graph of f, then so is the point $(-x, -y)$.

Even function A function f is even if $f(-x) = f(x)$ for every x in the domain of f. When a function f is even, its graph is symmetric with respect to the y-axis. That is, if the point (x, y) is on the graph of f, then so is the point $(-x, y)$.

Example 1 Part of the graph of a function f having period 2 is shown at the right. Graph f in the interval $-3 \le x \le 3$ and tell whether f is even, odd, or neither.

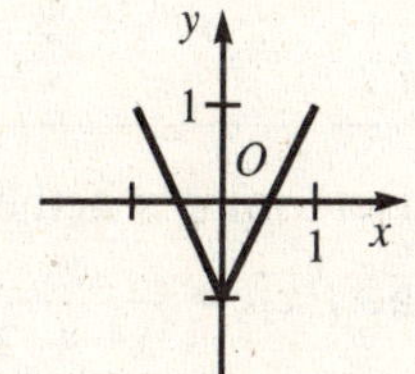

Solution Since the given part of the graph covers one period, repeat it to the right until you reach 3 and to the left until you reach -3.

The graph is symmetric with respect to the y-axis, so f is an even function.

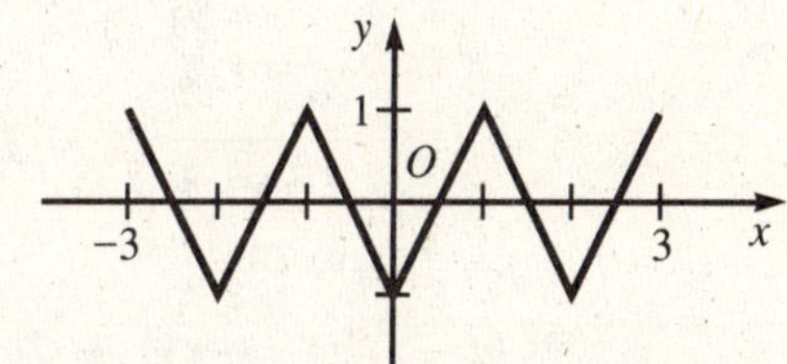

Example 2 Part of the graph of a function f is shown at the right. Graph f in the interval $-2 \le x \le 2$ assuming that f is (a) even and (b) odd.

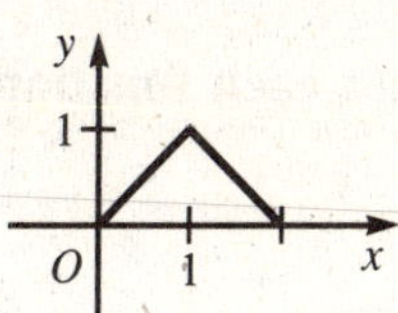

Solution

a. Reflect the given part of the graph in the y-axis.

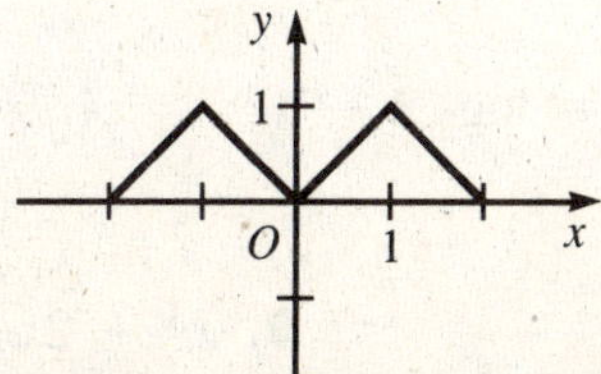

b. Reflect the given part of the graph in the origin.

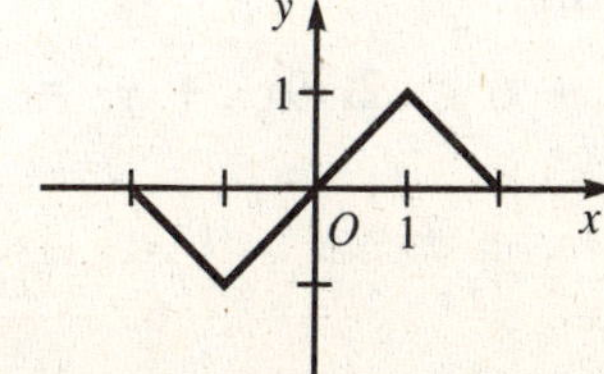

NAME ______________________ DATE ______________

13–3 Periodicity and Symmetry (continued)

Part of the graph of a function f having the given period p is shown. Graph f in the indicated interval and tell whether f is even, odd, or neither.

1. 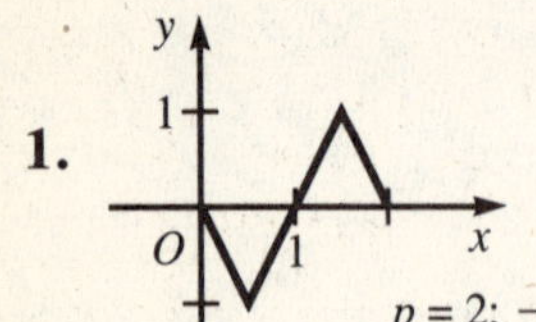
$p = 2$; $-4 \le x \le 4$

2. 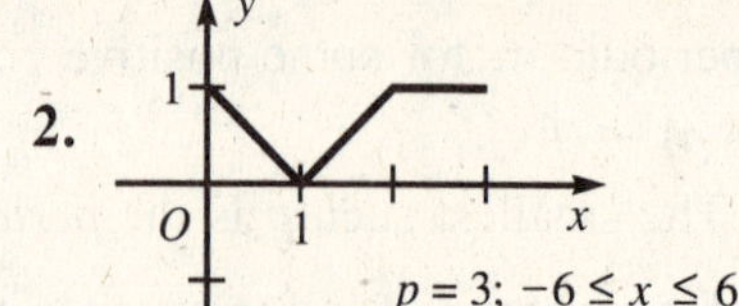
$p = 3$; $-6 \le x \le 6$

3. 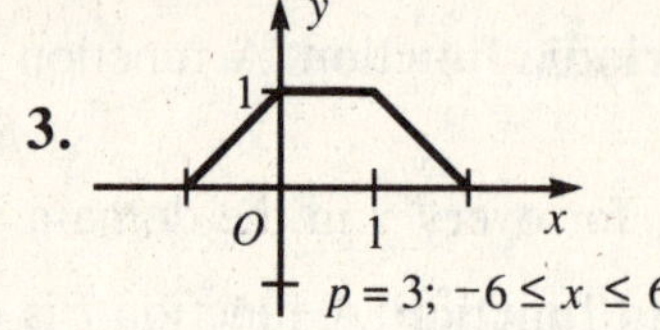
$p = 3$; $-6 \le x \le 6$

Part of the graph of a function f is given. Graph f in the interval $-2 \le x \le 2$ assuming f is (a) even and (b) odd.

4. 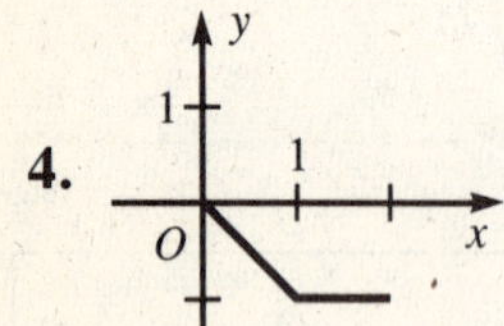

5. 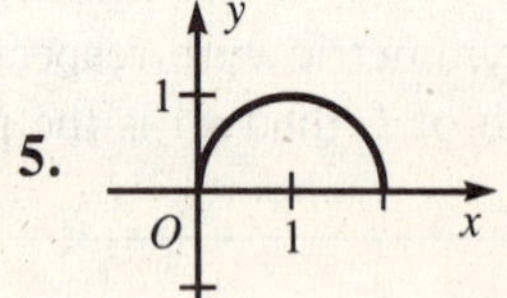

6. 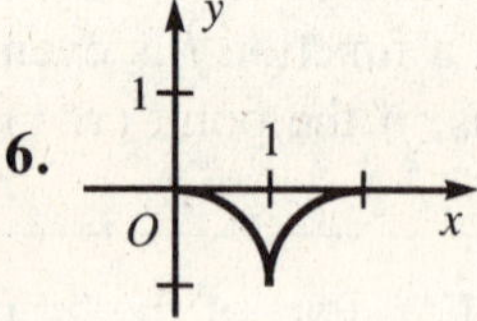

Example 3 Determine whether each function is even, odd, or neither.

a. $f(x) = \dfrac{x^2}{x^2 + 2}$ **b.** $g(x) = x^3 - x$ **c.** $h(x) = x^2 + 2x$

Solution

a. $f(-x) = \dfrac{(-x)^2}{(-x)^2 + 2} = \dfrac{x^2}{x^2 + 2} = f(x)$

$\therefore f$ is even.

b. $g(-x) = (-x)^3 - (-x) = -x^3 + x = -g(x)$

$\therefore g$ is odd.

c. $h(-x) = (-x)^2 + 2(-x) = x^2 - 2x$

Notice that $x^2 - 2x \ne x^2 + 2x$ and $x^2 - 2x \ne -(x^2 + 2x)$.

$\therefore h$ is neither even nor odd.

Determine whether each function f is even, odd, or neither.

7. $f(x) = x^2 + 3x$
8. $f(x) = x^3 + 4x$
9. $f(x) = x^4 + x^2$
10. $f(x) = \dfrac{x^4}{x^2 + 2}$
11. $f(x) = x\sqrt[3]{x^3 + 1}$
12. $f(x) = x + \dfrac{1}{x}$

Mixed Review Exercises

Graph each equation.

1. $y = x^2 + 4x + 6$
2. $9x^2 + y^2 = 9$
3. $y = \log_2 x - 1$
4. $x^2 + y^2 + 2x = 3$

Solve.

5. $\sqrt{3x - 2} = x$
6. $\dfrac{2x}{x + 1} - \dfrac{1}{x} = \dfrac{2}{x^2 + x}$
7. $x^3 + 2x^2 + 2x - 5 = 0$

NAME ______________________________ DATE ______________

13–4 Graphs of the Sine and Cosine

Objective: To graph the sine, cosine, and related functions.

Vocabulary

Graph of sine The graph of $y = \sin x$, called a *sine curve,* is shown at the right. Notice that the graph repeats itself every 2π units along the x-axis (because sine has period 2π), is symmetric with respect to the origin (because sine is an odd function), and has a maximum height of 1 and a minimum height of -1 (because the range of sine is $\{y: -1 \leq y \leq 1\}$).

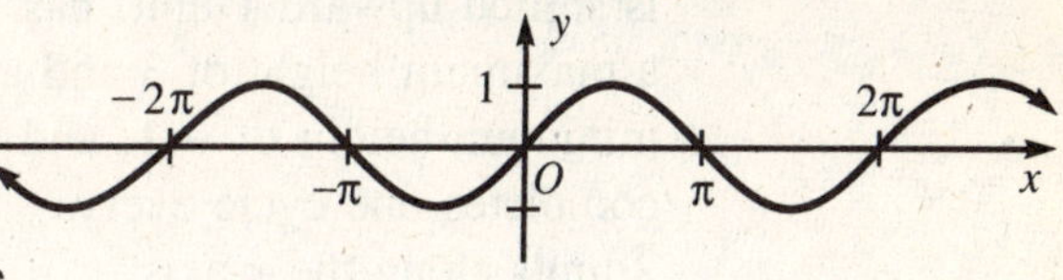

Graph of cosine The graph of $y = \cos x$, called a *cosine curve,* is shown at the right. Notice that the graph repeats itself every 2π units along the x-axis (because cosine has period 2π), is symmetric with respect to the y-axis (because cosine is an even function), and has a maximum height of 1 and a minimum height of -1 (because the range of cosine is $\{y: -1 \leq y \leq 1\}$). Notice also that if the cosine curve were shifted $\frac{\pi}{2}$ units to the right, it would coincide with the sine curve.

Vertical shift The graph of a function of the form $y = c + \sin x$ (or $y = c + \cos x$) can be obtained from the graph of $y = \sin x$ (or $y = \cos x$) by shifting the graph upward c units if $c > 0$ or downward $|c|$ units if $c < 0$.

Amplitude For a function of the form $y = a \sin x$ (or $y = a \cos x$) where $a > 0$, the number a is called the amplitude of the function. Such a function has maximum value a and minimum value $-a$. That is, the constant a causes a vertical stretch or shrinking of the sine (or cosine) curve.

Maximum and minimum values The maximum value M and minimum value m of a function of the form $y = c + a \sin x$ (or $y = c + a \cos x$) where $a > 0$ are given by

$$M = c + a \quad \text{and} \quad m = c - a.$$

These equations can be rewritten to obtain the vertical shift and amplitude from M and m:

$$c = \frac{M + m}{2} \quad \text{and} \quad a = \frac{M - m}{2}.$$

Period The period of a function of the form $y = \sin bx$ (or $y = \cos bx$) where $b > 0$ is $\frac{2\pi}{b}$. The constant b causes a horizontal stretch or shrinking of the sine (or cosine) curve.

Example 1 For the function $y = 1 + 2 \sin \pi x$, find (a) the amplitude, (b) the maximum and minimum values, and (c) the period.

Solution The function has the form $y = c + a \sin bx$.

a. Amplitude: $a = 2$

b. Maximum value: $M = c + a = 1 + 2 = 3$
Minimum value: $m = c - a = 1 - 2 = -1$

c. Period: $\frac{2\pi}{b} = \frac{2\pi}{\pi} = 2$

NAME ______________________ DATE ______________

13–4 Graphs of the Sine and Cosine *(continued)*

Example 2 Graph the function given in Example 1. Show at least two periods.

Solution The graph is a sine curve that is shifted upward 1 unit, has a maximum height of 3 and a minimum height of -1, and completes one cycle every 2 units along the x-axis.

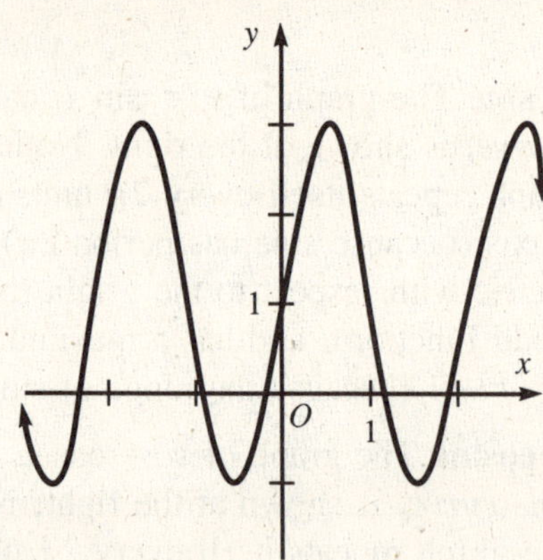

In Exercises 1–12:
a. Find the amplitude of each function.
b. Find the maximum and minimum values.
c. Find the period.

1. $y = 3 \sin x$
2. $y = \frac{1}{2} \cos x$
3. $y = \sin 2x$
4. $y = \cos 4x$
5. $y = 2 \sin 4x$
6. $y = \frac{1}{2} \cos 2x$
7. $y = \frac{1}{2} \sin 4x$
8. $y = 3 \cos \pi x$
9. $y = 2 \sin \frac{\pi}{2} x$
10. $y = 2 + 4 \cos \frac{\pi}{3} x$
11. $y = 3 \sin \pi x - 2$
12. $y = \frac{1}{2} \cos 2\pi x + 3$

13–16. Graph each function in Exercises 1, 4, 7, and 10. Show at least two periods.

Example 3 Find an equation of the form $y = c + a \sin bx$ that satisfies the following conditions: $M = 4$, $m = -2$, period 4π.

Solution

1. $a = \frac{M - m}{2} = \frac{4 - (-2)}{2} = 3$

2. $c = \frac{M + m}{2} = \frac{4 + (-2)}{2} = 1$

3. $\frac{2\pi}{b} = 4\pi$, so $b = \frac{2\pi}{4\pi} = \frac{1}{2}$

$\therefore y = 1 + 3 \sin \frac{1}{2} x$

Find an equation of the form $y = c + a \sin bx$ that satisfies the given conditions.

17. $M = 3$, $m = -1$, period π
18. $M = 5$, $m = 1$, period 4
19. $M = 2$, $m = -2$, period 4π

NAME ____________________ DATE ____________

13–5 Graphs of the Other Functions

Objective: To graph the tangent, cotangent, secant, cosecant, and related functions.

Vocabulary

Asymptote A line which a graph approaches without intersecting. Example: The lines $y = x$ and $y = -x$ are asymptotes for the graph of the hyperbola $x^2 - y^2 = 1$.

Graph of tangent The graph of $y = \tan x$ is shown at the right. Notice that the graph repeats itself every π units along the x-axis (because tangent has period π), is symmetric with respect to the origin (because tangent is an odd function), and has vertical asymptotes at $x = \frac{(2k + 1)\pi}{2}$ where k is an integer (because tangent is undefined for odd multiples of $\frac{\pi}{2}$).

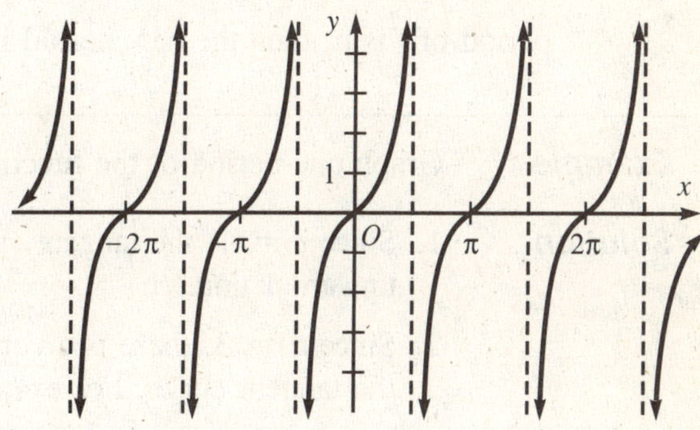

Graph of cotangent The graph of $y = \cot x$ is shown at the right. Notice that the graph repeats itself every π units along the x-axis (because cotangent has period π), is symmetric with respect to the origin (because cotangent is an odd function), and has vertical asymptotes at $x = k\pi$ where k is an integer (because cotangent is undefined for multiples of π).

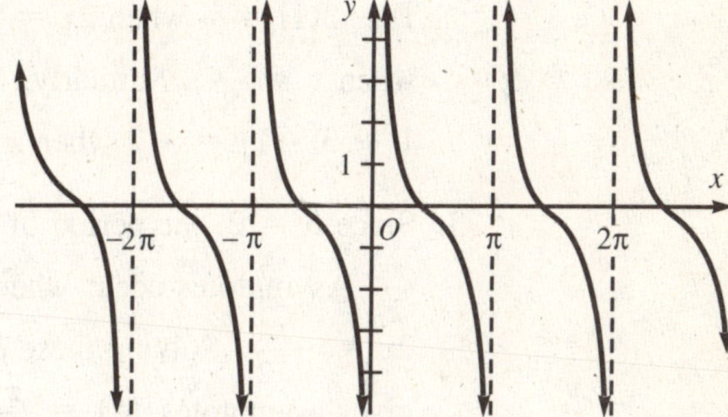

Graph of secant The graph of $y = \sec x$ is shown at the right. Notice that the graph repeats itself every 2π units along the x-axis (because secant has period 2π), is symmetric with respect to the y-axis (because secant is an even function), and has vertical asymptotes at $x = \frac{(2k + 1)\pi}{2}$ where k is an integer (because secant is undefined for odd multiples of $\frac{\pi}{2}$).

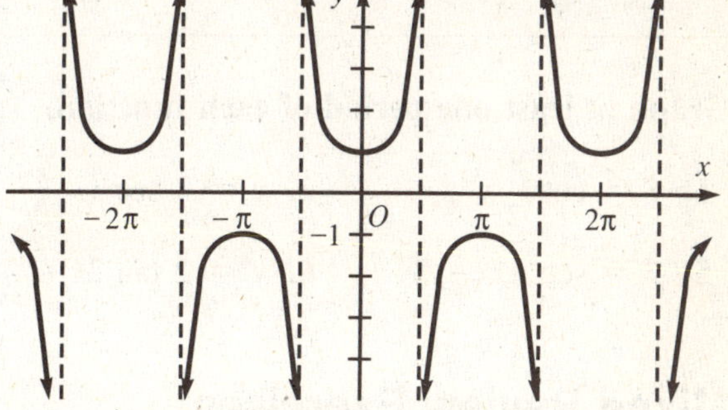

Graph of cosecant The graph of $y = \csc x$ is shown at the right. Notice that the graph repeats itself every 2π units along the x-axis (because cosecant has period 2π), is symmetric with respect to the origin (because cosecant is an odd function), and has vertical asymptotes at $x = k\pi$ where k is an integer (because cosecant is undefined for multiples of π).

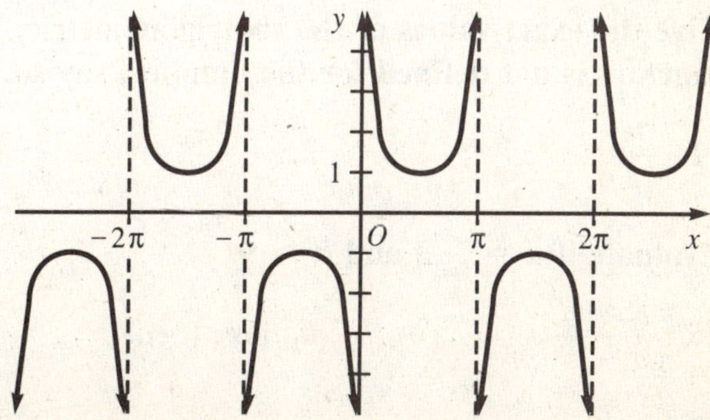

NAME ______________________ DATE ______________

Effect of constants on the graphs of the trigonometric functions For an equation of the form $y = c + a[f(bx)]$ where f is one of the six trigonometric functions, the constants c, a, and b ($a, b > 0$) have the following effects on the graph of f:

1. c produces a vertical shift.
2. a produces a vertical stretching or shrinking.
3. b produces a horizontal stretching or shrinking. (Specifically, if the period of f is p, then the new period is $\frac{p}{b}$.)

Example Graph one period of the function $y = 1 + 3 \tan 2x$.

Solution

1. Since $c = 1$, the tangent curve is shifted upward 1 unit.
2. Since $a = 3$, there is a vertical stretch of the tangent curve. For example, since $\tan x = 1$ when $x = \frac{\pi}{4}$, $1 + 3 \tan 2x = 1 + 3(1) = 4$ when $2x = \frac{\pi}{4}$ $\left(\text{that is, when } x = \frac{\pi}{8}\right)$. Similarly, $1 + 3 \tan 2x = 1 + 3(-1) = -2$ when $x = -\frac{\pi}{8}$.
3. Since $b = 2$, the period of the function is $\frac{\pi}{2}$. Asymptotes occur when $2x = \frac{\pi}{2}$ and $2x = -\frac{\pi}{2}$. Solving these equations gives asymptotes at $x = \frac{\pi}{4}$ and $x = -\frac{\pi}{4}$.

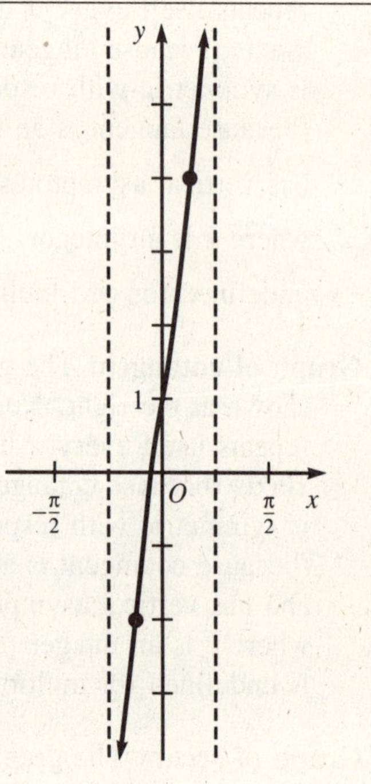

Graph at least one period of each function.

1. $y = \cot \frac{1}{2}x$
2. $y = 3 \sec x$
3. $y = \frac{3}{2} \tan x + 2$
4. $y = \frac{1}{2} \csc \frac{\pi}{2}x$
5. $y = \sec \pi x - 3$
6. $y = 2 \tan 2x$
7. $y = \csc 2x + 1$
8. $y = 3 \cot x - 2$

Mixed Review Exercises

Give the exact values of the six trigonometric functions of each number. If a function is not defined for the number, say so.

1. $\frac{7\pi}{2}$
2. $-\frac{2\pi}{3}$
3. $\frac{11\pi}{6}$
4. $-\frac{7\pi}{4}$

Evaluate if $a = -3$ and $b = 9$.

5. $\frac{a + b}{a - b}$
6. $\log_b(-a)$
7. $\sum_{n=1}^{4} b \cdot a^{n-1}$
8. $b^{-a/2}$

NAME ______________________ DATE ______________

13–6 The Fundamental Identities

Objective: To simplify trigonometric expressions and to prove identities.

Vocabulary

The Reciprocal Identities

$\sin\alpha = \dfrac{1}{\csc\alpha}$ $\qquad \sin\alpha\csc\alpha = 1 \qquad$ $\csc\alpha = \dfrac{1}{\sin\alpha}$

$\cos\alpha = \dfrac{1}{\sec\alpha}$ $\qquad \cos\alpha\sec\alpha = 1 \qquad$ $\sec\alpha = \dfrac{1}{\cos\alpha}$

$\tan\alpha = \dfrac{1}{\cot\alpha}$ $\qquad$ $\cot\alpha = \dfrac{1}{\tan\alpha}$

$\tan\alpha\cot\alpha = 1$

$\tan\alpha = \dfrac{\sin\alpha}{\cos\alpha}$ $\qquad$ $\cot\alpha = \dfrac{\cos\alpha}{\sin\alpha}$

The Cofunction Identities

$\sin\theta = \cos(90° - \theta)$ $\qquad \cos\theta = \sin(90° - \theta)$ $\qquad \tan\theta = \cot(90° - \theta)$

$\csc\theta = \sec(90° - \theta)$ $\qquad \sec\theta = \csc(90° - \theta)$ $\qquad \cot\theta = \tan(90° - \theta)$

The Pythagorean Identities

$\sin^2\alpha + \cos^2\alpha = 1$ $\qquad 1 + \tan^2\alpha = \sec^2\alpha$ $\qquad 1 + \cot^2\alpha = \csc^2\alpha$

General strategies for proving identities You can't prove an identity by "working across the = sign" (for example, by cross multiplying), because doing so assumes that what is being proved is true. Instead, you can usually use one of the following strategies.

1. Simplify the more complicated side of the identity until it is identical to the other side.
2. Transform both sides of the identity into the same expression.

Special strategies for proving identities There are often several ways of proving identities. Here are some special strategies that can be helpful.

1. Express functions in terms of sine and cosine.
2. Look for expressions to which the Pythagorean identities can be applied. Example: $\cot^2 x - \csc^2 x = -1$
3. Use factoring. Example: $1 - \sin^2 x = (1 + \sin x)(1 - \sin x)$
4. Combine terms on each side of the identity into a single fraction.
5. Multiply one side of the identity by an expression equal to 1.

Example 1 Simplify: **a.** $\csc\theta\tan\theta$ $\qquad$ **b.** $\dfrac{1 - \sin^2 x}{\sin x\cos x}$

Solution **a.** Write each function in terms of sine and cosine. Then simplify.

$$\csc\theta\tan\theta = \frac{1}{\sin\theta}\cdot\frac{\sin\theta}{\cos\theta} = \frac{1}{\cos\theta} = \sec\theta$$

(*Solution continues on the next page.*)

13-6 The Fundamental Identities *(continued)*

b. From the first Pythagorean identity you have $1 - \sin^2 x = \cos^2 x$.

$$\frac{1 - \sin^2 x}{\sin x \cos x} = \frac{\cos^2 x}{\sin x \cos x} = \frac{\cos x}{\sin x} = \cot x$$

Simplify.

1. $\sin \alpha \sec \alpha \cot \alpha$
2. $\cos^2 45° + \sin^2 45°$
3. $(\csc x - 1)(\csc x + 1)$
4. $\dfrac{\cot x}{\cos x}$
5. $\csc \alpha (\sec \alpha - \cos \alpha)$
6. $\dfrac{1 + \cot^2 \theta}{1 + \tan^2 \theta}$
7. $\dfrac{\cos \alpha}{\csc \alpha} - \dfrac{\sin \alpha}{\sec \alpha}$
8. $\dfrac{\csc x - \sin x}{\cos x}$
9. $\cot^2 (90° - \theta) - \sec^2 \theta$

Example 2 Write $\dfrac{\tan^2 x}{1 + \sec x}$ in terms of $\cos x$.

Solution From the second Pythagorean identity you have $\tan^2 x = \sec^2 x - 1$.

$$\frac{\tan^2 x}{1 + \sec x} = \frac{\sec^2 x - 1}{1 + \sec x} = \frac{(\sec x + 1)(\sec x - 1)}{1 + \sec x}$$

$$= \sec x - 1 = \frac{1}{\cos x} - 1 = \frac{1 - \cos x}{\cos x}$$

Write the first function in terms of the second.

10. $\cos \alpha \cot \alpha$; $\sin \alpha$
11. $\csc^2 x - 1$; $\cos x$
12. $\dfrac{\cot^2 \theta}{1 - \sin^2 \theta}$; $\cos \theta$

Example 3 Prove that $\dfrac{\cos x}{1 + \sin x} = \dfrac{1 - \sin x}{\cos x}$.

Solution Although each side is as "complicated" as the other, work only with the left side.

$$\frac{\cos x}{1 + \sin x} \stackrel{?}{=} \frac{1 - \sin x}{\cos x}$$

$$\frac{1 - \sin x}{1 - \sin x} \cdot \frac{\cos x}{1 + \sin x}$$ Use special strategy 5.

$$\frac{(1 - \sin x)\cos x}{1 - \sin^2 x}$$

$$\frac{(1 - \sin x)\cos x}{\cos^2 x}$$ Use special strategy 2.

$$\frac{1 - \sin x}{\cos x} = \frac{1 - \sin x}{\cos x}$$

Prove each identity.

13. $\tan x (\cot x + \tan x) = \sec^2 x$
14. $1 - 2 \sin^2 \theta = 2 \cos^2 \theta - 1$
15. $\sec \theta - \cos \theta = \sin \theta \tan \theta$
16. $\sec^4 \alpha - \tan^4 \alpha = 2 \sec^2 \alpha - 1$

NAME ______________________ DATE ______________

13–7 Trigonometric Addition Formulas

Objective: To use formulas for the sine and cosine of a sum or difference.

Vocabulary

Addition Formulas for the Sine and Cosine

$\sin(\alpha + \beta) = \sin\alpha\cos\beta + \cos\alpha\sin\beta$ $\qquad$ $\cos(\alpha + \beta) = \cos\alpha\cos\beta - \sin\alpha\sin\beta$

$\sin(\alpha - \beta) = \sin\alpha\cos\beta - \cos\alpha\sin\beta$ $\qquad$ $\cos(\alpha - \beta) = \cos\alpha\cos\beta + \sin\alpha\sin\beta$

CAUTION $\sin(\alpha + \beta) \neq \sin\alpha + \sin\beta$. Here is a counterexample:

$$\sin(30° + 60°) = \sin 90° = 1 \quad \text{but} \quad \sin 30° + \sin 60° = \frac{1}{2} + \frac{\sqrt{3}}{2} = \frac{1 + \sqrt{3}}{2}$$

Also, $\cos(\alpha + \beta) \neq \cos\alpha + \cos\beta$; $\sin(\alpha - \beta) \neq \sin\alpha - \sin\beta$; and $\cos(\alpha - \beta) \neq \cos\alpha - \cos\beta$.

Example 1 Find the exact value: **a.** $\cos 165°$ **b.** $\sin 195°$

Solution

a.
$$\begin{aligned}\cos 165° &= \cos(120° + 45°)\\ &= \cos 120°\cos 45° - \sin 120°\sin 45°\\ &= \left(-\frac{1}{2}\right)\left(\frac{\sqrt{2}}{2}\right) - \left(\frac{\sqrt{3}}{2}\right)\left(\frac{\sqrt{2}}{2}\right)\\ &= -\frac{\sqrt{2}}{4} - \frac{\sqrt{6}}{4}\\ &= \frac{-\sqrt{2} - \sqrt{6}}{4}, \text{ or } -\frac{\sqrt{2} + \sqrt{6}}{4}\end{aligned}$$

Note: There are often several ways to evaluate an expression like $\cos 165°$, such as $\cos(120° + 45°)$, $\cos(210° - 45°)$, and $\cos(135° + 30°)$. You should get the same value in each case.

b.
$$\begin{aligned}\sin 195° &= \sin(225° - 30°)\\ &= \sin 225°\cos 30° - \cos 225°\sin 30°\\ &= \left(-\frac{\sqrt{2}}{2}\right)\left(\frac{\sqrt{3}}{2}\right) - \left(-\frac{\sqrt{2}}{2}\right)\left(\frac{1}{2}\right)\\ &= -\frac{\sqrt{6}}{4} + \frac{\sqrt{2}}{4} = \frac{\sqrt{2} - \sqrt{6}}{4}\end{aligned}$$

Find the exact value of each of the following.

1. $\sin 105°$ **2.** $\cos 15°$ **3.** $\cos 255°$

4. $\sin 285°$ **5.** $\sin 345°$ **6.** $\cos 75°$

Example 2 Simplify to a trigonometric function of a single angle. Then give the exact value if possible.

a. $\cos 50°\cos 20° + \sin 50°\sin 20°$ **b.** $\sin 70°\cos 20° + \cos 70°\sin 20°$

Solution

a.
$$\begin{aligned}\cos 50°\cos 20° + \sin 50°\sin 20° &= \cos(50° - 20°)\\ &= \cos 30° = \frac{\sqrt{3}}{2}\end{aligned}$$

(*Solution continues on the next page.*)

13–7 Trigonometric Addition Formulas *(continued)*

b. $\sin 70° \cos 20° + \cos 70° \sin 20° = \sin(70° + 20°)$
$= \sin 90° = 1$

Simplify to a trigonometric function of a single angle. Then give the exact value if possible.

7. $\sin 65° \cos 5° - \cos 65° \sin 5°$

8. $\sin 75° \cos 15° + \cos 75° \sin 15°$

9. $\cos 25° \cos 20° - \sin 25° \sin 20°$

10. $\cos 110° \cos 50° + \sin 110° \sin 50°$

11. $\sin \frac{\pi}{6} \cos \frac{\pi}{12} + \cos \frac{\pi}{6} \sin \frac{\pi}{12}$

12. $\cos \frac{5\pi}{12} \cos \frac{\pi}{12} - \sin \frac{5\pi}{12} \sin \frac{\pi}{12}$

13. $\cos 2\alpha \cos 3\alpha + \sin 2\alpha \sin 3\alpha$

14. $\sin 5\theta \cos \theta - \cos 5\theta \sin \theta$

Example 3 Prove the identity $\cos\left(\frac{\pi}{2} - x\right) = \sin x$.

Solution

$$\cos\left(\frac{\pi}{2} - x\right) \stackrel{?}{=} \sin x$$

$$\cos \frac{\pi}{2} \cos x + \sin \frac{\pi}{2} \sin x$$

$$0 \cdot \cos x + 1 \cdot \sin x$$

$$\sin x = \sin x$$

Prove each identity.

15. $\cos(x - \pi) = -\cos x$

16. $\sin\left(\frac{\pi}{2} - \theta\right) = \cos \theta$

17. $\sin\left(\frac{3\pi}{2} + \theta\right) = -\cos \theta$

18. $\sin(\pi - x) + \sin(\pi + x) = 0$

19. $\cos\left(\frac{3\pi}{2} - x\right) + \cos\left(\frac{3\pi}{2} + x\right) = 0$

20. $\cos\left(\frac{\pi}{3} + \theta\right) + \cos\left(\frac{\pi}{3} - \theta\right) = \cos \theta$

Mixed Review Exercises

Graph each function. Show at least two periods.

1. $y = 2 \cos \frac{1}{2}x$

2. $y = \tan \frac{1}{2}x$

3. $y = 1 + \sin \pi x$

Simplify.

4. $\cos^2 x(1 + \tan^2 x)$

5. $x^8(x^{\pi-2})^4$

6. $\sqrt[5]{64x^{12}}$

7. $\log_3 \frac{1}{27}$

8. $\frac{3 + \sqrt{3}}{1 + \sqrt{3}}$

9. $\sqrt{\frac{1 + \frac{3}{5}}{2}}$

10. $\left(\frac{x^{3/4}}{x^{1/4}}\right)^8$

11. $\frac{2x}{x^2 - 1} - \frac{1}{x - 1}$

12. $\sec x - \sin x \tan x$

NAME ______________________________ DATE ______________

13–8 Double-Angle and Half-Angle Formulas

Objective: To use the double-angle and half-angle formulas for the sine and cosine.

Vocabulary

Double-Angle Formulas for Sine and Cosine

$$\sin 2\theta = 2 \sin \theta \cos \theta$$

$$\cos 2\theta = \cos^2 \theta - \sin^2 \theta \qquad \cos 2\theta = 1 - 2 \sin^2 \theta \qquad \cos 2\theta = 2 \cos^2 \theta - 1$$

Half-Angle Formulas for Sine and Cosine

$$\sin \frac{\theta}{2} = \pm\sqrt{\frac{1 - \cos \theta}{2}} \qquad \cos \frac{\theta}{2} = \pm\sqrt{\frac{1 + \cos \theta}{2}}$$

Note: The quadrant in which $\frac{\theta}{2}$ is located determines the sign (+ or −) you use.

Example 1 Simplify to a trigonometric function of a single angle. Do not evaluate.

a. $1 - 2 \sin^2 20°$ **b.** $\sqrt{\frac{1 - \cos 110°}{2}}$

Solution

a. $1 - 2 \sin^2 20° = \cos 2(20°)$
$= \cos 40°$

b. $\sqrt{\frac{1 - \cos 110°}{2}} = \sin \frac{110°}{2}$
$= \sin 55°$

Example 2 Find the exact value: **a.** $\sin \frac{3\pi}{8}$ **b.** $\cos 105°$

Solution

a. Since $\frac{3\pi}{8}$ is a first-quadrant angle, $\sin \frac{3\pi}{8}$ is positive.

$$\sin \frac{3\pi}{8} = \sin \frac{1}{2}\left(\frac{3\pi}{4}\right) = +\sqrt{\frac{1 - \cos\frac{3\pi}{4}}{2}} = \sqrt{\frac{1 - \left(-\frac{\sqrt{2}}{2}\right)}{2}} = \sqrt{\frac{2 + \sqrt{2}}{4}} = \frac{1}{2}\sqrt{2 + \sqrt{2}}$$

b. Since 105° is a second-quadrant angle, cos 105° is negative.

$$\cos 105° = \cos \frac{1}{2}(210°) = -\sqrt{\frac{1 + \cos 210°}{2}} = -\sqrt{\frac{1 + \left(-\frac{\sqrt{3}}{2}\right)}{2}} = -\sqrt{\frac{2 - \sqrt{3}}{4}} = -\frac{1}{2}\sqrt{2 - \sqrt{3}}$$

Simplify to a trigonometric function of a single angle. Do not evaluate.

1. $2 \sin 20° \cos 20°$ **2.** $\cos^2 25° - \sin^2 25°$ **3.** $1 - 2 \sin^2 65°$

4. $2 \cos^2 35° - 1$ **5.** $\sqrt{\frac{1 - \cos 70°}{2}}$ **6.** $\sqrt{\frac{1 + \cos 50°}{2}}$

NAME ______________________ DATE ______________

13–8 Double-Angle and Half-Angle Formulas *(continued)*

Simplify to a trigonometric function of a single angle. Do not evaluate.

7. $\sqrt{\frac{1 - \cos 62°}{2}}$ **8.** $\sqrt{\frac{1 + \cos 40°}{2}}$ **9.** $2 \sin 4t \cos 4t$

10. $\cos^2 3t - \sin^2 3t$ **11.** $1 - 2 \sin^2 \frac{1}{2}\theta$ **12.** $2 \cos^2 2\theta - 1$

Use a half-angle or double-angle formula to find the exact value of each of the following.

13. $\cos^2 15° - \sin^2 15°$ **14.** $1 - 2 \sin^2 22.5°$ **15.** $2 \sin 75° \cos 75°$

16. $2 \sin \frac{\pi}{12} \cos \frac{\pi}{12}$ **17.** $2 \cos^2 165° - 1$ **18.** $1 - 2 \sin^2 \frac{\pi}{8}$

19. $\cos 67.5°$ **20.** $\cos 75°$ **21.** $\sin 22.5°$

22. $\sin \frac{5\pi}{8}$ **23.** $\cos \frac{11\pi}{12}$ **24.** $\sin 75°$

Example 3 If $\cos \theta = -\frac{4}{5}$ and $0° < \theta < 180°$, find (a) $\sin 2\theta$ and (b) $\cos \frac{\theta}{2}$.

Solution If $0° < \theta < 180°$, θ is either a first- or a second-quadrant angle. Since $\cos \theta$ is negative, θ must be a second-quadrant angle. You can find the exact value of $\sin \theta$ in one of two ways:

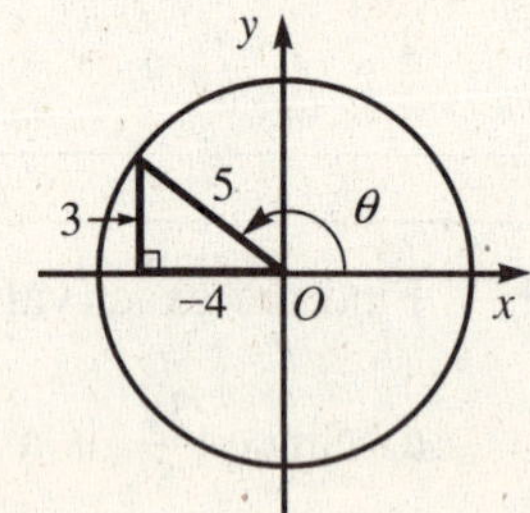

1. Make a sketch of θ as shown at the right. You can see from the sketch that if $\cos \theta = -\frac{4}{5}$, then $\sin \theta = \frac{3}{5}$.
2. In the second quadrant, $\sin \theta > 0$.

$\therefore \sin \theta = \sqrt{1 - \cos^2 \theta} = \sqrt{1 - \left(-\frac{4}{5}\right)^2} = \frac{3}{5}$

a. $\sin 2\theta = 2 \sin \theta \cos \theta = 2\left(\frac{3}{5}\right)\left(-\frac{4}{5}\right) = -\frac{24}{25}$

b. Since $0° < \theta < 180°$, $0° < \frac{\theta}{2} < 90°$. Therefore $\cos \frac{\theta}{2}$ is positive.

$$\cos \frac{\theta}{2} = +\sqrt{\frac{1 + \cos \theta}{2}} = \sqrt{\frac{1 + \left(-\frac{4}{5}\right)}{2}} = \sqrt{\frac{1}{10}} = \frac{\sqrt{10}}{10}$$

In Exercises 25–28, $0° < \theta < 180°$ and $\cos \theta = -\frac{5}{13}$. Find the following.

25. $\cos 2\theta$ **26.** $\sin 2\theta$ **27.** $\cos \frac{\theta}{2}$ **28.** $\sin \frac{\theta}{2}$

In Exercises 29–32, $180° < \theta < 360°$ and $\cos \theta = \frac{12}{13}$. Find the following.

29. $\cos 2\theta$ **30.** $\sin 2\theta$ **31.** $\cos \frac{\theta}{2}$ **32.** $\sin \frac{\theta}{2}$

NAME ______________________ DATE ____________

13–9 Formulas for the Tangent

Objective: To use the addition, double-angle, and half-angle formulas for the tangent.

Vocabulary

Addition Formulas for the Tangent

$$\tan(\alpha + \beta) = \frac{\tan\alpha + \tan\beta}{1 - \tan\alpha\tan\beta} \qquad \tan(\alpha - \beta) = \frac{\tan\alpha - \tan\beta}{1 + \tan\alpha\tan\beta}$$

Double-Angle Formula for the Tangent

$$\tan 2\alpha = \frac{2\tan\alpha}{1 - \tan^2\alpha}$$

Half-Angle Formulas for the Tangent

$$\tan\frac{\theta}{2} = \pm\sqrt{\frac{1-\cos\theta}{1+\cos\theta}} \qquad \tan\frac{\theta}{2} = \frac{\sin\theta}{1+\cos\theta} \qquad \tan\frac{\theta}{2} = \frac{1-\cos\theta}{\sin\theta}$$

Example 1 Simplify to a trigonometric function of a single angle. Do not evaluate.

a. $\dfrac{\tan 27° + \tan 31°}{1 - \tan 27° \tan 31°}$ **b.** $\dfrac{1 - \cos 43°}{\sin 43°}$

Solution **a.** $\dfrac{\tan 27° + \tan 31°}{1 - \tan 27° \tan 31°} = \tan(27° + 31°) = \tan 58°$

b. $\dfrac{1 - \cos 43°}{\sin 43°} = \tan\dfrac{43°}{2} = \tan 21.5°$

Simplify to a trigonometric function of a single angle. Do not evaluate.

1. $\dfrac{\tan 52° + \tan 37°}{1 - \tan 52° \tan 37°}$
2. $\dfrac{1 - \cos 330°}{\sin 330°}$
3. $\dfrac{2\tan 35°}{1 - \tan^2 35°}$
4. $\sqrt{\dfrac{1 - \cos 42°}{1 + \cos 42°}}$
5. $\dfrac{\tan 230° - \tan 40°}{1 + \tan 230° \tan 40°}$
6. $\dfrac{\sin 110°}{1 + \cos 110°}$
7. $\dfrac{\tan 76° - \tan 22°}{1 + \tan 76° \tan 22°}$
8. $\dfrac{2\tan\pi}{1 - \tan^2\pi}$
9. $\dfrac{\tan 3\pi + \tan\pi}{1 - \tan 3\pi \tan\pi}$

Example 2 Find the exact value: **a.** $\tan 165°$ **b.** $\tan\dfrac{3\pi}{8}$

Solution **a.** $\tan 165° = \tan(120° + 45°) = \dfrac{\tan 120° + \tan 45°}{1 - \tan 120° \tan 45°}$

$$= \frac{-\sqrt{3} + 1}{1 - (-\sqrt{3})(1)}$$

$$= \frac{-\sqrt{3} + 1}{1 + \sqrt{3}} \cdot \frac{1 - \sqrt{3}}{1 - \sqrt{3}}$$

$$= \frac{4 - 2\sqrt{3}}{-2} = -2 + \sqrt{3}$$

(Solution continues on the next page.)

13–9 Formulas for the Tangent *(continued)*

b. $\tan \frac{3\pi}{8} = \tan \frac{1}{2}\left(\frac{3\pi}{4}\right) = \dfrac{1 - \cos \frac{3\pi}{4}}{\sin \frac{3\pi}{4}}$

$$= \frac{1 - \left(-\frac{\sqrt{2}}{2}\right)}{\frac{\sqrt{2}}{2}} = \frac{2 + \sqrt{2}}{\sqrt{2}} = \sqrt{2} + 1$$

Find the exact value of each of the following.

10. $\tan 105°$ **11.** $\tan 247.5°$ **12.** $\tan 345°$ **13.** $\tan 195°$

14. $\tan \frac{7\pi}{8}$ **15.** $\tan \frac{19\pi}{12}$ **16.** $\tan 292.5°$ **17.** $\tan 375°$

Example 3 If $\tan \alpha = -\frac{3}{4}$, $\tan \beta = \frac{5}{12}$, and α is a second-quadrant angle, find:

a. $\tan (\alpha - \beta)$ **b.** $\tan 2\alpha$

Solution **a.** $\tan (\alpha - \beta) = \dfrac{\tan \alpha - \tan \beta}{1 + \tan \alpha \tan \beta}$

$$= \frac{-\frac{3}{4} - \frac{5}{12}}{1 + \left(-\frac{3}{4}\right)\left(\frac{5}{12}\right)}$$

$$= \frac{-\frac{14}{12}}{\frac{33}{48}} = -\frac{56}{33}$$

b. $\tan 2\alpha = \dfrac{2 \tan \alpha}{1 - \tan^2 \alpha}$

$$= \frac{2\left(-\frac{3}{4}\right)}{1 - \left(-\frac{3}{4}\right)^2}$$

$$= \frac{-\frac{3}{2}}{\frac{7}{16}} = -\frac{24}{7}$$

In Exercises 18–21, $\tan \alpha = -\frac{12}{5}$, $\tan \beta = \frac{4}{3}$, and α is a second-quadrant angle. Find the value of each expression.

18. $\tan (\alpha + \beta)$ **19.** $\tan (\alpha - \beta)$ **20.** $\tan 2\alpha$ **21.** $\tan \frac{\alpha}{2}$

Mixed Review Exercises

Prove each identity.

1. $\csc^2 \theta(1 - \sin^2 \theta) = \cot^2 \theta$ **2.** $\tan \theta(\sin \theta - \csc \theta) = -\cos \theta$

3. $(\csc x - \cot x)(\csc x + \cot x) = 1$ **4.** $\csc 2\alpha = \frac{1}{2}\sec \alpha \csc \alpha$

Solve each inequality and graph the solution set.

5. $|x + 1| > 2$ **6.** $x^2 < 4x - 3$ **7.** $3 \le 4x - 1 < 7$

8. $|3x + 2| \le 1$ **9.** $5 - 2x < -3$ **10.** $x^3 - 4x > 0$

NAME ______________________ DATE ______________

14 Trigonometric Applications

14–1 Vector Operations

Objective: To define vector operations and apply the resultant of two vectors.

Vocabulary

Vector quantity Any quantity that has both magnitude (size) and direction.

Vector An arrow representing a vector quantity. The length of the arrow indicates the magnitude of the vector quantity and the direction of the arrow indicates the direction of the vector. $\overrightarrow{AB}$ denotes the vector extending from point A, the *initial point*, to point B, the *terminal point.*

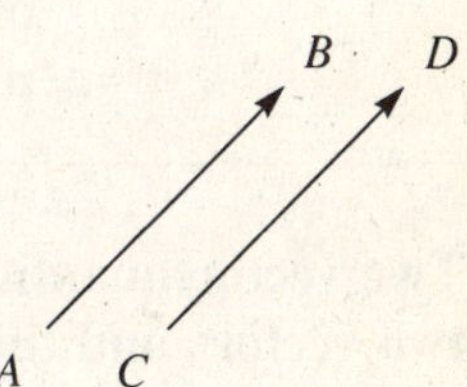

Equivalent vectors Two vectors that have the same magnitude and same direction. Example: $\overrightarrow{AB}$ and $\overrightarrow{CD}$, shown at the right, are equivalent. You write $\overrightarrow{AB} = \overrightarrow{CD}$. Equivalent vectors are often denoted by the same boldface lowercase letter, such as **v**.

Vector addition Given two vectors **u** and **v**, you can find their *sum*, or *resultant*, by using either of the following two methods. These methods show that vector addition is commutative: **u** + **v** = **v** + **u**.

The Triangle Method

Place the initial point of **v** at the terminal point of **u**. Then **u** + **v** is the vector extending from the initial point of **u** to the terminal point of **v**.

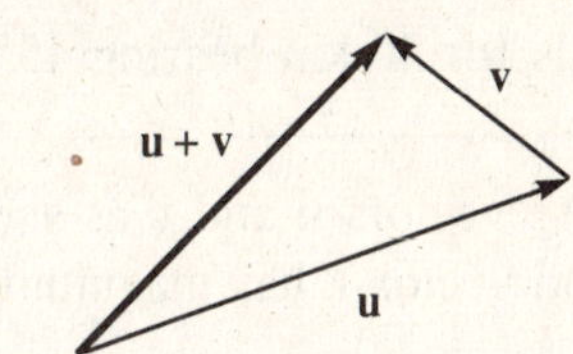

The Parallelogram Method

Form a parallelogram with **u** and **v** as adjacent sides starting from a common point. Then **u** + **v** extends from that point to the opposite vertex of the parallelogram.

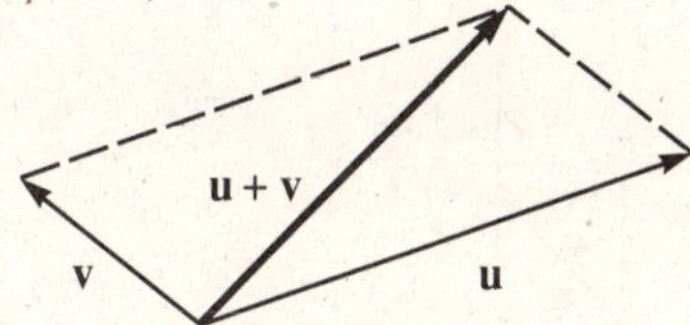

Scalar multiplication To multiply the vector **v** by the real number t (called a *scalar*), multiply the length of **v** by $|t|$ and reverse the direction if $t < 0$.

Zero vector A vector of length 0, represented by a point and denoted by **0**.

Difference of vectors The difference **u** − **v** is defined to be **u** + (−**v**) where −**v** is the product of **v** and the scalar −1. (*Note*: **u** − **u** = **0** for any vector **u**.)

Norm of a vector The length or magnitude of the vector. ‖**v**‖ denotes the norm of vector **v**.

Unit vector A vector with a norm of 1.

Bearing of a vector The angle measured clockwise from due north around to the vector.

Symbols

$\overrightarrow{AB}$ (vector AB) ‖**v**‖ (norm of vector **v**) **0** (zero vector)

NAME ______________________________ DATE ______________

14–1 Vector Operations *(continued)*

Example 1 For the following two vector quantities, make a scale drawing showing the two vectors and their sum: a 6 km trip northwest followed by a 3 km trip due east.

Solution Let **u** be the vector representing a 6 km trip northwest and **v** be the vector representing a 3 km trip due east. These vectors are shown at the right. (Notice that the vectors point in the proper directions and that the length of **u** is twice the length of **v**.) The sum **u** + **v** is also shown.

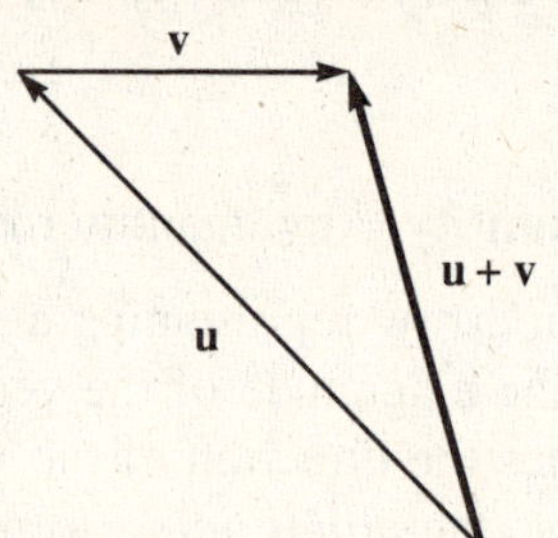

Two vector quantities are given. Make a scale drawing showing the two vectors and their sum.

1. A 7 km trip west followed by a 5 km trip northeast.
2. An 8 km trip southeast followed by a 10 km trip north.
3. A plane flies northeast at 400 km/h, while a wind of 40 km/h blows from the east.
4. A catamaran sails for 20 km bearing 45° and then sails for 15 km bearing 150°.

Example 2 Use vectors **u** and **v** as shown below. Vector **u** has magnitude 6 and bearing 45°, and vector **v** has magnitude 10 and bearing 135°. Sketch (a) **u** + **v** and (b) **u** − **v**.

Solution **a.** **u** + **v**

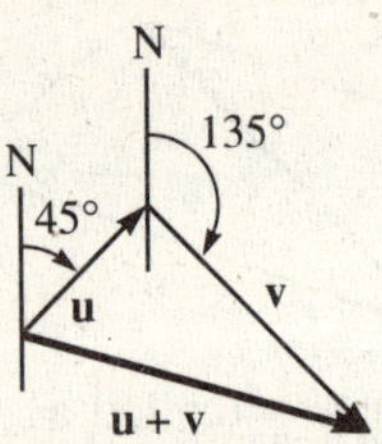

b. Use the fact that **u** − **v** = **u** + (−**v**).

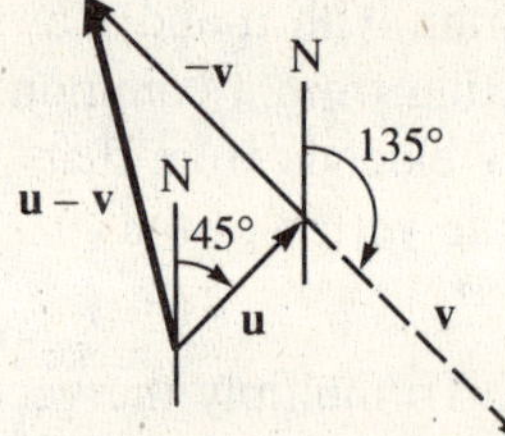

In the diagram at the right, vector u has magnitude 4 and bearing 270°, vector v has magnitude 2 and bearing 225°, and vector w has magnitude 3 and bearing 180°. Use these vectors to sketch the following vectors.

u v w

5. **u** + **v** 6. **u** + **w** 7. **u** − **v** 8. **u** − **w**

9. **u** + 2**v** 10. 2**u** + **w** 11. 2**u** + 3**v** 12. 2**v** − **w**

Mixed Review Exercises

Solve.

1. $x^2 + 5x = 7$ 2. $|4z - 9| = 3$ 3. $\log_4 u = \frac{5}{2}$ 4. $\sqrt{3t + 1} = 7$

NAME ______________________ DATE ______________

14–2 Vectors in the Plane

Objective: To find vectors in component form and to apply the dot product.

Vocabulary

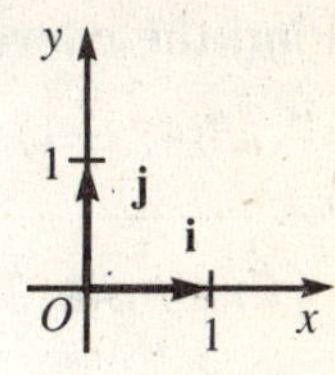
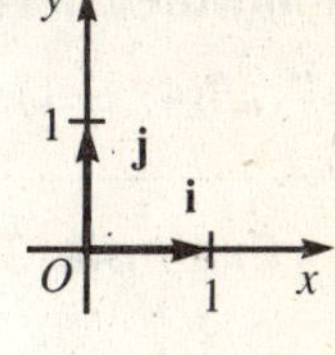

i and j The vector **i** has initial point (0, 0) and terminal point (1, 0). The vector **j** has initial point (0, 0) and terminal point (0, 1). If point P has coordinates (a, b), then $\overrightarrow{OP}$ can be written in terms of **i** and **j**: $\overrightarrow{OP} = a\mathbf{i} + b\mathbf{j}$.

Component form When a vector **u** is written in the form $\mathbf{u} = a\mathbf{i} + b\mathbf{j}$, vector **u** is said to be in component form. The number a is the *x-component* of **u** and the number b is the *y-component* of **u**.

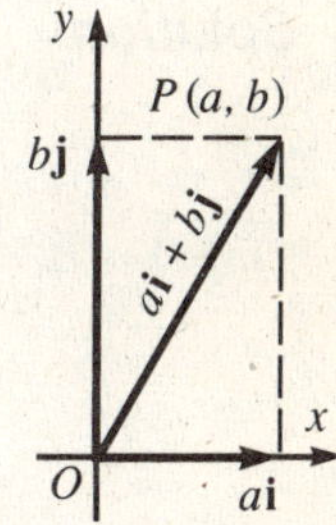

Properties of vectors in component form If $\mathbf{u} = a\mathbf{i} + b\mathbf{j}$, $\mathbf{v} = c\mathbf{i} + d\mathbf{j}$, then:

1. $\mathbf{u} = \mathbf{v}$ if and only if $a = c$ and $b = d$
2. $\mathbf{u} + \mathbf{v} = (a + c)\mathbf{i} + (b + d)\mathbf{j}$
3. $t\mathbf{u} = ta\mathbf{i} + tb\mathbf{j}$ (t a scalar)
4. $\|\mathbf{u}\| = \sqrt{a^2 + b^2}$

Dot product The dot product of two nonzero vectors **u** and **v** is $\mathbf{u} \cdot \mathbf{v} = \|\mathbf{u}\| \, \|\mathbf{v}\| \cos \theta$ where θ is the angle between **u** and **v**.
Note: The dot product is a real number, not a vector.

Theorem If $\mathbf{u} = a\mathbf{i} + b\mathbf{j}$ and $\mathbf{v} = c\mathbf{i} + d\mathbf{j}$, then $\mathbf{u} \cdot \mathbf{v} = ac + bd$.

Orthogonal Two vectors are orthogonal if either is **0** or if they are perpendicular. This means that vectors **u** and **v** are orthogonal if and only if $\mathbf{u} \cdot \mathbf{v} = 0$.

Example 1 Given the points $A(-4, 2)$ and $B(1, 1)$, express $\overrightarrow{AB}$ in component form.

Solution 1 Draw an equivalent vector, say $\overrightarrow{OP}$, with its initial point at the origin, as shown at the right. Then P has coordinates $(5, -1)$, and $\overrightarrow{AB} = \overrightarrow{OP} = 5\mathbf{i} - \mathbf{j}$.

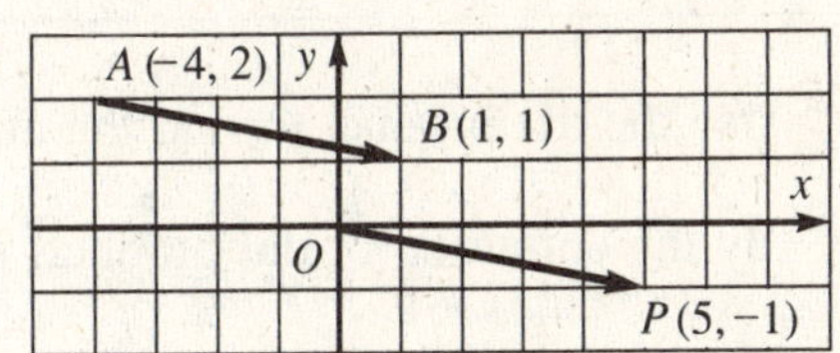

Solution 2 Notice that the x-component of $\overrightarrow{AB}$ is the difference of the x-coordinates of B and A. The y-component is the difference in the y-coordinates of B and A.
$\therefore \overrightarrow{AB} = [1 - (-4)]\mathbf{i} + [1 - 2]\mathbf{j} = 5\mathbf{i} - \mathbf{j}$.

Given the points A and B, express $\overrightarrow{AB}$ in component form.

1. $A(2, 1)$, $B(6, 3)$ **2.** $A(3, -1)$, $B(1, 4)$ **3.** $A(-3, 2)$, $B(1, 2)$ **4.** $A(1, 3)$, $B(-2, -2)$

Example 2 Find the coordinates of B if A has coordinates $(4, -1)$ and $\overrightarrow{AB} = 3\mathbf{i} + 2\mathbf{j}$.

Solution Draw the vector $\overrightarrow{OP} = 3\mathbf{i} + 2\mathbf{j}$. Point P is 3 units to the right and 2 units above point O. If $\overrightarrow{AB} = \overrightarrow{OP}$, then B must be 3 units to the right and 2 units above point A.
$\therefore$ B has coordinates $(7, 1)$.

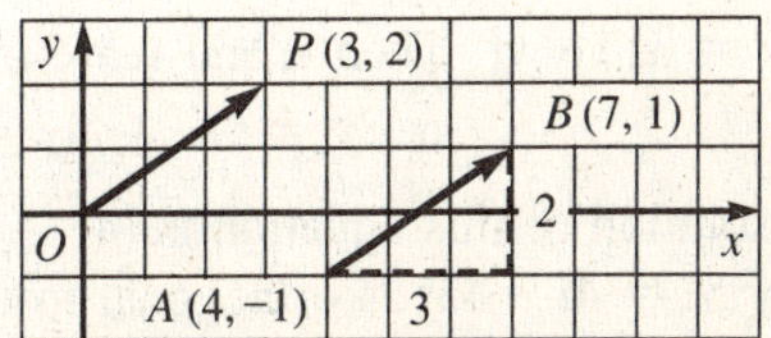

NAME ____________________ DATE ____________

14–2 Vectors in the Plane *(continued)*

Find the coordinates of B, given the following.

5. $A(-3, -4)$, $\overrightarrow{AB} = 3\mathbf{i} + 4\mathbf{j}$

6. $A(-4, 2)$, $\overrightarrow{AB} = \mathbf{i} - 3\mathbf{j}$

Find the coordinates of A, given the following. (*Hint*: Use the vector $\overrightarrow{BA} = -\overrightarrow{AB}$.)

7. $B(4, -1)$, $\overrightarrow{AB} = -\mathbf{i} + 2\mathbf{j}$

8. $B(0, -5)$, $\overrightarrow{AB} = 3\mathbf{i} - \mathbf{j}$

Example 3 Let $\mathbf{u} = \mathbf{i} - 4\mathbf{j}$, $\mathbf{v} = -3\mathbf{i} + \mathbf{j}$, and $\mathbf{w} = \mathbf{u} + 2\mathbf{v}$. (a) Find $\mathbf{w}$ in component form. (b) Draw a diagram showing $\mathbf{u}$, $\mathbf{v}$, and $\mathbf{w}$. (c) Find $\|\mathbf{w}\|$. (d) Find $\mathbf{u} \cdot \mathbf{v}$.

Solution

a. $\mathbf{w} = \mathbf{u} + 2\mathbf{v} = \mathbf{i} - 4\mathbf{j} + 2(-3\mathbf{i} + \mathbf{j})$
$= \mathbf{i} - 4\mathbf{j} - 6\mathbf{i} + 2\mathbf{j} = -5\mathbf{i} - 2\mathbf{j}$

b.

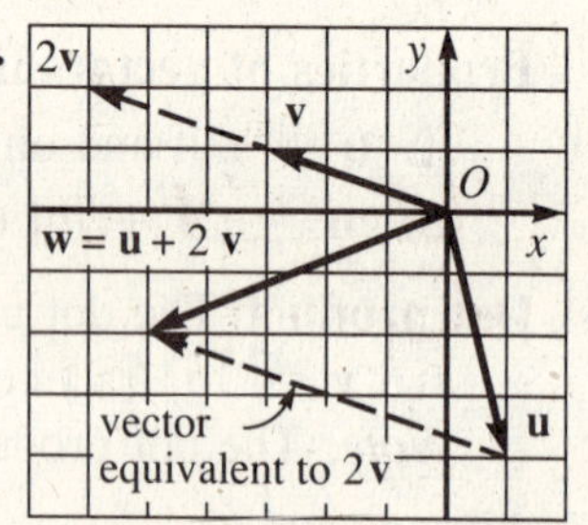

c. $\|\mathbf{w}\| = \sqrt{(-5)^2 + (-2)^2} = \sqrt{29}$

d. To find $\mathbf{u} \cdot \mathbf{v}$, multiply the x-components of $\mathbf{u}$ and $\mathbf{v}$, multiply the y-components, and add the products:

$\mathbf{u} \cdot \mathbf{v} = 1(-3) + (-4)(1) = -7$

For Exercises 9–10: **(a) Find w in component form. (b) Draw a diagram showing u, v, and w. (c) Find $\|\mathbf{w}\|$. (d) Find u · v.**

9. $\mathbf{u} = -4\mathbf{i} - 3\mathbf{j}$; $\mathbf{v} = \mathbf{i} + 2\mathbf{j}$; $\mathbf{w} = \mathbf{u} + \mathbf{v}$

10. $\mathbf{u} = 2\mathbf{i} - \mathbf{j}$; $\mathbf{v} = -\mathbf{i} + 3\mathbf{j}$; $\mathbf{w} = 2\mathbf{u} - \mathbf{v}$

Example 4 Use the dot product to find the angle between $\mathbf{u}$ and $\mathbf{v}$ if $\mathbf{u} = 2\mathbf{i} + 3\mathbf{j}$ and $\mathbf{v} = \mathbf{i} - 4\mathbf{j}$.

Solution By the definition of dot product, $\mathbf{u} \cdot \mathbf{v} = \|\mathbf{u}\|\,\|\mathbf{v}\| \cos\theta$. Solving for $\cos\theta$ gives:

$$\cos\theta = \frac{\mathbf{u} \cdot \mathbf{v}}{\|\mathbf{u}\|\,\|\mathbf{v}\|} = \frac{2(1) + 3(-4)}{\sqrt{2^2 + 3^2}\sqrt{1^2 + (-4)^2}} = \frac{-10}{\sqrt{13}\sqrt{17}} \approx -0.6727$$

$\therefore \theta = \cos^{-1}(-0.6727) = 132.3°$.

Use the dot product to find the angle between u and v.

11. $\mathbf{u} = 2\mathbf{i} - 3\mathbf{j}$, $\mathbf{v} = \mathbf{i} + 6\mathbf{j}$

12. $\mathbf{u} = 4\mathbf{i} + \mathbf{j}$, $\mathbf{v} = \mathbf{i} + 4\mathbf{j}$

13. $\mathbf{u} = 6\mathbf{i} - 3\mathbf{j}$, $\mathbf{v} = 4\mathbf{i} + 8\mathbf{j}$

Example 5 Find a unit vector orthogonal to $\mathbf{u} = 3\mathbf{i} + 4\mathbf{j}$.

Solution

Let $\mathbf{v} = x\mathbf{i} + y\mathbf{j}$. Then $\mathbf{u}$ and $\mathbf{v}$ are orthogonal if and only if

$$\mathbf{u} \cdot \mathbf{v} = 3x + 4y = 0.$$

One solution of this equation is $(4, -3)$, since $3(4) + 4(-3) = 0$. So the vector $\mathbf{v} = 4\mathbf{i} - 3\mathbf{j}$ is orthogonal to $\mathbf{u}$. To find a *unit* vector orthogonal to $\mathbf{u}$, use the fact that any nonzero vector divided by its norm has length 1.

$\|\mathbf{v}\| = \sqrt{4^2 + (-3)^2} = 5$, so $\frac{4}{5}\mathbf{i} - \frac{3}{5}\mathbf{j}$ is a unit vector orthogonal to $\mathbf{u}$.

Find a unit vector orthogonal to u.

14. $\mathbf{u} = 12\mathbf{i} + 5\mathbf{j}$

15. $\mathbf{u} = -8\mathbf{i} + 15\mathbf{j}$

16. $\mathbf{u} = 3\mathbf{i} - \mathbf{j}$

NAME ______________________ DATE ______________

14–3 Polar Coordinates

Objective: To define polar coordinates and graph polar equations.

Vocabulary

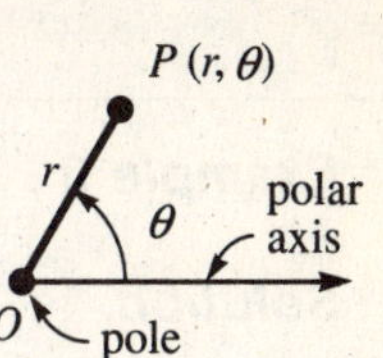

Polar coordinate system A system consisting of a point O called the *pole*, and a ray, called the *polar axis*, having O as its endpoint.

Polar coordinates of a point P An ordered pair of the form (r, θ) where $r = OP$ and θ is the measure of an angle from the polar axis to the segment $\overline{OP}$.

Coordinate-system conversion formulas

From polar to rectangular

$x = r\cos\theta$ and $y = r\sin\theta$

From rectangular to polar

$r = \pm\sqrt{x^2 + y^2}$, $\cos\theta = \frac{x}{r}$, and $\sin\theta = \frac{y}{r}$

Example 1 Plot each point in a polar coordinate system and find its rectangular coordinates.

a. $P(3, 30°)$ **b.** $R(-2, -45°)$

Solution **a.** The graph of P is at the right.

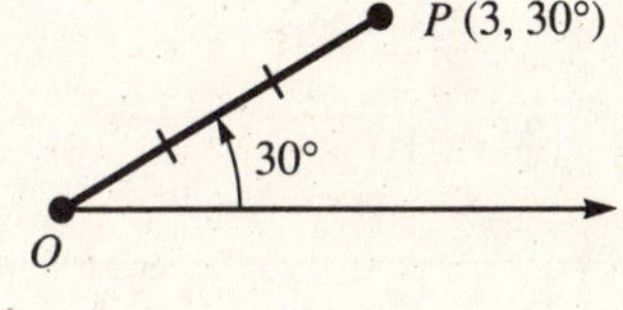

$x = 3\cos 30° = 3\left(\frac{\sqrt{3}}{2}\right) = \frac{3\sqrt{3}}{2}$

$y = 3\sin 30° = 3\left(\frac{1}{2}\right) = \frac{3}{2}$

$\therefore$ P has rectangular coordinates $\left(\frac{3\sqrt{3}}{2}, \frac{3}{2}\right)$.

b. The graph of R is at the right.

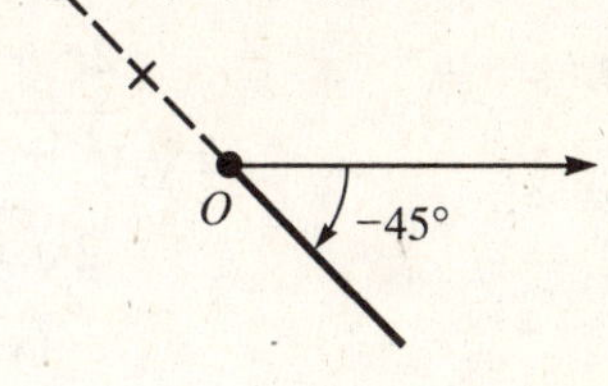

$x = -2\cos(-45°) = -2\left(\frac{\sqrt{2}}{2}\right) = -\sqrt{2}$

$y = -2\sin(-45°) = -2\left(-\frac{\sqrt{2}}{2}\right) = \sqrt{2}$

$\therefore$ R has rectangular coordinates $(-\sqrt{2}, \sqrt{2})$.

Plot the point whose polar coordinates are given and find its rectangular coordinates. Give answers in simplest radical form.

1. $(2, 60°)$ **2.** $(5, -150°)$ **3.** $(-1, 120°)$ **4.** $(-4, -135°)$

Example 2 Convert $A(3, -3)$ to polar coordinates.

Solution $r = \pm\sqrt{x^2 + y^2} = \pm\sqrt{3^2 + (-3)^2} = \pm\sqrt{9 + 9} = \pm\sqrt{18} = \pm 3\sqrt{2}$

If you choose the positive value of r, then the terminal side of θ passes through A and θ is a fourth-quadrant angle. Since $\sin\theta = \frac{y}{r} = \frac{-3}{3\sqrt{2}} = -\frac{\sqrt{2}}{2}$, you can use 315° for θ. $\therefore$ A has polar coordinates $(3\sqrt{2}, 315°)$.

Choosing the negative value of r would result in answers like $(-3\sqrt{2}, 135°)$ or $(-3\sqrt{2}, -225°)$, which are also correct.

NAME ______________________ DATE ______________

14–3 Polar Coordinates *(continued)*

Find a pair of polar coordinates for each pair of rectangular coordinates.

5. $(2, 0)$ **6.** $(0, -1)$ **7.** $(2, -2)$ **8.** $(-2, 2\sqrt{3})$

9. $(\sqrt{5}, -\sqrt{5})$ **10.** $(\sqrt{3}, 1)$ **11.** $(\sqrt{2}, \sqrt{6})$ **12.** $(-1, -1)$

Example 3 Find a polar equation for $x^2 + y^2 = 4x$.

Solution

$$x^2 + y^2 = 4x$$

$$r^2 \cos^2 \theta + r^2 \sin^2 \theta = 4r \cos \theta$$ Substitute $r \cos \theta$ for x and $r \sin \theta$ for y.

$$r^2(\cos^2 \theta + \sin^2 \theta) = 4r \cos \theta$$

$$r^2(1) = 4r \cos \theta$$ Use the Pythagorean identity $\cos^2\theta + \sin^2 \theta = 1$.

$$r^2 = 4r \cos \theta$$

$$r = 4 \cos \theta$$

Find a polar equation for each rectangular-coordinate equation.

13. $y = 3$ **14.** $x = -4$ **15.** $x + y = 0$

16. $x^2 + y^2 = 16$ **17.** $x^2 + y^2 = 2x$ **18.** $x^2 + y^2 - y = 0$

Example 4 Find a rectangular-coordinate equation for $r(2 + \sin \theta) = 1$.

Solution

$$r(2 + \sin \theta) = 1$$

$$2r + r \sin \theta = 1$$

$$2\sqrt{x^2 + y^2} + y = 1$$ Substitute $\sqrt{x^2 + y^2}$ for r and y for $r \sin \theta$.

$$2\sqrt{x^2 + y^2} = 1 - y$$ Isolate the radical.

$$4(x^2 + y^2) = 1 - 2y + y^2$$ Square both sides.

$$4x^2 + 3y^2 + 2y - 1 = 0$$

Find a rectangular-coordinate equation for each polar equation.

19. $r = 3$ **20.** $\theta = 135°$ **21.** $r \cos \theta = -2$

22. $r \sin \theta = 4$ **23.** $r(1 + \cos \theta) = 2$ **24.** $r(1 + \sin \theta) = 1$

Mixed Review Exercises

Find the quotient and remainder.

1. $\dfrac{x^4 - 3x^2 + 2}{x + 2}$ **2.** $\dfrac{4x^3 + 6x^2 + 2}{2x + 1}$ **3.** $\dfrac{8x^3 + 4x^2 - 3x + 1}{x - 1}$

Graph each equation.

4. $x^2 + 4x - y = 3$ **5.** $5x + 6y = 8$ **6.** $x^2 + y^2 + 6x - 2y = 15$

7. $y = \log_3 x$ **8.** $y = -2$ **9.** $4x^2 + y^2 = 16$

NAME ______________________ DATE ____________

14–4 The Geometry of Complex Numbers

Objective: To plot complex numbers in the complex plane and to use the polar form of complex numbers.

Vocabulary

Complex plane The rectangular coordinate system when it is used to represent complex numbers. The horizontal axis is called the *real axis*, and the vertical axis is called the *imaginary axis*. The point (x, y) corresponds to the complex number $x + yi$.

Polar (or **trigonometric**) **form** The *polar*, or *trigonometric*, form of a nonzero complex number $z = x + yi$ is $z = r(\cos\theta + i\sin\theta)$ where $r = \sqrt{x^2 + y^2}$ and θ is an angle such that $\cos\theta = \frac{x}{r}$ and $\sin\theta = \frac{y}{r}$. The number r is called the *absolute value* of z (that is, $r = |z|$) and sometimes the *modulus* of z. Angle θ, chosen so that $0° \le \theta < 360°$, is called the *amplitude*, or *argument*, of z.

Multiplication and division of complex numbers in polar form If $w = a(\cos\alpha + i\sin\alpha)$ and $z = b(\cos\beta + i\sin\beta)$, then:

1. $wz = ab[\cos(\alpha + \beta) + i\sin(\alpha + \beta)]$
2. $\frac{w}{z} = \frac{a}{b}[\cos(\alpha - \beta) + i\sin(\alpha - \beta)]$

Example 1 Plot $w = 2 + 3i$, $z = 3 - i$, $w + z$, and $w - z$ in the complex plane.

Solution

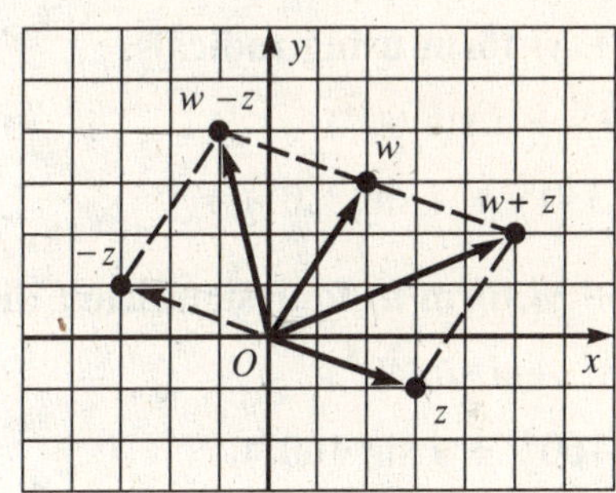

Plot w, z, $w + z$, and $w - z$ in the complex plane.

1. $w = 2 + i$, $z = 3$
2. $w = 4 + 2i$, $z = 4 - 2i$
3. $w = -1 + i$, $z = 3 - i$
4. $w = 3i$, $z = 3 + 2i$
5. $w = 3 + i$, $z = -2 - 2i$
6. $w = -3 - 2i$, $z = -1 + 3i$

Example 2 Let $w = 4(\cos 25° + i\sin 25°)$ and $z = 2(\cos 110° + i\sin 110°)$. Find (a) wz and (b) $\frac{w}{z}$ in polar form with $0° \le \theta < 360°$.

Solution

a. $wz = (4)(2)[\cos(25° + 110°) + i\sin(25° + 110°)]$
$= 8(\cos 135° + i\sin 135°)$

b. $\frac{w}{z} = \frac{4}{2}[\cos(25° - 110°) + i\sin(25° - 110°)]$
$= 2[\cos(-85°) + i\sin(-85°)]$
$= 2(\cos 275° + i\sin 275°)$

NAME ______________________________ DATE ______________

14–4 The Geometry of Complex Numbers *(continued)*

Find wz and $\frac{w}{z}$ in polar form with $0° \le \theta < 360°$.

7. $w = 3(\cos 50° + i \sin 50°)$, $z = 2(\cos 95° + i \sin 95°)$

8. $w = 5(\cos 200° + i \sin 200°)$, $z = 3(\cos 25° + i \sin 25°)$

9. $w = 6(\cos 140° + i \sin 140°)$, $z = 7.2(\cos 70° + i \sin 70°)$

10. $w = 2.5(\cos 280° + i \sin 280°)$, $z = 0.5(\cos 125° + i \sin 125°)$

Example 3 **a.** Write $2(\cos 210° + i \sin 210°)$ in $x + yi$ form using radicals.
b. Write $5(\cos 50° + i \sin 50°)$ in $x + yi$ form to four significant digits.

Solution **a.** $\cos 210° = -\frac{\sqrt{3}}{2}$ and $\sin 210° = -\frac{1}{2}$

$$\therefore\ 2(\cos 210° + i \sin 210°) = 2\left(-\frac{\sqrt{3}}{2} - \frac{1}{2}i\right) = -\sqrt{3} - i$$

b. $\cos 50° = 0.6428$ and $\sin 50° = 0.7660$

$$\therefore\ 5(\cos 50° + i \sin 50°) = 5(0.6428 + 0.7660i) = 3.214 + 3.830i$$

Write in $x + yi$ form using radicals.

11. $2(\cos 60° + i \sin 60°)$

12. $4(\cos 135° + i \sin 135°)$

13. $3.5(\cos 330° + i \sin 330°)$

14. $5.2(\cos 225° + i \sin 225°)$

Write in $x + yi$ form to four significant digits. You may wish to use a calculator.

15. $2(\cos 48° + i \sin 48°)$

16. $7(\cos 80° + i \sin 80°)$

17. $3.1(\cos 160° + i \sin 160°)$

18. $0.4(\cos 310° + i \sin 310°)$

Example 4 Write $-2 + 3i$ in polar form with absolute value in simplest radical form and amplitude to the nearest tenth of a degree.

Solution $r = \sqrt{(-2)^2 + 3^2} = \sqrt{4 + 9} = \sqrt{13}$

$\sin \theta = \frac{3}{\sqrt{13}} = 0.8321$

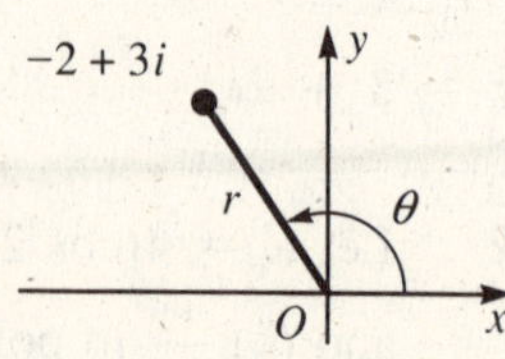

So the reference angle θ is 56.3°.

Since θ is a second-quadrant angle,
$\theta = 180° - 56.3° = 123.7°$

$\therefore\ -2 + 3i = \sqrt{13}(\cos 123.7° + i \sin 123.7°)$

Write in polar form with absolute value in simplest radical form and amplitude to the nearest tenth of a degree.

19. $\sqrt{2} - i\sqrt{2}$

20. $12 - 5i$

21. $3 + 4i$

22. $-6 - i$

NAME ______________________ DATE ____________

14–5 De Moivre's Theorem

Objective: To use De Moivre's theorem.

Vocabulary

De Moivre's theorem If $z = r(\cos\theta + i\sin\theta)$ and n is a positive integer, then $z^n = r^n(\cos n\theta + i\sin n\theta)$.

***n*th roots of unity** The nth roots of the number 1. There are n nth roots of unity, which are given by the formula $\cos\frac{k \cdot 360°}{n} + i\sin\frac{k \cdot 360°}{n}$ $(k = 0, 1, 2, \ldots, n - 1)$.

Example 1 Use De Moivre's theorem to find $(\sqrt{3} - i)^6$. Give the answer in $x + yi$ form.

Solution Let $z = \sqrt{3} - i$. Then write z in polar form.

$|z| = \sqrt{(\sqrt{3})^2 + (-1)^2} = \sqrt{3 + 1} = 2$, so $z = 2\left(\frac{\sqrt{3}}{2} - \frac{1}{2}i\right)$.

Since $\frac{\sqrt{3}}{2} = \cos\theta$ and $-\frac{1}{2} = \sin\theta$, you have $\theta = 330°$ and

$$z = 2(\cos 330° + i\sin 330°).$$

Now use De Moivre's theorem.

$z^6 = 2^6[\cos(6 \cdot 330°) + i\sin(6 \cdot 330°)]$

$= 64[\cos 1980° + i\sin 1980°]$

$= 64[\cos(180° + 5 \cdot 360°) + i\sin(180° + 5 \cdot 360°)]$

$= 64(\cos 180° + i\sin 180°)$ Simplify.

$= 64(-1 + 0i)$ Convert to $x + yi$ form.

$= -64$

$\therefore (\sqrt{3} - i)^6 = -64$

Use De Moivre's theorem to find each power. Give answers in $x + yi$ form.

1. $(1 + i\sqrt{3})^3$ **2.** $(\sqrt{3} - i)^7$ **3.** $(1 + i)^{10}$

4. $(-1 + i)^{12}$ **5.** $(-1 + i\sqrt{3})^6$ **6.** $(-1 - i\sqrt{3})^4$

Example 2 Find the cube roots of $4 - 4i\sqrt{3}$ in polar form.

Solution Write $4 - 4i\sqrt{3}$ in polar form:

$4 - 4i\sqrt{3} = 8\left(\frac{1}{2} - \frac{\sqrt{3}}{2}i\right) =$ Factor out $|4 - 4i\sqrt{3}|$, or 8.

$8(\cos 300° + i\sin 300°)$ $\frac{1}{2} = \cos\theta$ and $-\frac{\sqrt{3}}{2} = \sin\theta$, so $\theta = 300°$.

Let $z = r(\cos\theta + i\sin\theta)$ be a cube root of $4 - 4i\sqrt{3}$. Then $z^3 = 4 - 4i\sqrt{3}$ and by De Moivre's theorem:

$$z^3 = r^3(\cos 3\theta + i\sin 3\theta) = 8(\cos 300° + i\sin 300°)$$

From this equation, $r^3 = 8$, so $r = 2$.

(Solution continues on the next page.)

14–5 De Moivre's Theorem *(continued)*

Also, $\cos 3\theta = \cos 300°$ and $\sin 3\theta = \sin 300°$.
Therefore 3θ differs from 300° by an integral multiple of 360°.

$$3\theta = 300° + k \cdot 360° \text{ (}k\text{ is an integer)}$$

$$\theta = 100° + k \cdot 120° \text{ (}k\text{ is an integer)}$$

Substituting 0, 1, and 2 for k gives $\theta = 100°$, $\theta = 100° + 120° = 220°$, and $\theta = 100° + 240° = 340°$. (Other values of k give angles coterminal with one of these angles.)

$\therefore 4 - 4i\sqrt{3}$ has these three cube roots:

$$2(\cos 100° + i \sin 100°)$$
$$2(\cos 220° + i \sin 220°)$$
$$2(\cos 340° + i \sin 340°)$$

Example 3 Find the cube roots of unity in polar form.

Solution Use the formula $\cos \dfrac{k \cdot 360°}{n} + i \sin \dfrac{k \cdot 360°}{n}$ where $n = 3$ and $k = 0, 1,$ and 2.

$$\cos \frac{0 \cdot 360°}{3} + i \sin \frac{0 \cdot 360°}{3} = \cos 0° + i \sin 0°$$

$$\cos \frac{1 \cdot 360°}{3} + i \sin \frac{1 \cdot 360°}{3} = \cos 120° + i \sin 120°$$

$$\cos \frac{2 \cdot 360°}{3} + i \sin \frac{2 \cdot 360°}{3} = \cos 240° + i \sin 240°$$

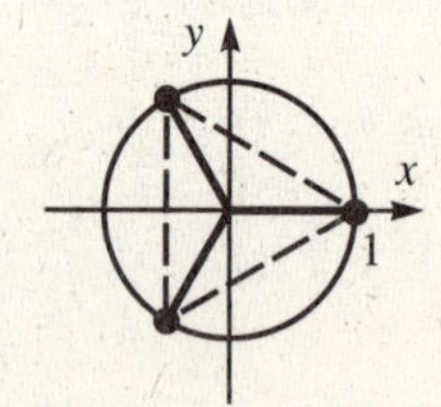

As shown in the diagram, the cube roots of unity are evenly spaced around the unit circle. Therefore, they are the vertices of an equilateral triangle inscribed in the circle.

Find the required roots in polar form.

7. The square roots of $-2\sqrt{3} + 2i$

8. The cube roots of $-1 - i$

9. The fourth roots of $8 + 8i\sqrt{3}$

10. The fifth roots of unity

11. The sixth roots of unity

12. The 12th roots of unity

Mixed Review Exercises

Let $\mathbf{u} = 2\mathbf{i} + 3\mathbf{j}$, $\mathbf{v} = \mathbf{i} - 4\mathbf{j}$, and $\mathbf{w} = \mathbf{u} + \mathbf{v}$.

1. Find **w** in component form.

2. Draw a diagram showing **u, v,** and **w**.

3. Find $\|\mathbf{w}\|$.

4. Find $\mathbf{u} \cdot \mathbf{v}$.

Plot each pair of polar coordinates and find the rectangular coordinates.

5. (2, 30°) **6.** (3, −45°) **7.** (−1, 150°) **8.** (−4, −120°)

NAME ______________________ DATE ______________

14–6 The Inverse Cosine and Inverse Sine

Objective: To evaluate expressions involving the inverse cosine and inverse sine.

Vocabulary

Inverse cosine Because the cosine function is not one-to-one $\left(\text{for example, } \cos\frac{\pi}{3} = \frac{1}{2} = \cos\left(-\frac{\pi}{3}\right)\right)$, cosine does *not* have an inverse. By restricting the domain of cosine to the interval $0 \le x \le \pi$, however, a new function called Cosine—which *does* have an inverse—is created. The inverse of Cosine is called *inverse cosine* and is denoted by Cos^{-1}. (It is also sometimes called *Arc cosine*, denoted by Arccos.) The definition of inverse cosine can be stated as follows:

$$y = \text{Cos}^{-1} x \text{ if and only if } \cos y = x \text{ and } 0 \le y \le \pi$$

The graph of inverse cosine is shown at the right.

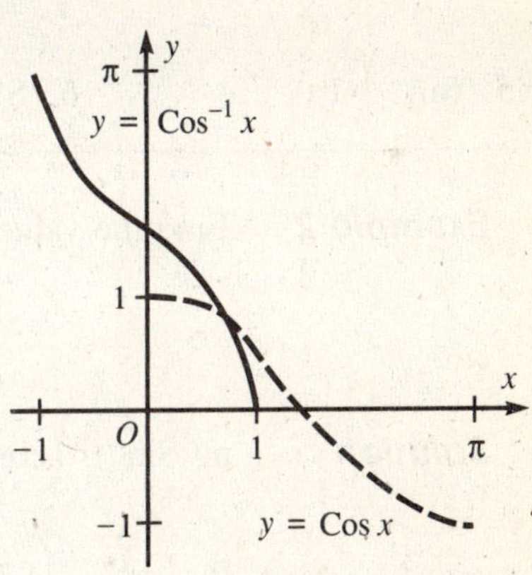

Inverse sine Because the sine function is not one-to-one $\left(\text{for example, } \sin\frac{\pi}{6} = \frac{1}{2} = \sin\frac{5\pi}{6}\right)$, sine does *not* have an inverse. By restricting the domain of sine to the interval $-\frac{\pi}{2} \le x \le \frac{\pi}{2}$, however, a new function called Sine—which *does* have an inverse—is created. The inverse of Sine is called *inverse sine* and is denoted by Sin^{-1}. (It is also sometimes called *Arc sine*, denoted by Arcsin.) The definition of inverse sine can be stated as follows:

$$y = \text{Sin}^{-1} x \text{ if and only if } \sin y = x \text{ and } -\frac{\pi}{2} \le y \le \frac{\pi}{2}$$

The graph of inverse sine is shown at the right.

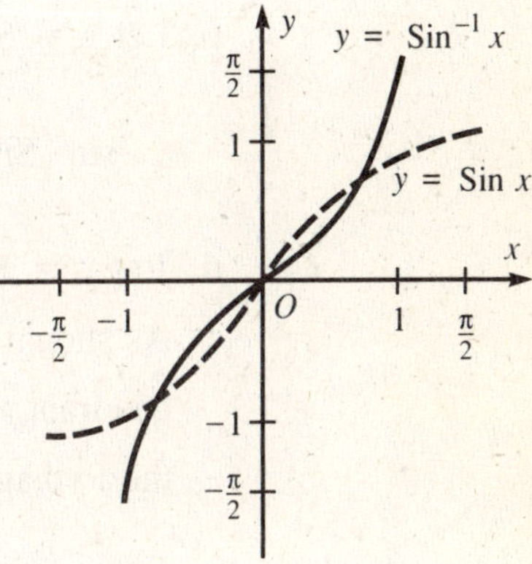

CAUTION $\text{Cos}^{-1} x$ does *not* mean $\frac{1}{\text{Cos } x}$, and $\text{Sin}^{-1} x$ does *not* mean $\frac{1}{\text{Sin } x}$.

Example 1 Find the value: **a.** $\text{Cos}^{-1}\frac{\sqrt{2}}{2}$ **b.** $\text{Cos}^{-1}\left(-\frac{1}{2}\right)$ **c.** $\text{Sin}^{-1}\left(-\frac{\sqrt{3}}{2}\right)$

Solution

a. Let $y = \text{Cos}^{-1}\frac{\sqrt{2}}{2}$. Then $\cos y = \frac{\sqrt{2}}{2}$ and $0 \le y \le \pi$.

So $y = \frac{\pi}{4}$ and therefore $\text{Cos}^{-1}\frac{\sqrt{2}}{2} = \frac{\pi}{4}$.

b. Let $y = \text{Cos}^{-1}\left(-\frac{1}{2}\right)$. Then $\cos y = -\frac{1}{2}$ and $0 \le y \le \pi$.

So $y = \frac{2\pi}{3}$ and therefore $\text{Cos}^{-1}\left(-\frac{1}{2}\right) = \frac{2\pi}{3}$.

c. Let $y = \text{Sin}^{-1}\left(-\frac{\sqrt{3}}{2}\right)$. Then $\sin y = -\frac{\sqrt{3}}{2}$ and $-\frac{\pi}{2} \le y \le \frac{\pi}{2}$.

So $y = -\frac{\pi}{3}$ and therefore $\text{Sin}^{-1}\left(-\frac{\sqrt{3}}{2}\right) = -\frac{\pi}{3}$.

NAME ______________________ DATE ______________

14–6 The Inverse Cosine and Inverse Sine *(continued)*

Find the value of each of the following.

1. $\text{Cos}^{-1}\frac{\sqrt{3}}{2}$ **2.** $\text{Sin}^{-1}\frac{1}{2}$ **3.** $\text{Cos}^{-1}\frac{\sqrt{2}}{2}$ **4.** $\text{Cos}^{-1} 0$

5. $\text{Sin}^{-1} 1$ **6.** $\text{Sin}^{-1}\left(-\frac{\sqrt{2}}{2}\right)$ **7.** $\text{Cos}^{-1}\frac{1}{2}$ **8.** $\text{Sin}^{-1}(-1)$

Example 2 Find the value: **a.** $\text{Sin}^{-1}\left(\cos\frac{7\pi}{6}\right)$ **b.** $\text{Cos}^{-1}\left(\cos\frac{5\pi}{4}\right)$

c. $\sin\left(\text{Sin}^{-1}\frac{2}{5}\right)$ **d.** $\cos\left[\text{Sin}^{-1}\left(-\frac{1}{3}\right)\right]$

Solution **a.** $\text{Sin}^{-1}\left(\cos\frac{7\pi}{6}\right) = \text{Sin}^{-1}\left(-\frac{\sqrt{3}}{2}\right) = -\frac{\pi}{3}$

b. $\text{Cos}^{-1}\left(\cos\frac{5\pi}{4}\right) = \text{Cos}^{-1}\left(-\frac{\sqrt{2}}{2}\right) = \frac{3\pi}{4}$

c. Let $y = \text{Sin}^{-1}\frac{2}{5}$. Then by the definition of the inverse sine, $\sin y = \frac{2}{5}$.

$\therefore \sin\left(\text{Sin}^{-1}\frac{2}{5}\right) = \frac{2}{5}$

d. Let $y = \text{Sin}^{-1}\left(-\frac{1}{3}\right)$. Then $\sin y = -\frac{1}{3}$ and $-\frac{\pi}{2} \le y \le \frac{\pi}{2}$. As shown at the right below, you might make a sketch of y as a fourth-quadrant angle having a sine of $-\frac{1}{3}$. Since you want to find $\cos y$, use the Pythagorean identity $\sin^2 y + \cos^2 y = 1$:

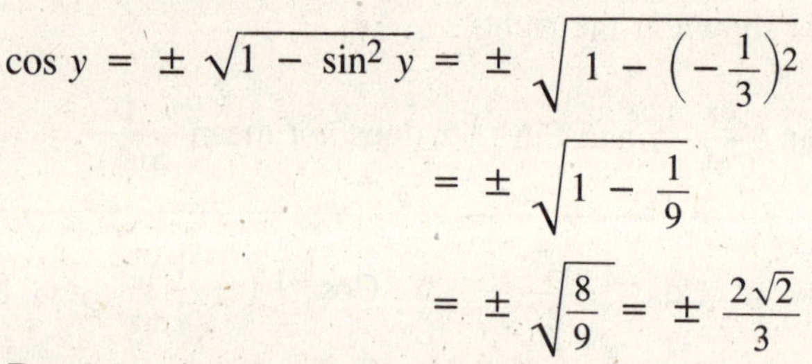

$$\cos y = \pm\sqrt{1 - \sin^2 y} = \pm\sqrt{1 - \left(-\frac{1}{3}\right)^2}$$
$$= \pm\sqrt{1 - \frac{1}{9}}$$
$$= \pm\sqrt{\frac{8}{9}} = \pm\frac{2\sqrt{2}}{3}$$

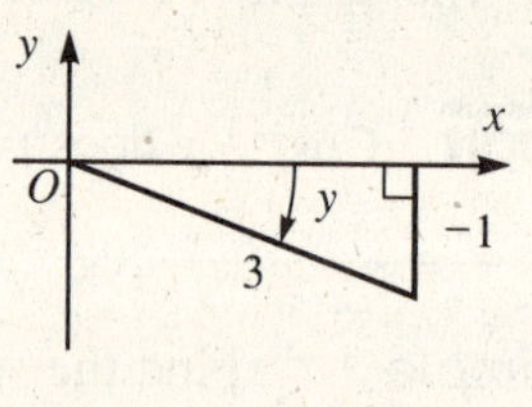

Because y is a fourth-quadrant angle, $\cos y$ is positive.

$\therefore \cos\left[\text{Sin}^{-1}\left(-\frac{1}{3}\right)\right] = \frac{2\sqrt{2}}{3}$

Find the value of each of the following.

9. $\text{Sin}^{-1}\left(\sin\frac{5\pi}{4}\right)$ **10.** $\sin\left[\text{Cos}^{-1}\left(-\frac{1}{2}\right)\right]$ **11.** $\text{Cos}^{-1}\left[\cos\left(-\frac{2\pi}{3}\right)\right]$

12. $\cos\left[\text{Sin}^{-1}\left(-\frac{\sqrt{2}}{2}\right)\right]$ **13.** $\text{Sin}^{-1}\left(\cos\frac{7\pi}{6}\right)$ **14.** $\text{Cos}^{-1}\left(\sin\frac{\pi}{6}\right)$

15. $\text{Sin}^{-1}\left(\cos\frac{5\pi}{3}\right)$ **16.** $\text{Cos}^{-1}\left(\cos\frac{\pi}{3}\right)$ **17.** $\sin\left(\text{Cos}^{-1}\frac{12}{13}\right)$

18. $\cos\left[\text{Sin}^{-1}\left(-\frac{4}{5}\right)\right]$ **19.** $\cos\left(\text{Sin}^{-1}\frac{8}{17}\right)$ **20.** $\sin\left[\text{Cos}^{-1}\left(-\frac{3}{4}\right)\right]$

NAME ______________________ DATE ______________

14–7 Other Inverse Functions

Objective: To evaluate expressions involving the inverse trigonometric functions.

Vocabulary

Inverse tangent $y = \text{Tan}^{-1} x$ if and only if $\tan y = x$ and $-\frac{\pi}{2} < y < \frac{\pi}{2}$.

Inverse cotangent $y = \text{Cot}^{-1} x$ if and only if $\cot y = x$ and $0 < y < \pi$.

Relationship between inverse tangent and inverse cotangent If $x > 0$, then $\text{Cot}^{-1} x = \text{Tan}^{-1} \frac{1}{x}$.

Inverse secant If $|x| \geq 1$, then $\text{Sec}^{-1} x = \text{Cos}^{-1} \frac{1}{x}$.

Inverse cosecant If $|x| \geq 1$, then $\text{Csc}^{-1} x = \text{Sin}^{-1} \frac{1}{x}$.

Example 1 Find the value: **a.** $\text{Tan}^{-1}(-\sqrt{3})$ **b.** $\text{Cot}^{-1} 1$ **c.** $\text{Sec}^{-1}(-2)$

Solution

a. Let $y = \text{Tan}^{-1}(-\sqrt{3})$. Then $\tan y = -\sqrt{3}$ and $-\frac{\pi}{2} < y < \frac{\pi}{2}$.

So $y = -\frac{\pi}{3}$ and therefore $\text{Tan}^{-1}(-\sqrt{3}) = -\frac{\pi}{3}$.

b. Let $y = \text{Cot}^{-1} 1$. Then $\cot y = 1$ and $0 < y < \pi$.

So $y = \frac{\pi}{4}$ and therefore $\text{Cot}^{-1} 1 = \frac{\pi}{4}$.

c. Use $\text{Sec}^{-1} x = \text{Cos}^{-1} \frac{1}{x}$: $\text{Sec}^{-1}(-2) = \text{Cos}^{-1}\left(-\frac{1}{2}\right)$

Let $y = \text{Cos}^{-1}\left(-\frac{1}{2}\right)$. Then $\cos y = -\frac{1}{2}$ and $0 \leq y \leq \pi$.

So $y = \frac{2\pi}{3}$ and therefore $\text{Sec}^{-1}(-2) = \frac{2\pi}{3}$.

Find the value of each of the following.

1. $\text{Tan}^{-1} 0$
2. $\text{Csc}^{-1} 1$
3. $\text{Sec}^{-1}\left(-\frac{2\sqrt{3}}{3}\right)$
4. $\text{Csc}^{-1} \frac{2\sqrt{3}}{3}$
5. $\text{Sec}^{-1} \sqrt{2}$
6. $\text{Tan}^{-1} \sqrt{3}$
7. $\text{Sec}^{-1}(-\sqrt{2})$
8. $\text{Cot}^{-1}\left(-\frac{\sqrt{3}}{3}\right)$
9. $\text{Csc}^{-1}(-\sqrt{2})$

Example 2 Find the value: **a.** $\cos(\text{Tan}^{-1} 1)$ **b.** $\sin(\text{Sec}^{-1} 2)$ **c.** $\text{Cot}^{-1}\left(\cot \frac{5\pi}{3}\right)$ **d.** $\cos\left(\text{Csc}^{-1} \frac{5}{2}\right)$

Solution **a.** $\cos(\text{Tan}^{-1} 1) = \cos \frac{\pi}{4} = \frac{\sqrt{2}}{2}$

(Solution continues on the next page.)

NAME ______________________ DATE ____________

14–7 Other Inverse Functions *(continued)*

b. Use $\text{Sec}^{-1} x = \text{Cos}^{-1} \frac{1}{x}$: $\text{Sec}^{-1} 2 = \text{Cos}^{-1} \frac{1}{2} = \frac{\pi}{3}$

$\therefore \sin(\text{Sec}^{-1} 2) = \sin \frac{\pi}{3} = \frac{\sqrt{3}}{2}$

c. $\text{Cot}^{-1}\left(\cot \frac{5\pi}{3}\right) = \text{Cot}^{-1}\left(-\frac{\sqrt{3}}{3}\right)$

Let $y = \text{Cot}^{-1}\left(-\frac{\sqrt{3}}{3}\right)$. Then $\cot y = -\frac{\sqrt{3}}{3}$ and $0 < y < \pi$. So $y = \frac{2\pi}{3}$.

$\therefore \text{Cot}^{-1}\left(\cot \frac{5\pi}{3}\right) = \frac{2\pi}{3}$

d. Use $\text{Csc}^{-1} x = \text{Sin}^{-1} \frac{1}{x}$: $\text{Csc}^{-1}\left(\frac{5}{2}\right) = \text{Sin}^{-1}\left(\frac{2}{5}\right)$

Let $y = \text{Sin}^{-1}\left(\frac{2}{5}\right)$. Then $\sin y = \frac{2}{5}$ and $-\frac{\pi}{2} \le y \le \frac{\pi}{2}$.

As shown at the right below, you might make a sketch of y as a first-quadrant angle having a sine of $\frac{2}{5}$. Since you want to find $\cos y$, use the Pythagorean identity $\sin^2 y + \cos^2 y = 1$:

$$\cos y = \pm\sqrt{1 - \sin^2 y} = \pm\sqrt{1 - \left(\frac{2}{5}\right)^2}$$
$$= \pm\sqrt{1 - \frac{4}{25}}$$
$$= \pm\sqrt{\frac{21}{25}} = \pm\frac{\sqrt{21}}{5}$$

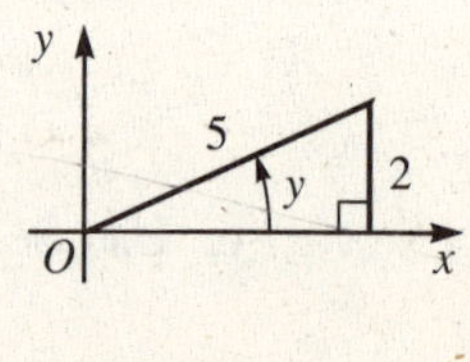

Because y is a first-quadrant angle, $\cos y$ is positive.

$\therefore \cos\left(\text{Csc}^{-1} \frac{5}{2}\right) = \frac{\sqrt{21}}{5}$

Find the value of each of the following.

10. $\tan\left(\text{Sec}^{-1} \frac{2\sqrt{3}}{3}\right)$ **11.** $\csc(\text{Csc}^{-1} 1)$ **12.** $\csc[\text{Cot}^{-1}(-1)]$

13. $\text{Sec}^{-1}\left[\sec\left(-\frac{\pi}{6}\right)\right]$ **14.** $\text{Tan}^{-1}\left(\cot \frac{3\pi}{4}\right)$ **15.** $\text{Csc}^{-1}\left(\tan \frac{\pi}{4}\right)$

16. $\tan\left(\text{Cot}^{-1} \frac{4}{3}\right)$ **17.** $\sin\left[\text{Sec}^{-1}\left(-\frac{5}{3}\right)\right]$ **18.** $\cos\left[\text{Tan}^{-1}\left(-\frac{5}{12}\right)\right]$

Mixed Review Exercises

Determine whether each sequence is arithmetic or geometric. Then find the specified term.

1. 5, 9, 13, 17, . . . ; t_{20} **2.** $-4, -2, -1, -\frac{1}{2}, \ldots; t_9$ **3.** $3, -8, -19, -30, \ldots; t_{18}$

Write each series using sigma notation. Then find the sum.

4. $3 + 6 + 12 + \cdots + 384$ **5.** $3 + 7 + 11 + \cdots + 99$ **6.** $27 + 9 + 3 + \cdots$

NAME ______________________ DATE ______________

14–8 Trigonometric Equations

Objective: To solve trigonometric equations.

Vocabulary

Strategies for solving trigonometric equations

1. Use trigonometric identities so that there is only one trigonometric function in the equation.
2. Use algebraic transformations (such as adding the same number or expression to both sides of an equation) and algebraic methods (such as factoring).

General solutions Formulas that name all possible solutions of a trigonometric equation. Example: The general solution of $\sin x = 0$ is $x = k\pi$ where k is an integer.

Primary solutions Solutions of trigonometric equations in a specified interval, usually $0 \le x < 2\pi$ or $0° \le \theta < 360°$. Example: The primary solutions of $\sin x = 0$ in the interval $0 \le x < 2\pi$ are $x = 0, \pi$.

Example 1 Solve $\cos 2\theta = -\cos\theta$ for θ, $0° \le \theta < 360°$.

Solution Of the three formulas for $\cos 2\theta$, it is best to use the identity $\cos 2\theta = 2\cos^2\theta - 1$ so that $\cos\theta$ is the only function in the transformed equation.

$$\cos 2\theta = -\cos\theta$$
$$2\cos^2\theta - 1 = -\cos\theta$$
$$2\cos^2\theta + \cos\theta - 1 = 0 \quad \text{Make one side equal 0.}$$
$$(2\cos\theta - 1)(\cos\theta + 1) = 0 \quad \text{Factor and solve.}$$

$$2\cos\theta - 1 = 0 \quad \text{or} \quad \cos\theta + 1 = 0$$
$$\cos\theta = \frac{1}{2} \qquad\qquad \cos\theta = -1$$
$$\theta = 60°, 300° \qquad\qquad \theta = 180°$$

$\therefore$ the solution set over the interval $0° \le \theta < 360°$ is $\{60°, 180°, 300°\}$.

Example 2 Solve $\tan\theta = 2\sin\theta$. Find (a) the solution in the interval $0° \le \theta < 360°$ and (b) the formulas giving general solutions.

Solution **a.**

$$\tan\theta = 2\sin\theta$$
$$\frac{\sin\theta}{\cos\theta} = 2\sin\theta \quad \text{Write tangent in terms of sine and cosine.}$$
$$\sin\theta = 2\sin\theta\cos\theta \quad \text{Multiply both sides by } \cos\theta.$$
$$\sin\theta - 2\sin\theta\cos\theta = 0 \quad \text{Make one side equal 0.}$$
$$\sin\theta\,(1 - 2\cos\theta) = 0 \quad \text{Factor and solve.}$$

$$\sin\theta = 0 \quad \text{or} \quad 1 - 2\cos\theta = 0$$
$$\theta = 0°, 180° \qquad\qquad \cos\theta = \frac{1}{2}$$
$$\theta = 60°, 300°$$

$\therefore$ the solution set over the interval $0° \le \theta < 360°$ is $\{0°, 60°, 180°, 300°\}$.

(Solution continues on the next page.)

14–8 Trigonometric Equations *(continued)*

b. To find formulas giving general solutions, you should recognize that all other solutions differ from the primary solutions by integral multiples of 360°, the period of the sine and cosine functions. Therefore, all solutions are of this form:

$\theta = 0° + k \cdot 360°$, k is an integer
$\theta = 180° + k \cdot 360°$, k is an integer
} or $\theta = 0° + k \cdot 180°$, k is an integer

$\theta = 60° + k \cdot 360°$, k is an integer

$\theta = 300° + k \cdot 360°$, k is an integer

$\therefore$ the set of general solutions is:

$\{\theta: \theta = 0° + k \cdot 180°, \theta = 60° + k \cdot 360°, \text{ or } \theta = 300° + k \cdot 360°\}$.

Example 3 Solve $3 \sin x \sec x = \sec x$ for x, $0 \le x < 2\pi$.

Solution

$3 \sin x \sec x = \sec x$

$3 \sin x \sec x - \sec x = 0$ Make one side equal 0.

$(3 \sin x - 1) \sec x = 0$ Factor and solve.

$3 \sin x - 1 = 0$ or $\sec x = 0$

$3 \sin x = 1$

$\sin x = \frac{1}{3} \approx 0.3333$

To the nearest hundredth of a radian, a calculator or table gives

$x = 0.34$.

Since sine takes on a value of $\frac{1}{3}$ in the *second* (as well as the first) quadrant, another solution in the given interval is $x = \pi - 0.34$ or $x = 3.14 - 0.34 = 2.80$.

Since the secant function never takes on a value of 0 (that is, 0 is not in the range of the secant function), the equation $\sec x = 0$ has no solution.

$\therefore$ the solution set over the interval $0 \le x < 2\pi$ is $\{0.34, 2.80\}$.

Find the primary solutions of each equation for $0° \le \theta < 360°$ or $0 \le x < 2\pi$. Then find the formulas giving the general solution.

Give your answers to the nearest tenth of a degree if the variable is θ. Give your answers to the nearest hundredth of a radian if the variable is x, unless you can express the answers in terms of π.

1. $\sec x = 1$

2. $2 \sin \theta = -\sqrt{3}$

3. $2 \cos x - \sqrt{2} = 0$

4. $4 \sin^2 x = 1$

5. $\sin \theta \cos \theta = 0$

6. $\cot \theta = \cos \theta$

7. $\tan\left(x + \frac{\pi}{4}\right) = 0$

8. $\sqrt{3} \sec (\theta - 10°) = 2$

9. $\tan x = 4 \cot x$

10. $\sin 2\theta = -\sin \theta$

11. $\cos^2 x + \cos x = 0$

12. $\sin^2 \theta + 3 \sin \theta + 2 = 0$

13. $2 + \tan^2 x = 2 \sec x$

14. $5 \cos x = 6 \sin^2 x$

15. $2 \cos x = -\sin x$

NAME ______________________ DATE ______________

15 Statistics and Probability

15-1 Presenting Statistical Data

Objective: To display data using frequency distributions, histograms, and stem-and-leaf plots, and to compute measures of central tendency.

Vocabulary

Frequency distribution An organized listing of data that shows how many times each item or group of items occurs.

Histogram A method of displaying a frequency distribution using bars.

Stem-and-leaf plot A way of displaying data and including the data in the display.

Mode The number or numbers that occur most frequently in a set of data.

Median The middle number (when the count of the numbers is odd) or the arithmetic average of the two middle numbers (when the count is even) in a set of *ordered* data.

Mean Denoted by M, the mean is the arithmetic average of all the numbers in a set of data.

Example 1 Thirty high school students with after-school or weekend jobs were asked to give the number of hours per week they work. The results of this survey appear in the histogram below. Use the histogram to answer the questions.

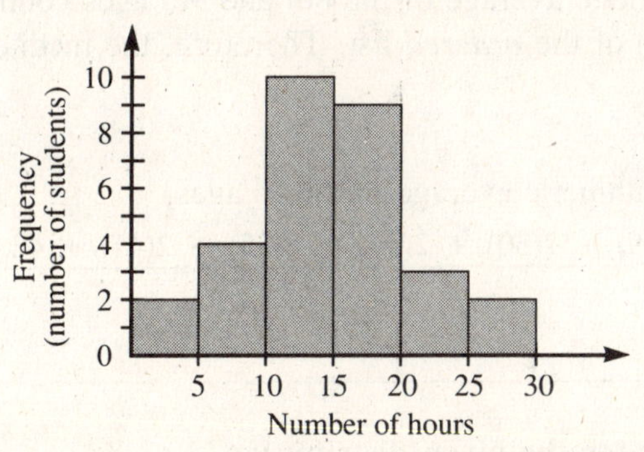

Note: In this histogram, each boundary number on the horizontal axis belongs to the bar on the right. For example, the bar over the interval 10–15 includes those students who work from 10 to 15 hours, including 10 but *not* including 15.

a. How many students work at least 10 but fewer than 15 hours per week?

b. How many students work at least 20 hours per week?

Solution **a.** The height of the bar over the interval 10–15 gives the number of students.
$\therefore$ 10 students work at least 10 but fewer than 15 hours per week.

b. Add the frequencies for the intervals 20–25 and 25–30: $3 + 2 = 5$
$\therefore$ 5 students work at least 20 hours per week.

Use the histogram of Example 1 to answer the following questions.

1. How many students work at least 15 but fewer than 20 hours per week?

2. How many students work less than 10 hours per week?

15–1 Presenting Statistical Data *(continued)*

Example 2 Draw a stem-and-leaf plot for the following distribution of ages:
26, 30, 29, 41, 35, 26, 34, 29, 35, 30, 25, 42, 26, 34, 41, 35

Solution You first obtain the *stem* by using only the tens' digit of each number. These are written to the left of a vertical line. Then, for each age, you record the *leaf*, or units' digit, to the right of the stem. After making an unordered stem-and-leaf plot (as shown at the left below), you should reorder the elements of each row to complete the plot (as shown at the right below).

Stem	Leaf
2	6, 9, 6, 9, 5, 6
3	0, 5, 4, 5, 0, 4, 5
4	1, 2, 1

Stem	Leaf
2	5, 6, 6, 6, 9, 9
3	0, 0, 4, 4, 5, 5, 5
4	1, 1, 2

Example 3 Use the distribution of ages given in Example 2 to find (a) the mode, (b) the median, and (c) the mean.

Solution

a. Look for the numbers that occur most frequently. The ages 26 and 35 both occur three times, more often than any other age. Therefore, the distribution has two modes, 26 and 35.

b. Since there are 16 numbers in the distribution, there is no middle number. The median is the arithmetic average of the 8th and 9th ages counting from the top (or bottom) of the *ordered* list. Therefore, the median is $\frac{30 + 34}{2}$, or 32.

c. Since the mean is the arithmetic average of all 16 ages,

$$M = \frac{25 + 3(26) + 2(29) + 2(30) + 2(34) + 3(35) + 2(41) + 42}{16}$$
$$\approx 32.4$$

In Exercises 3–6, draw a stem-and-leaf plot for the given distribution.

3. 37, 41, 41, 42, 38, 38, 40, 41, 39, 39, 42, 37, 36

4. 61, 66, 92, 98, 72, 76, 79, 83, 92, 91, 85, 86, 76

5. 3, 7, 11, 5, 21, 27, 4, 8, 7, 21, 15, 17, 19, 21, 21, 8

6. 117, 115, 136, 142, 140, 141, 135, 135, 110, 153, 156, 147, 156, 115

7–10. Find (a) the mode, (b) the median, and (c) the mean for each distribution given in Exercises 3–6.

Mixed Review Exercises

Write in simplest form without negative exponents.

1. $\left(\frac{x^4y^{-2}}{x^{-3}y}\right)^{-3}\left(\frac{xy^{-1}}{x^2y}\right)$ **2.** $\frac{x^2 - 7x + 10}{x^3 - 125}$ **3.** $(x^{-1} + y^{-1})^{-1}(x + y)$ **4.** $9^{\sqrt{2}} \cdot 3^{\sqrt{2}}$

NAME ______________________ DATE ____________

15–2 Analyzing Statistical Data

Objective: To compute measures of dispersion and, together with measures of central tendency, to describe and compare distributions using these statistics.

Vocabulary

First quartile The median of the lower half of the data in a set of ordered data.

Third quartile The median of the upper half of the data in a set of ordered data.

Range The difference between the largest and smallest numbers in a set of data.

Box-and-whisker plot A diagram used to show the median, the first and third quartiles, and the range of a distribution.

Measures of central tendency The mean, the median, and the mode, each of which indicates a center of a distribution.

Variance The arithmetic average of the squares of the differences between the n numbers $x_1, x_2, \ldots, x_n$ and the mean M of a distribution:

$$\text{variance} = \frac{\text{sum of the squares of the deviations from the mean}}{\text{number of elements in the distribution}}$$

$$= \frac{(x_1 - M)^2 + (x_2 - M)^2 + \cdots + (x_n - M)^2}{n}$$

Standard deviation Denoted by the Greek letter σ (sigma), the standard deviation is the principal square root of the variance: $\sigma = \sqrt{\text{variance}}$.

Measures of dispersion The range, the first and third quartiles, the variance, and the standard deviation, each of which indicates how scattered a distribution is.

Example 1 For the distribution of children's weights (in pounds) shown in the stem-and-leaf plot at the right, find (a) the median, (b) the first quartile, (c) the third quartile, and (d) the range.

	Stem	Leaves	
low end →	3	2, 4, 6	
	4	5, 7, 8, 9	
	5	1, 2, 2, 8, 9, 9	
	6	0, 5, 7, 8	
	7	0, 1, 2, 5	← *high end*

Solution

a. With 21 weights in the distribution, the median is the 11th weight counting from either end: 58.

b. With 11 weights (including the median) in the lower half of the distribution, the first quartile is the 6th weight counting from the low end: 48.

c. With 11 weights (including the median) in the upper half of the distribution, the third quartile is the 6th weight counting from the high end: 67.

d. The difference between the largest and smallest weights is 75 − 32, or 43.

For each stem-and-leaf plot find (a) the median, (b) the first quartile, (c) the third quartile, and (d) the range.

1.

Stem	Leaves
1	1, 5, 6
2	5, 5, 7, 7, 8
3	1, 2, 3, 4, 5, 6
4	0, 2, 3

2.

Stem	Leaves
5	1, 8, 9
6	2, 2, 3, 4, 7, 9
7	3, 3, 4, 6, 8
8	6, 7

NAME ______________________ DATE ______________

15–2 Analyzing Statistical Data (continued)

Example 2 Draw a box-and-whisker plot for the distribution in Example 1.

Solution 1. First, using dots below a number line, show the median (Q_2), the first and third quartiles (Q_1 and Q_3), and the least and greatest weights (L and G).

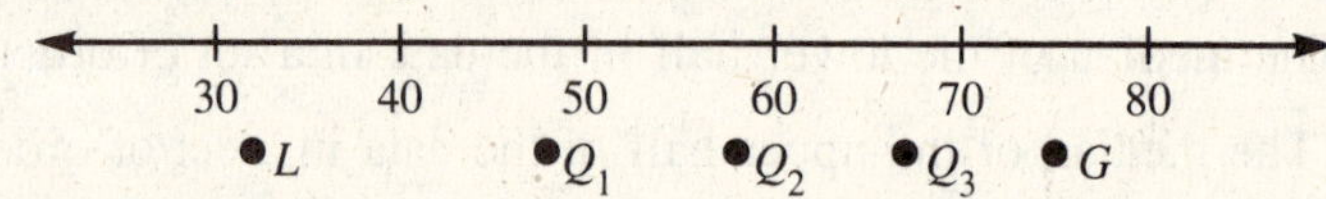

2. Next draw a narrow rectangular box with its shorter sides containing the two quartile dots. Then draw a line segment through the median dot parallel to the shorter sides. Finally, draw line segments, or "whiskers," from the first and third quartile dots to the least- and greatest-weight dots, respectively.

3–4. Draw a box-and-whisker plot for each of the distributions in Exercises 1 and 2 on page 241.

Example 3 The stem-and-leaf plot at the right shows the distribution of scores on an algebra test. Find (a) the mean, (b) the variance, and (c) the standard deviation for the distribution.

7	2, 3, 3, 3
8	0, 0, 0, 5, 5
9	3, 5, 5

Solution Use a table like the one shown below to organize your computations.

Score	Frequency	Score × Frequency	Deviation from the mean M	(Deviation)^2	(Deviation)^2 × Frequency
72	1	72	$72 - 82 = -10$	100	100
73	3	219	$73 - 82 = -9$	81	243
80	3	240	$80 - 82 = -2$	4	12
85	2	170	$85 - 82 = 3$	9	18
93	1	93	$93 - 82 = 11$	121	121
95	2	190	$95 - 82 = 13$	169	338
Sums	12	984			832

a. After completing the first three columns, you can compute the mean: $M = \frac{984}{12} = 82$

b. After completing the last three columns, you can compute the variance: variance $= \frac{832}{12} \approx 69.3$

c. Use the variance to compute the standard deviation: $\sigma \approx \sqrt{69.3} \approx 8.3$

For each distribution in Exercises 5–7 find (a) the mean, (b) the variance, and (c) the standard deviation to the nearest tenth.

5. 2, 3, 5, 5, 7, 8, 8, 10 **6.** 2, 3, 3, 5, 5, 5, 7, 10 **7.** 20, 22, 23, 23, 23, 24, 24, 25

NAME ______________________ DATE ______________

15–3 The Normal Distribution

Objective: To recognize and analyze normal distributions.

Vocabulary

Normal distribution A distribution for which the bulk of the data is clustered about the mean, with the remainder tapering off away from the mean. A histogram of a normal distribution has a characteristic "bell-shaped" appearance.

Standard normal distribution A normal distribution with mean equal to zero and standard deviation equal to one.

Standard normal curve The bell-shaped curve that represents standard normal distribution. The standard normal curve has the following properties:

1. It is symmetric with respect to the y-axis.
2. As $|x|$ increases, the curve gets closer and closer to the x-axis.
3. The total area under the curve and above the x-axis is equal to 1.

Area under the Standard Normal Curve for $0 \le x \le 4.0$

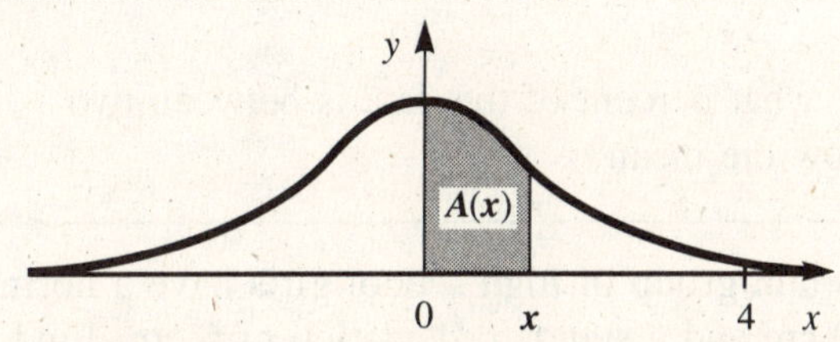

x	Area, $A(x)$	x	Area, $A(x)$	x	Area, $A(x)$
0.0	0.0000	1.4	0.4192	2.8	0.4974
0.2	0.0793	1.6	0.4452	3.0	0.4987
0.4	0.1554	1.8	0.4641	3.2	0.4993
0.6	0.2257	2.0	0.4772	3.4	0.4997
0.8	0.2881	2.2	0.4861	3.6	0.4998
1.0	0.3413	2.4	0.4918	3.8	0.4999
1.2	0.3849	2.6	0.4953	4.0	0.5000

Standardized value The number x obtained from the number z, a given value from a normal distribution with mean M and standard deviation σ, using the formula:

$$x = \frac{z - M}{\sigma}$$

Example 1 In a standard normal distribution, what fraction of the data is between 2.0 and 2.4?

Solution The required fraction is equal to the area under the standard normal curve bounded by $x = 2.0$ and $x = 2.4$. This is the difference between the area from 0 to 2.4 and the area from 0 to 2.0. Therefore:

$$\begin{aligned} A(2.0 \le x \le 2.4) &= A(2.4) - A(2.0) \\ &= 0.4918 - 0.4772 \\ &= 0.0146 \end{aligned}$$

15–3 The Normal Distribution *(continued)*

Example 2 In a standard normal distribution, what percent of the data is within one standard deviation of the mean?

Solution The mean is 0 and the standard deviation is 1. Since the standard normal curve is symmetric with respect to the y-axis, the area under the curve between -1 and 1 is twice the area between 0 and 1. Therefore:

$$\begin{aligned} A(-1 \le x \le 1) &= 2 \cdot A(1) \\ &= 2(0.3413) \\ &= 0.6826, \text{ or } 68.26\% \end{aligned}$$

In Exercises 1–3, use the table on page 243.

1. In a standard normal distribution, what fraction of the data is between the mean and 0.6?
2. In a standard normal distribution, what percent of the data is between -0.4 and 0.4?
3. In a standard normal distribution, what percent of the data is between two and three standard deviations below the mean?

Example 3 The heights of a certain group of high school girls have a normal distribution with a mean of 150 cm and a standard deviation of 5 cm. Find the percent of the group having a height greater than 158 cm.

Solution The standardized value corresponding to 158 is:

$$x = \frac{158 - M}{\sigma} = \frac{158 - 150}{5} = 1.6$$

The area under the standard normal curve to the right of $x = 1.6$ is the difference between the area to the right of $x = 0$ and the area between $x = 0$ and $x = 1.6$:

$$0.5 - 0.4452 = 0.0548$$

$\therefore$ about 5.5% of the group has a height greater than 158.

For Exercise 4, use the table on page 243.

4. The weights of a certain group of high school boys have a normal distribution with a mean of 146 lb and a standard deviation of 10 lb. Find the percent of the group having a weight (a) greater than 154 lb and (b) less than 132 lb.

Mixed Review Exercises

For the line containing the given points, find (a) the slope and (b) an equation in standard form.

1. $(2, -3), (5, 1)$
2. $(4, -3), (-3, 4)$
3. $(4, 7), (5, -2)$
4. $(-3, 4), (5, 4)$
5. $(-2, -3), (-4, -5)$
6. $(-1, -1), (-2, -2)$

5–4 Correlation *(continued)*

Draw a scatter plot. Then describe the nature of the correlation.

1.

Manufacturers' Shipments of Televisions (five consecutive years)	
Color TVs (in millions)	Black and White TVs (in millions)
11.4	5.8
13.9	5.8
15.9	5.1
16.9	3.7
18.9	3.7

2.

Retail Sales of Passenger Cars in U.S. (five consecutive years)	
Domestic Cars (in millions)	Foreign Cars (in millions)
6.2	2.3
5.8	2.2
6.8	2.4
8.0	2.4
8.2	2.8

Example 2 For the table in Example 1, suppose x represents the number of students and y represents the number of teachers. Given that $M_{xy} \approx 87.616$, $M_x = 40.402$, $M_y = 2.168$, $\sigma_x \approx 0.879$, and $\sigma_y \approx 0.028$, determine (a) the correlation coefficient and (b) an equation of the regression line. Then (c) use the equation of the regression line to find the expected number of teachers when the number of students is 39 million.

Solution **a.** The correlation coefficient of the ordered pairs is:

$$r = \frac{M_{xy} - M_x \cdot M_y}{\sigma_x \cdot \sigma_y} \approx \frac{87.616 - (40.402)(2.168)}{(0.879)(0.028)} \approx 0.99$$

b. The regression line contains the point $(M_x, M_y) = (40.402, 2.168)$.

The slope of the regression line is: $r\left(\frac{\sigma_y}{\sigma_x}\right) \approx 0.99\left(\frac{0.028}{0.879}\right) \approx 0.032.$

To obtain an equation of the regression line, use the point-slope form of the equation of a line:

$$y - 2.168 = 0.032(x - 40.402)$$
$$y = 0.032x + 0.875$$

c. Substituting 39 for x in the equation from part (b), you get a value of $0.032(39) + 0.875$, or 2.123, for y.

$\therefore$ the expected number of teachers is about 2.12 million.

3. For the table in Exercise 1, suppose x represents the number of color televisions and y represents the number of black and white televisions. Given that $M_{xy} \approx 72.06$, $M_x = 15.4$, $M_y = 4.82$, $\sigma_x \approx 2.57$, and $\sigma_y \approx 0.95$, determine (a) the correlation coefficient and (b) an equation of the regression line. Then (c) determine the expected number of black and white televisions when the number of color televisions is 20 million.

4. For the table in Exercise 2, suppose x represents the number of domestic cars and y represents the number of foreign cars. Given that $M_{xy} = 17.1$, $M_x = 7$, $M_y = 2.42$, $\sigma_x \approx 0.95$, and $\sigma_y \approx 0.20$, determine (a) the correlation coefficient and (b) an equation of the regression line. Then (c) determine the expected number of foreign cars when the number of domestic cars is 10 million.

NAME ______________________ DATE ____________

15–5 Fundamental Counting Principles

Objective: To apply fundamental counting principles.

Vocabulary

Mutually exclusive cases Two situations such that when one occurs, the other cannot. Example: A number can be less than 100 or greater than 100, but not both.

Fundamental counting principles There are two such principles, the first of which involves multiplication and the second, addition:

1. If one selection can be made in m ways, and for each of these a second selection can be made in n ways, then the number of ways the two selections can be made is $m \times n$.
2. If the possibilities being counted can be grouped into *mutually exclusive* cases, then the total number of possibilities is the sum of the number of possibilities in each case.

Example 1 How many *even* 2-digit positive integers less than 80 are there?

Solution A diagram such as $\boxed{}\boxed{}$ is useful to help analyze the problem.

Since the integers have 2 digits and are less than 80, there are *seven* possible selections for the tens' digit: 1, 2, 3, 4, 5, 6, and 7. Write 7 in the first box: $\boxed{7}\boxed{}$

Since the integers are even, there are *five* possible selections for the units' digit: 0, 2, 4, 6, and 8. Write 5 in the second box: $\boxed{7}\ \boxed{5}$

∴ by the first fundamental counting principle there are $7 \cdot 5$, or 35, even 2-digit positive integers less than 80.

Example 2 How many positive integers less than 100 can be written using the digits 1, 3, 5, 7, and 9?

Solution Consider two mutually exclusive cases: (1) the 1-digit integers and (2) the 2-digit integers.

	Tens	Units	Number of integers	
1-digit integers	$\boxed{-}$	$\boxed{5}$	5	
2-digit integers	$\boxed{5}$	$\boxed{5}$	$5 \cdot 5 = 25$	by the first fundamental counting principle

∴ by the second fundamental counting principle there are $5 + 25$, or 30, positive integers less than 100 that can be written using the digits 1, 3, 5, 7, and 9.

NAME ______________________ DATE ______________

15–5 Fundamental Counting Principles *(continued)*

Example 3 In how many ways can a 5-question true-false section of a test be answered:

a. if every question must be answered?

b. if it is all right to leave questions unanswered?

Solution **a.** Each of the 5 questions has *two* possible answers: true or false.

Use a diagram where each box represents a question: [2] [2] [2] [2] [2]

$\therefore$ there are 2^5, or 32, possible ways to answer the true-false section of the test if every question must be answered.

b. Each of the 5 questions has *three* possible answers: true, false, or blank.

Use a diagram where each box represents a question: [3] [3] [3] [3] [3]

$\therefore$ there are 3^5, or 243, possible ways to answer the true-false section of the test if it is all right to leave questions unanswered.

1. How many odd 2-digit positive integers less than 50 are there?

2. How many even 2-digit positive integers greater than 60 are there?

3. How many positive integers less than 1000 can be written using the digits 3, 4, 5, and 6?

4. How many even 3-digit positive integers can be written using the digits 1, 3, 4, 5, and 6?

5. In how many different ways can an 8-question true-false test be answered:
a. if every question must be answered?
b. if it is all right to leave questions unanswered?

6. A bookshelf contains 6 different algebra books, 5 different geometry books, and 3 different trigonometry books. In how many ways can you select:
a. any one of the math books?
b. one of each of the three types of math books?

7. In how many ways can you select 4 cards, one after another, from a 52-card deck:
a. if the cards are returned to the deck after being selected?
b. if the cards are *not* returned to the deck after being selected?

Mixed Review Exercises

Evaluate if $x = -3$ and $y = 4$.

1. $\frac{y^3}{x^4}$ **2.** $|y - x|$ **3.** y^x **4.** $\sqrt{xy}$

Use the given root to solve each equation completely.

5. $x^3 - x^2 - x - 2 = 0$; 2

6. $x^3 - 3x^2 + 4x - 12 = 0$; $2i$

7. $2x^3 - x^2 + 18x - 9 = 0$; $3i$

8. $2x^3 + 3x^2 - 4x - 4 = 0$; -2

NAME ______________________ DATE ____________

15–6 Permutations

Objective: To find the number of permutations of the elements of a set.

Vocabulary

Factorial of a positive integer The product of the given integer and all positive integers less than the given integer. Example: The factorial of 5 (written 5!) is $5 \cdot 4 \cdot 3 \cdot 2 \cdot 1$, or 120. (*Note*: 0! is defined to be 1.)

Permutation An arrangement of the elements of a set *in a definite order*. Example: Six permutations of the elements of $\{m, n, o\}$ are possible: *mno, mon, nmo, nom, omn,* and *onm*. Permutations have the following properties:

1. The number of permutations of n objects is $n!$.
2. The number of permutations of n objects taken r at a time, denoted ${}_nP_r$, is given by the formula ${}_nP_r = \dfrac{n!}{(n-r)!}$.
3. If a set of n elements has n_1 elements of one kind alike, n_2 of another kind alike, and so on, then the number of permutations, P, of the n elements taken n at a time is given by $P = \dfrac{n!}{n_1!n_2!\cdots}$.

Symbols

! (factorial) ${}_nP_r$ (the number of permutations of n objects taken r at a time)

Example 1 Evaluate: **a.** 4! 6! **b.** $\dfrac{6!}{(6-2)!}$

Solution **a.** $4!\,6! = (4 \cdot 3 \cdot 2 \cdot 1) \cdot (6 \cdot 5 \cdot 4 \cdot 3 \cdot 2 \cdot 1) = 17{,}280$

b. $\dfrac{6!}{(6-2)!} = \dfrac{6!}{4!} = \dfrac{6 \cdot 5 \cdot 4 \cdot 3 \cdot 2 \cdot 1}{4 \cdot 3 \cdot 2 \cdot 1} = 30$

Evaluate.

1. 3! 4! **2.** 2(6!) **3.** $\dfrac{7!}{(7-3)!}$ **4.** $\dfrac{7!}{2!\,5!}$

Example 2 Find the number of permutations of 5 objects taken:
a. 5 at a time. **b.** 3 at a time.

Solution **a.** Use the second property of permutations:

$${}_5P_5 = \frac{5!}{(5-5)!} = \frac{5!}{0!} = \frac{5!}{1} = 5! = 120$$

(Notice that this result agrees with the first property of permutations.)

b. ${}_5P_3 = \dfrac{5!}{(5-3)!} = \dfrac{5!}{2!} = 5 \cdot 4 \cdot 3 = 60$

Find ${}_nP_r$ for the given values of n and r.

5. $n = 8, r = 8$ **6.** $n = 6, r = 2$ **7.** $n = 9, r = 1$ **8.** $n = 10, r = 5$

NAME ______________________________ DATE ______________

15–6 Permutations *(continued)*

Example 3 In how many ways can 6 potted plants be arranged on a windowsill if there is room: **a.** for all 6 pots? **b.** for only 5 pots?

Solution **a.** You are looking for the number of permutations of 6 objects, which is just 6!, or 720.

∴ there are 720 ways to arrange 6 potted plants on a windowsill.

b. You are looking for the number of permutations of 6 objects taken 5 at a time:

$$_6P_5 = \frac{6!}{(6-5)!} = \frac{6!}{1!} = 6 \cdot 5 \cdot 4 \cdot 3 \cdot 2 = 720$$

∴ there are 720 ways to arrange any 5 of 6 potted plants on a windowsill.

Example 4 In how many ways can the letters in the word BASE be arranged using: **a.** all 4 letters? **b.** only 2 letters at a time?

Solution **a.** $4! = 24$

∴ there are 24 ways to arrange the letters in the word BASE.

b. $_4P_2 = \frac{4!}{(4-2)!} = \frac{4!}{2!} = 4 \cdot 3 = 12$

∴ there are 12 ways to arrange the letters in the word BASE using only 2 letters at a time.

9. In how many ways can 8 different cars parallel park along one side of a street if there is room: **a.** for all 8 cars? **b.** for only 5 cars?

10. In how many ways can 10 different books be arranged on a shelf if there is room: **a.** for all 10 books? **b.** for only 7 books?

11. In how many ways can the letters of the word UMPIRE be arranged using: **a.** all 6 letters? **b.** only 4 letters?

12. In how many ways can the letters of the word PITCHER be arranged using: **a.** all 7 letters? **b.** only 3 letters?

Example 5 In how many ways can the letters in the word BASEBALL be arranged?

Solution There are 8 letters, of which 2 are B's, 2 are A's, and 2 are L's. Use the third property of permutations:

$$\frac{8!}{2!\,2!\,2!} = \frac{8 \cdot 7 \cdot 6 \cdot 5 \cdot 4 \cdot 3 \cdot 2 \cdot 1}{8} = 5040$$

∴ there are 5040 ways to arrange the letters in the word BASEBALL.

Find the number of ways the letters of each word can be arranged.

13. SOCCER **14.** RUNNING **15.** FOOTBALL **16.** VOLLEYBALL

NAME ______________________ DATE ______________

15–7 Combinations

Objective: To find the combinations of a set of elements.

Vocabulary

Subset Set B is a subset of set A if each member of B is also a member of A. Example: If $C = \{1, 2, 3\}$ and $D = \{1, 2\}$, then D is a subset of C, but C is not a subset of D. (*Note*: The empty set, $\emptyset$, is a subset of every set.)

Combination An r-element subset of a set of n elements is called a combination of n elements taken r at a time. The number of combinations of n elements taken r at a time, denoted ${}_nC_r$, is given by:

$$_nC_r = \frac{_nP_r}{_rP_r} = \frac{n!}{r!\,(n-r)!}$$

Symbol

${}_nC_r$ (the combination of n elements taken r at a time)

CAUTION Recall that the order of the elements matters for permutations: MN and NM, for example, are different permutations. The order of the elements is *not* important for combinations, however: {M, N} and {N, M} are the same combination.

Example 1 For the 3-digit set $\{2, 4, 6\}$, find:
a. all the subsets. **b.** the 2-digit subsets.

Solution **a.** $\{2, 4, 6\}, \{2, 4\}, \{2, 6\}, \{4, 6\}, \{2\}, \{4\}, \{6\}, \emptyset$
b. $\{2, 4\}, \{2, 6\}, \{4, 6\}$

1. For the two-digit set $\{1, 3\}$, find:
a. all the subsets. **b.** the subsets containing at least one digit.

2. For the four-letter set {A, B, C, D}, find:
a. the 3-letter subsets. **b.** the subsets containing fewer than three letters.

Example 2 Evaluate: **a.** ${}_4C_3$ **b.** ${}_{10}C_6$

Solution **a.** $_4C_3 = \frac{4!}{3!\,1!} = \frac{4 \cdot 3!}{3! \cdot 1} = \frac{4}{1} = 4$

b. $_{10}C_6 = \frac{10!}{6!\,4!} = \frac{10 \cdot 9 \cdot 8 \cdot 7 \cdot 6!}{6! \cdot 4 \cdot 3 \cdot 2 \cdot 1} = \frac{10 \cdot 9 \cdot 8 \cdot 7}{4 \cdot 3 \cdot 2 \cdot 1} = 210$

Evaluate.

3. ${}_6C_2$ **4.** ${}_5C_1$ **5.** ${}_8C_5$ **6.** ${}_7C_3$

7. ${}_{11}C_9$ **8.** ${}_{20}C_4$ **9.** ${}_{52}C_3$ **10.** ${}_{96}C_2$

15–7 Combinations *(continued)*

Example 3 Find the number of combinations of the letters in the word GRADE, taking them (a) 5 at a time and (b) 2 at a time. List each combination.

Solution **a.** ${}_5C_5 = \frac{5!}{5!\,0!} = 1$ {G, R, A, D, E}

b. ${}_5C_2 = \frac{5!}{2!\,3!} = 10$ {G, R}, {G, A}, {G, D}, {G, E}, {R, A} {R, D}, {R, E}, {A, D}, {A, E}, {D, E}

11. How many combinations can be formed from the letters in NUMBER, taking them:
a. 5 at a time? **b.** 3 at a time? **c.** 2 at a time?

12. How many combinations can be formed from the letters in EQUATION, taking them:
a. 7 at a time? **b.** 5 at a time? **c.** 2 at a time?

Example 4 In how many ways can a basketball team having 5 players be chosen from 6 seniors and 4 juniors:
a. if all are equally eligible? **b.** if the team must have 3 seniors and 2 juniors?

Solution **a.** There are 10 people eligible for the team: ${}_{10}C_5 = \frac{10!}{5!\,5!} = 252$

b. There are ${}_6C_3$ ways to choose seniors and ${}_4C_2$ ways to choose juniors. Use the first fundamental counting principle to find the number of ways

to select the team: ${}_6C_3 \cdot {}_4C_2 = \frac{6!}{3!\,3!} \cdot \frac{4!}{2!\,2!} = 20 \cdot 6 = 120$

13. A sample of 4 light bulbs taken from a group of 50 light bulbs is to be inspected. How many different samples could be selected?

14. Seven boys and 8 girls want to participate in a school play. In how many ways can a cast be selected if it must have 4 boys and 5 girls?

Mixed Review Exercises

Find the value of each function if $x = 3$.

1. $f(x) = \frac{2x + 5}{3x - 2}$ **2.** $g(x) = |5 - 2x|$ **3.** $h(x) = (3 - x)^2$

4. $F(x) = -5$ **5.** $G(x) = -\frac{2}{3}x + 4$ **6.** $H(x) = x^3 - 27$

Simplify.

7. $\sqrt{48} - \sqrt{8} + \sqrt{2}$ **8.** $\log_5 100 - 2 \log_5 2$ **9.** $\frac{1}{x + 3} + \frac{2}{x^2 + x - 6}$

10. $(5 - 2i)(3 + i)$ **11.** $\frac{4x^2 + 6x + 9}{8x^3 - 27}$ **12.** $\left(\frac{\sqrt{11} + 1}{4}\right)\left(\frac{\sqrt{11} - 1}{4}\right)$

NAME ______________________ DATE __________

15–8 Sample Spaces and Events

Objective: To specify sample spaces and events for random experiments.

Vocabulary

Random experiment An experiment in which you do not necessarily get the same outcome when you repeat it under the same conditions. Example: Tossing a coin is a random experiment.

Sample space The set of all possible outcomes of a random experiment. Example: If H represents heads and T tails, then $\{H, T\}$ is the sample space for a coin tossing experiment.

Event Any subset of possible outcomes for an experiment.

Simple event An event that represents a single element of a random experiment's sample space. Example: $\{H\}$ and $\{T\}$ are each simple events for a coin tossing experiment.

Example 1 Each letter of the word RANDOM is written on a separate card. The cards are then shuffled, and one card is drawn.

a. Specify the sample space for the experiment.
b. Specify the event that the letter on the card that is drawn is a vowel.
c. Specify the event that the letter on the card that is drawn is neither M nor N.

Solution

a. {R, A, N, D, O, M}
b. {A, O}
c. {R, A, D, O}

Example 2 Suppose you roll two dice, one red and one green.

a. Specify the sample space for the experiment.
b. Specify the event that both dice turn up the same number.

Solution

a. The outcome can be represented by the ordered pair (r, g) where r is the number on the red die and g is the number on the green die.

$\{(1, 1), (1, 2), (1, 3), (1, 4), (1, 5), (1, 6),$
$(2, 1), (2, 2), (2, 3), (2, 4), (2, 5), (2, 6),$
$(3, 1), (3, 2), (3, 3), (3, 4), (3, 5), (3, 6),$
$(4, 1), (4, 2), (4, 3), (4, 4), (4, 5), (4, 6),$
$(5, 1), (5, 2), (5, 3), (5, 4), (5, 5), (5, 6),$
$(6, 1), (6, 2), (6, 3), (6, 4), (6, 5), (6, 6)\}$

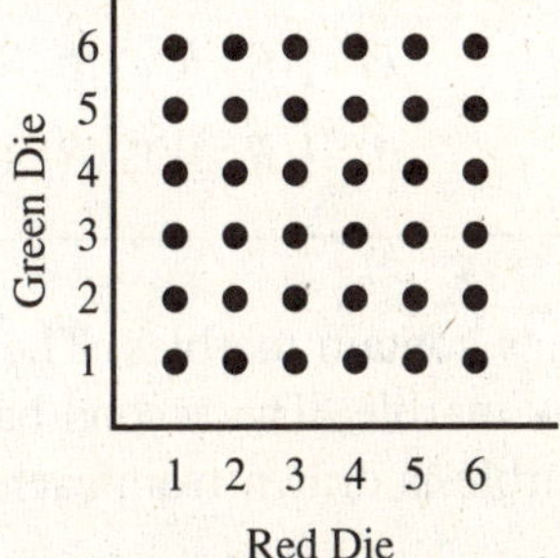

Another way of specifying this sample space is shown at the right. The dot in the upper left corner, for example, represents the simple event "red die shows 1 and green die shows 6."

b. $\{(1, 1), (2, 2), (3, 3), (4, 4), (5, 5), (6, 6)\}$

15–8 Sample Spaces and Events *(continued)*

Exercises 1–3 refer to an experiment in which each letter of the word SAMPLE is written on a separate card. The cards are then shuffled, and one card is drawn.

1. Specify the sample space for the experiment.

2. Specify the event that the letter on the card that is drawn is a consonant.

3. Specify the event that the letter on the card that is drawn comes before N in the alphabet.

For Exercises 4–9, refer to the two-dice experiment described in Example 2. Specify each event.

4. The sum of the numbers showing on the two dice is 4.

5. The product of the numbers showing on the two dice is 6.

6. The sum of the numbers showing on the two dice is greater than 9.

7. The product of the numbers showing on the two dice is greater than 20.

8. The number on the red die is one more than the number on the green die.

9. The number on the green die is 5, and the number on the red die is greater than 3.

Example 3 One box contains a red ball and a blue ball. A second box contains a red ball, a white ball, and a green ball. One ball is drawn from each box. For the experiment specify:

a. the sample space.
b. the event that one ball is green.
c. the event that at least one ball is red.
d. the event that exactly one ball is red.

Solution

a. An outcome of the experiment can be represented by the ordered pair (b_1, b_2) where b_1 is the ball drawn from the first box and b_2 is the ball drawn from the second box. If R represents a red ball, B a blue ball, W a white ball, and G a green ball, then the sample space for the experiment is $\{(R, R), (R, W), (R, G), (B, R), (B, W), (B, G)\}$.

b. $\{(R, G), (B, G)\}$

c. $\{(R, R), (R, W), (R, G), (B, R)\}$

d. $\{(R, W), (R, G), (B, R)\}$

10. Two bags contain marbles. The first bag contains a white, an orange, and a yellow marble. The second bag contains a white and an orange marble. One marble is drawn from each bag. For the experiment specify:

a. the sample space.
b. the event that one marble is yellow.
c. the event that at most one marble is white.
d. the event that neither marble is orange.

NAME ______________________ DATE ______________

15–9 Probability

Objective: To find the probability that an event will occur.

Vocabulary

Probability A number (between 0 and 1, inclusive) that expresses the likelihood of an event occurring. Example: The probability of obtaining a head on the toss of a coin is $\frac{1}{2}$; this is written $P(H) = \frac{1}{2}$. Some properties of probability are:

1. If $\{a_1, a_2, a_3, \ldots, a_n\}$ is a sample space containing n equally likely outcomes, then the probability of each simple event is $\frac{1}{n}$:

$$P(a_1) = P(a_2) = P(a_3) = \cdots = P(a_n) = \frac{1}{n}$$

2. If the sample space for an experiment consists of n equally likely outcomes, and if k of them are in event E, then:

$$P(E) = \underbrace{\frac{1}{n} + \frac{1}{n} + \frac{1}{n} + \cdots + \frac{1}{n}}_{k \text{ addends}} = \frac{k}{n}$$

3. If event E contains all elements of the sample space, then $P(E) = \frac{n}{n} = 1$. That is, the event is certain to occur.

4. If event E contains no elements of the sample space, then $P(E) = \frac{0}{n} = 0$. That is, the event is certain *not* to occur.

Symbol

$P(E)$ (the probability of event E)

Example 1 A die is rolled. Find the probability of each event.
a. Event A: The number showing is even.
b. Event B: The number showing is 5 or 6.

Solution The sample space is $\{1, 2, 3, 4, 5, 6\}$. Since the six outcomes are equally likely, the probability of each simple event is $\frac{1}{6}$.

a. Event $A = \{2, 4, 6\}$

$P(A) = \frac{3}{6} = \frac{1}{2}$

b. Event $B = \{5, 6\}$

$P(B) = \frac{2}{6} = \frac{1}{3}$

1. A number wheel is divided into 12 congruent sectors, numbered 1–12. You spin the wheel and record the number. Find the probability of each event.

a. It is a 5.
b. It is a 2-digit number.
c. It is a factor of 12.
d. It is an odd number.
e. It is a negative number.
f. It is a factor of 27,720.

NAME ______________________ DATE ______________

15–9 Probability *(continued)*

2. A letter is selected at random from those in the word RECTANGLE. Find the probability of each event.

a. It is a vowel.
b. It is a consonant.
c. It is between D and M in the alphabet.
d. It is either C or G.

3. One marble is drawn at random from a bag containing 2 red, 4 yellow, and 6 blue marbles. Find the probability of each event.

a. It is yellow.
b. It is blue.
c. It is not yellow.
d. It is red or blue.

4. One card is drawn at random from a 52-card deck. Find the probability of each event.

a. It is a king.
b. It is a heart.
c. It is red.
d. It is the ace of spades.
e. It is a red jack.
f. It is red or black.

Example 2 Two coins are tossed. Find the probability of each event.

a. Event A: Both come up heads.
b. Event B: Each coin comes up either heads or tails.
c. Event C: Neither coin comes up heads.
d. Event D: One coin comes up heads and the other coin comes up tails.

Solution The sample space is $\{(H, T), (H, H), (T, H), (T, T)\}$.

a. Event $A = \{(H, H)\}$

$P(A) = \frac{1}{4}$

b. Event $B = \{(H, T), (H, H), (T, H), (T, T)\}$

$P(B) = \frac{4}{4} = 1$

c. Event $C = \{(T, T)\}$

$P(C) = \frac{1}{4}$

d. Event $D = \{(H, T), (T, H)\}$

$P(D) = \frac{2}{4} = \frac{1}{2}$

5. Two dice are rolled. Find the probability of each event.

a. The sum of the numbers showing on the dice is 7.
b. The sum of the numbers showing on the dice is less than 6.
c. Exactly one die shows 3.
d. At least one die shows 3.

Mixed Review Exercises

Solve each inequality and graph the solution set.

1. $|x + 3| > 5$
2. $y^2 + y \leq 6$
3. $3 - 2w > -1$
4. $-2 \leq 3m + 4 \leq 4$
5. $8u^2 > 32$
6. $|1 - 2k| < 3$

7. How many permutations are there of all the letters in AXIOM?

8. In how many ways can a committee of 4 be selected from a club having 16 members?

NAME ____________________ DATE ____________

15–10 Mutually Exclusive and Independent Events

Objective: To identify mutually exclusive and independent events and find the probability of such events.

Vocabulary

Intersection of sets The intersection of sets A and B is the set whose members are elements of both A and B. Example: If $A = \{1, 2, 3\}$ and $B = \{3, 4\}$, then the intersection of A and B (written $A \cap B$) is $\{3\}$.

Disjoint sets Sets that have no members in common. (*Note*: A and B are disjoint sets if and only if $A \cap B = \emptyset$.)

Union of sets The union of sets A and B is the set whose members are the elements belonging to either A or B (or both). Example: If $A = \{1, 2, 3\}$ and $B = \{3, 4\}$, then the union of A and B (written $A \cup B$) is $\{1, 2, 3, 4\}$.

Mutually exclusive events Two events that have no simple events in common. Example: Getting two tails and getting at least one head are mutually exclusive events when two coins are tossed.

Independent events Two events such that the occurrence of one has no effect on the probability of the other.

Complement of an event If event A is a subset of sample space S, the complement of A (written $\overline{A}$) consists of the elements of S that are *not* members of A.

Probability Here are some additional properties of probability:

1. For any sample space S, $P(S) = 1$ and $P(\emptyset) = 0$.
2. For any two events A and B in a sample space, $P(A \cup B) = P(A) + P(B) - P(A \cap B)$.
3. If A and B are mutually exclusive events, $P(A \cup B) = P(A) + P(B)$.
4. Two events A and B are independent if and only if $P(A \cap B) = P(A) \cdot P(B)$.
5. For any event A, $P(A) + P(\overline{A}) = 1$, or $P(\overline{A}) = 1 - P(A)$.

Symbols

$\cap$ (intersection) $\cup$ (union) $\overline{A}$ (the complement of event A)

Example 1 Specify each of the following sets by listing its elements. Use the *Venn diagram* (a diagram used to picture sets and set relationships) shown at the right.

a. $A \cap B$ **b.** $B \cap C$
c. $A \cup B$ **d.** $B \cup C$
e. $(A \cup B) \cap C$ **f.** $A \cap \overline{C}$

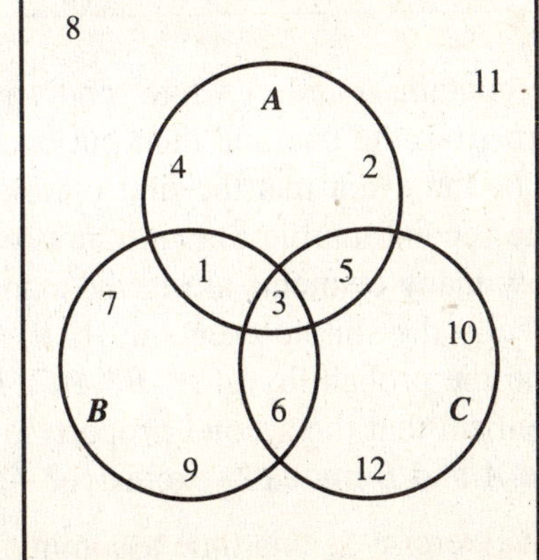

Solution **a.** $\{1, 3\}$ **b.** $\{3, 6\}$
c. $\{1, 2, 3, 4, 5, 6, 7, 9\}$
d. $\{1, 3, 5, 6, 7, 9, 10, 12\}$
e. $\{3, 5, 6\}$ **f.** $\{1, 2, 4\}$

NAME ______________________ DATE ______________

15–10 Mutually Exclusive and Independent Events *(continued)*

1. Use the Venn diagram given in Example 1 to specify each of the following sets by listing its elements.
a. $A \cap C$ **b.** $A \cup C$ **c.** $(A \cap B) \cup C$ **d.** $\bar{A} \cap B$

Example 2 A letter is selected at random from those in the word FOR. A second letter is selected at random from those in the word SOME. Let A be the event that each letter is an O, and let B be the event that neither letter is an O.

a. How many elements are in the sample space of the experiment?
b. Specify the simple events in A, B, $A \cap B$, and $A \cup B$.
c. Find the probability of A, B, $A \cap B$, and $A \cup B$.
d. Confirm that the second property of probability given on page 257 holds.
e. Are A and B mutually exclusive? Are they independent?

Solution

a. The word FOR contains 3 letters, and the word SOME contains 4. By the first fundamental counting principle, there are $3 \cdot 4$, or 12, ways of selecting a letter from FOR and then a letter from SOME.

b. Let the ordered pair (l_1, l_2) represent the selection of the letter l_1 from FOR followed by the selection of the letter l_2 from SOME.

$A = \{(O, O)\}$
$B = \{(F, S), (F, M), (F, E), (R, S), (R, M), (R, E)\}$
$A \cap B = \emptyset$
$A \cup B = \{(F, S), (F, M), (F, E), (O, O), (R, S), (R, M), (R, E)\}$

c. $P(A) = \frac{1}{12}$ $P(B) = \frac{6}{12} = \frac{1}{2}$

$P(A \cap B) = \frac{0}{12} = 0$ $P(A \cup B) = \frac{7}{12}$

d. $P(A) + P(B) - P(A \cap B) = \frac{1}{12} + \frac{6}{12} - 0 = \frac{7}{12} = P(A \cup B)$

e. Since $A \cap B$ is empty, A and B are mutually exclusive. Since $P(A) \cdot P(B) = \frac{1}{12} \cdot \frac{1}{2} = \frac{1}{24} \neq P(A \cap B)$, A and B are not independent.

2. A bag contains 1 red, 1 white, and 1 blue marble. A marble is drawn at random from the bag and then put back before a second marble is drawn. Let A be the event that the first marble drawn is red, and let B be the event that the second marble drawn is not red.

a. How many elements are in the sample space of the experiment?
b. Specify the simple events in A, B, $A \cap B$, and $A \cup B$.
c. Find the probability of A, B, $A \cap B$, and $A \cup B$.
d. Confirm that the second property of probability given on page 257 holds.
e. Are A and B mutually exclusive? Are they independent?

3. Repeat Exercise 2, this time assuming that the first marble is *not* put back before the second is drawn.

NAME ______________________ DATE ______________

16 Matrices and Determinants

16–1 Definition of Terms

Objective: To learn and apply matrix terminology.

Vocabulary

Matrix (plural, **matrices**) A rectangular array of numbers, called *elements,* enclosed by brackets and named with a capital letter. Example: $A = \begin{bmatrix} -1 & 3 & 0 \\ 2 & -4 & 1 \end{bmatrix}$

Dimensions of a matrix The number of rows (horizontal) and the number of columns (vertical) in the matrix. Subscripts are used to indicate dimensions. Example: $A_{2\times 3}$ denotes a matrix A with 2 rows and 3 columns.

Row matrix A matrix having exactly one row. Example: $[0 \quad -3 \quad 2]$

Column matrix A matrix having exactly one column. Example: $\begin{bmatrix} -2 \\ 5 \\ 4 \end{bmatrix}$

Square matrix A matrix for which the number of rows and the number of columns are equal (that is, an $n \times n$ matrix). Example: $\begin{bmatrix} -4 & 3 \\ 2 & 1 \end{bmatrix}$

Zero matrix A matrix in which all the elements are zero.

Example: $\begin{bmatrix} 0 & 0 \\ 0 & 0 \\ 0 & 0 \end{bmatrix}$ is a 3×2 zero matrix, denoted by $O_{3\times 2}$.

Equal matrices Two matrices are equal if and only if they have the same dimensions and the elements in all corresponding positions are equal.

Example: $\begin{bmatrix} \sqrt{8} & \frac{2}{4} \\ (-1)3 & -2+2 \end{bmatrix} = \begin{bmatrix} 2\sqrt{2} & \frac{1}{2} \\ -3 & 0 \end{bmatrix}$ but $\begin{bmatrix} -1 & 3 \\ 0 & 2 \end{bmatrix} \neq \begin{bmatrix} -1 & 3 \\ 2 & 0 \end{bmatrix}$

Symbols

$A_{m\times n}$ (the matrix A having m rows and n columns)

$O_{m\times n}$ (the zero matrix O having m rows and n columns)

Example 1 Write the zero matrix denoted by $O_{3\times 4}$.

Solution $O_{3\times 4}$ has 3 rows and 4 columns. Each element is zero.

$$\therefore O_{3\times 4} = \begin{bmatrix} 0 & 0 & 0 & 0 \\ 0 & 0 & 0 & 0 \\ 0 & 0 & 0 & 0 \end{bmatrix}$$

Write the zero matrix denoted by each of the following.

1. $O_{2\times 1}$ **2.** $O_{4\times 2}$ **3.** $O_{1\times 3}$ **4.** $O_{3\times 3}$

5. Of the matrices in Exercises 1–4, name (a) a row matrix, (b) a column matrix, and (c) a square matrix.

NAME ____________________ DATE ____________

16–1 Definition of Terms *(continued)*

Example 2 Find the value of each variable.

a. $O_{1\times 3} = [2x \quad 0 \quad y-1]$ **b.** $\begin{bmatrix} x+3 & -1 \\ 5 & 12 \end{bmatrix} = \begin{bmatrix} -3 & -1 \\ 5 & -6y \end{bmatrix}$

Solution **a.** Since $O_{1\times 3}$ is a zero matrix, each element is zero.

$2x = 0$ and $y - 1 = 0$

$\therefore$ $x = 0$ and $y = 1$

b. Since the matrices are equal, elements in corresponding positions are equal.

$x + 3 = -3$ and $12 = -6y$

$\therefore$ $x = -6$ and $y = -2$

Find the value of each variable.

6. $O_{2\times 2} = \begin{bmatrix} 0 & x \\ 0 & y \end{bmatrix}$

7. $O_{3\times 1} = \begin{bmatrix} x \\ y \\ z \end{bmatrix}$

8. $O_{2\times 3} = \begin{bmatrix} 2w & y+2 & 0 \\ -3x & 10-z & 0 \end{bmatrix}$

9. $O_{1\times 2} = [7 - x \quad 2y + 6]$

10. $[-3 \quad x] = [y \quad 1]$

11. $\begin{bmatrix} x & 0 \\ 5 & 6 \end{bmatrix} = \begin{bmatrix} -1 & 0 \\ 5 & y \end{bmatrix}$

12. $\begin{bmatrix} 3 & x \\ 4 & 2 \end{bmatrix} = \begin{bmatrix} 3 & 7 \\ y & 2 \end{bmatrix}$

13. $\begin{bmatrix} 2 \\ 0 \\ x \end{bmatrix} = \begin{bmatrix} y \\ 0 \\ 6 \end{bmatrix}$

14. $\begin{bmatrix} 4 & -1 \\ 2 & 3 \\ x & 5 \end{bmatrix} = \begin{bmatrix} 4 & -1 \\ y & 3 \\ -6 & 5 \end{bmatrix}$

15. $[3 \quad x \quad 5] = [y \quad 2 \quad z]$

16. $\begin{bmatrix} x-4 \\ 14 \\ 2 \end{bmatrix} = \begin{bmatrix} 10 \\ y \\ z-3 \end{bmatrix}$

17. $\begin{bmatrix} 3 & 4 \\ 2x & z \end{bmatrix} = \begin{bmatrix} y+3 & 4 \\ 2 & -6 \end{bmatrix}$

18. $\begin{bmatrix} -1 & x+2 & 0 & -y \\ 17 & 3 & 8 & 0 \end{bmatrix} = \begin{bmatrix} z & -2 & 0 & 5 \\ 17 & 3 & 8 & 0 \end{bmatrix}$

Mixed Review Exercises

Solve each triangle. Give lengths to three significant digits and angle measures to the nearest tenth of a degree.

1. $a = 8$, $b = 12$, $\angle C = 70°$

2. $a = 5$, $\angle C = 90°$, $\angle A = 60°$

3. $a = \dfrac{5\sqrt{2}}{2}$, $b = \dfrac{5\sqrt{2}}{2}$, $\angle C = 90°$

4. $\angle A = 30°$, $\angle C = 52°$, $b = 20$

5–8. Find the area of each triangle in Exercises 1–4 above. Give answers to three significant digits.

NAME ______________________ DATE ______________

16–2 Addition and Scalar Multiplication

Objective: To find sums and differences of matrices and products of a scalar and a matrix.

Vocabulary

Sum of matrices having the same dimensions The matrix whose elements are the sums of the corresponding elements of the matrices being added. (*Note*: Addition of matrices of different dimensions is not defined.)

Identity for addition of matrices For each set of $m \times n$ matrices, $O_{m \times n}$ is the identity for addition (that is, the sum of $A_{m \times n}$ and $O_{m \times n}$ is just $A_{m \times n}$).

Additive inverse of matrix *A* The matrix $-A$, where each element of $-A$ is the opposite of the corresponding element of A.

Example: If $A = \begin{bmatrix} 1 & -2 \\ -3 & 0 \end{bmatrix}$, then $-A = \begin{bmatrix} -1 & 2 \\ 3 & 0 \end{bmatrix}$.

Properties of addition of matrices Let A, B, and C be $m \times n$ matrices. Let $O_{m \times n}$ be the $m \times n$ zero matrix.

1. Closure property — $A + B$ is an $m \times n$ matrix.
2. Commutative property — $A + B = B + A$
3. Associative property — $(A + B) + C = A + (B + C)$
4. Identity property — $A + O_{m \times n} = O_{m \times n} + A = A$
5. Inverse property — $A + (-A) = -A + A = O_{m \times n}$

Subtraction of matrices For each set of $m \times n$ matrices, subtraction is defined as follows: $A_{m \times n} - B_{m \times n} = A_{m \times n} + (-B_{m \times n})$.

Scalar A real number.

Scalar product The matrix rA, which is the product of a scalar r and a matrix A. Each element of rA is determined by multiplying the corresponding element of A by r.

Properties of scalar multiplication Let A and B be $m \times n$ matrices. Let $O_{m \times n}$ be the $m \times n$ zero matrix, and let p and q be scalars.

1. Closure property — pA is an $m \times n$ matrix.
2. Commutative property — $pA = Ap$
3. Associative property — $p(qA) = (pq)A$
4. Distributive properties — $(p + q)A = pA + qA$; $p(A + B) = pA + pB$
5. Identity property — $1 \cdot A = A$
6. Multiplicative property of -1 — $(-1)A = -A$
7. Multiplicative property of 0 — $0 \cdot A = O_{m \times n}$

Example 1 Simplify: $\begin{bmatrix} 2 & 4 & 7 \\ -1 & 5 & -2 \end{bmatrix} + \begin{bmatrix} 5 & -3 & 8 \\ 6 & -4 & -6 \end{bmatrix}$

Solution Add the elements in corresponding positions.

$$\begin{bmatrix} 2 & 4 & 7 \\ -1 & 5 & -2 \end{bmatrix} + \begin{bmatrix} 5 & -3 & 8 \\ 6 & -4 & -6 \end{bmatrix} = \begin{bmatrix} 2+5 & 4+(-3) & 7+8 \\ -1+6 & 5+(-4) & -2+(-6) \end{bmatrix}$$

$$= \begin{bmatrix} 7 & 1 & 15 \\ 5 & 1 & -8 \end{bmatrix}$$

NAME ______________________ DATE ______________

16–2 Addition and Scalar Multiplication *(continued)*

Simplify.

1. $\begin{bmatrix} 5 & -1 \\ 3 & -2 \end{bmatrix} + \begin{bmatrix} 4 & -3 \\ -6 & 1 \end{bmatrix}$

2. $\begin{bmatrix} 6 \\ 1 \\ 3 \end{bmatrix} + \begin{bmatrix} -4 \\ -5 \\ 4 \end{bmatrix}$

3. $\begin{bmatrix} 5 & 2 \\ 3 & 5 \end{bmatrix} + \begin{bmatrix} -5 & 4 \\ 2 & -1 \end{bmatrix}$

4. $\begin{bmatrix} 4 & -1 & 0 \\ 5 & 6 & 2 \end{bmatrix} + \begin{bmatrix} -3 & -2 & 4 \\ 1 & -3 & 4 \end{bmatrix}$

Example 2 Simplify: $\begin{bmatrix} 4 & 1 & 3 \\ 2 & -2 & 4 \end{bmatrix} - \begin{bmatrix} 3 & -1 & -2 \\ 5 & 4 & 4 \end{bmatrix}$

Solution Use the definition of subtraction: $A_{m\times n} - B_{m\times n} = A_{m\times n} + (-B_{m\times n})$.

$$\begin{bmatrix} 4 & 1 & 3 \\ 2 & -2 & 4 \end{bmatrix} - \begin{bmatrix} 3 & -1 & -2 \\ 5 & 4 & 4 \end{bmatrix} = \begin{bmatrix} 4 & 1 & 3 \\ 2 & -2 & 4 \end{bmatrix} + \begin{bmatrix} -3 & 1 & 2 \\ -5 & -4 & -4 \end{bmatrix}$$
$$= \begin{bmatrix} 1 & 2 & 5 \\ -3 & -6 & 0 \end{bmatrix}$$

Simplify.

5. $\begin{bmatrix} 5 & 2 \\ 3 & -1 \\ -2 & 5 \end{bmatrix} - \begin{bmatrix} 6 & 2 \\ 4 & -2 \\ 2 & 4 \end{bmatrix}$

6. $\begin{bmatrix} 4 & 3 \\ 5 & -2 \\ -2 & 3 \end{bmatrix} - \begin{bmatrix} 8 & 2 \\ 4 & -1 \\ 3 & 4 \end{bmatrix}$

7. $\begin{bmatrix} 0 & 4 \\ 3 & -3 \end{bmatrix} - \begin{bmatrix} 4 & 6 \\ 3 & -1 \end{bmatrix}$

8. $\begin{bmatrix} 4 & 5 & 6 \\ 1 & -2 & 3 \end{bmatrix} - \begin{bmatrix} 3 & 2 & -6 \\ 5 & -1 & 4 \end{bmatrix}$

Example 3 Simplify: $4\begin{bmatrix} 1 & 2 & 3 \\ 5 & -1 & 2 \\ 0 & 3 & -1 \end{bmatrix}$

Solution To find the scalar product, multiply each element of the given matrix by the scalar 4.

$$4\begin{bmatrix} 1 & 2 & 3 \\ 5 & -1 & 2 \\ 0 & 3 & -1 \end{bmatrix} = \begin{bmatrix} 4(1) & 4(2) & 4(3) \\ 4(5) & 4(-1) & 4(2) \\ 4(0) & 4(3) & 4(-1) \end{bmatrix} = \begin{bmatrix} 4 & 8 & 12 \\ 20 & -4 & 8 \\ 0 & 12 & -4 \end{bmatrix}$$

Simplify.

9. $3\begin{bmatrix} 4 & -1 \\ 2 & 5 \end{bmatrix}$

10. $-2\begin{bmatrix} 4 & 1 & 3 \\ -2 & -1 & 2 \end{bmatrix}$

11. $5\begin{bmatrix} 6 & 1 \\ 2 & 4 \\ -5 & 3 \end{bmatrix}$

12. $6\begin{bmatrix} 3 & -1 \\ -1 & 3 \end{bmatrix}$

13. $O_{3\times 2} + \begin{bmatrix} 4 & 2 \\ 5 & -1 \\ 3 & 4 \end{bmatrix}$

14. $O_{2\times 2} - 2\begin{bmatrix} 6 & 5 \\ -2 & 3 \end{bmatrix}$

15. $5\begin{bmatrix} 1 & -2 \\ 4 & 3 \\ -2 & 1 \end{bmatrix} + \begin{bmatrix} 5 & 6 \\ -2 & 4 \\ 3 & -1 \end{bmatrix}$

16. $\begin{bmatrix} 4 & 3 \\ 1 & -2 \end{bmatrix} - 3\begin{bmatrix} 6 & -1 \\ -2 & 4 \end{bmatrix}$

NAME ______________________ DATE ______________

16–3 Matrix Multiplication

Objective: To find the product of two matrices.

Vocabulary

Product of matrices The product of matrices $A_{m\times n}$ and $B_{n\times p}$ is the $m\times p$ matrix whose element in the *a*th row and *b*th column is the sum of the products of corresponding elements of the *a*th row of A and *b*th column of B.

Identity matrix An $n\times n$ matrix, denoted by $I_{n\times n}$, whose *main diagonal* from upper left to lower right has all elements 1, while all other elements are 0.

Example: $I_{3\times 3} = \begin{bmatrix} 1 & 0 & 0 \\ 0 & 1 & 0 \\ 0 & 0 & 1 \end{bmatrix}$. The product of $I_{n\times n}$ and any $n\times n$ matrix A is A.

Properties of matrix multiplication Let A, B, and C be $n\times n$ matrices. Let $I_{n\times n}$ be the identity matrix and $O_{n\times n}$ be the zero matrix.

1. Associative property $\quad (AB)C = A(BC)$
2. Distributive properties $\quad A(B + C) = AB + AC;\ (B + C)A = BA + CA$
3. Identity property $\quad I_{n\times n} \cdot A = A \cdot I_{n\times n} = A$
4. Multiplicative property of $O_{n\times n}$ $\quad O_{n\times n} \cdot A = A \cdot O_{n\times n} = O_{n\times n}$

Symbols

AB (the product of matrices A and B) $\quad I_{n\times n}$ (the identity matrix I having n rows and n columns)

CAUTION 1 When you multiply two matrices, the number of *columns* of the first matrix must equal the number of *rows* of the second matrix. If this is not the case, the matrices cannot be multiplied.

CAUTION 2 Unlike real-number multiplication, matrix multiplication is in general *not* commutative. That is, if A and B are any two $n\times n$ matrices, the products AB and BA are not necessarily equal.

Example 1 Multiply: $[4\ \ 6\ \ -1]\begin{bmatrix} -4 \\ 3 \\ 4 \end{bmatrix}$

Solution A $\mathbf{1}\times 3$ matrix times a $3\times\mathbf{1}$ matrix gives a $\mathbf{1}\times\mathbf{1}$ matrix.

$$[4\ \ 6\ \ -1]\begin{bmatrix} -4 \\ 3 \\ 4 \end{bmatrix} = [4(-4) + 6(3) + (-1)(4)]$$
$$= [-16 + 18 - 4]$$
$$= [-2]$$

Multiply.

1. $[4\ \ 2]\begin{bmatrix} 3 \\ -1 \end{bmatrix}$

2. $[-2\ \ 3\ \ 1]\begin{bmatrix} 5 \\ 0 \\ -4 \end{bmatrix}$

3. $[3\ \ 0\ \ -1\ \ 2]\begin{bmatrix} -2 \\ 5 \\ -4 \\ 1 \end{bmatrix}$

NAME ______________________ DATE ______________

16-3 Matrix Multiplication *(continued)*

Example 2 Multiply: $\begin{bmatrix} 4 & 2 \\ -1 & 3 \end{bmatrix}\begin{bmatrix} -1 & 3 & 5 \\ -2 & 0 & -6 \end{bmatrix}$

Solution A 2×2 matrix times a 2×3 matrix gives a 2×3 matrix.

$\begin{bmatrix} 4 & 2 \\ -1 & 3 \end{bmatrix}\begin{bmatrix} -1 & 3 & 5 \\ -2 & 0 & -6 \end{bmatrix} = \begin{bmatrix} -8 & & \\ & & \end{bmatrix}$ row 1 × column 1: $4(-1) + 2(-2) = -8$

$\begin{bmatrix} 4 & 2 \\ -1 & 3 \end{bmatrix}\begin{bmatrix} -1 & 3 & 5 \\ -2 & 0 & -6 \end{bmatrix} = \begin{bmatrix} -8 & 12 & \\ & & \end{bmatrix}$ row 1 × column 2: $4(3) + 2(0) = 12$

$\begin{bmatrix} 4 & 2 \\ -1 & 3 \end{bmatrix}\begin{bmatrix} -1 & 3 & 5 \\ -2 & 0 & -6 \end{bmatrix} = \begin{bmatrix} -8 & 12 & 8 \\ & & \end{bmatrix}$ row 1 × column 3: $4(5) + 2(-6) = 8$

$\begin{bmatrix} 4 & 2 \\ -1 & 3 \end{bmatrix}\begin{bmatrix} -1 & 3 & 5 \\ -2 & 0 & -6 \end{bmatrix} = \begin{bmatrix} -8 & 12 & 8 \\ -5 & & \end{bmatrix}$ row 2 × column 1: $-1(-1) + 3(-2) = -5$

$\begin{bmatrix} 4 & 2 \\ -1 & 3 \end{bmatrix}\begin{bmatrix} -1 & 3 & 5 \\ -2 & 0 & -6 \end{bmatrix} = \begin{bmatrix} -8 & 12 & 8 \\ -5 & -3 & \end{bmatrix}$ row 2 × column 2: $-1(3) + 3(0) = -3$

$\begin{bmatrix} 4 & 2 \\ -1 & 3 \end{bmatrix}\begin{bmatrix} -1 & 3 & 5 \\ -2 & 0 & -6 \end{bmatrix} = \begin{bmatrix} -8 & 12 & 8 \\ -5 & -3 & -23 \end{bmatrix}$ row 2 × column 3: $-1(5) + 3(-6) = -23$

$\therefore$ the product is $\begin{bmatrix} -8 & 12 & 8 \\ -5 & -3 & -23 \end{bmatrix}$.

Give the dimensions of the product. Then multiply.

4. $\begin{bmatrix} 4 & -1 \\ 3 & 2 \end{bmatrix}\begin{bmatrix} 3 & -2 \\ 0 & 4 \end{bmatrix}$

5. $\begin{bmatrix} -2 & 0 \\ 1 & 4 \end{bmatrix}\begin{bmatrix} -1 & 0 & 2 \\ 4 & -3 & -5 \end{bmatrix}$

6. $\begin{bmatrix} 3 \\ -1 \end{bmatrix}\begin{bmatrix} 4 & -1 & 2 & 0 \end{bmatrix}$

7. $\begin{bmatrix} 2 & -5 \\ 3 & -1 \end{bmatrix}\begin{bmatrix} -4 \\ 1 \end{bmatrix}$

8. $\begin{bmatrix} 1 & 2 \\ 5 & 4 \\ 2 & -1 \end{bmatrix}\begin{bmatrix} 4 & -2 & 3 \\ 0 & 6 & -1 \end{bmatrix}$

9. $\begin{bmatrix} 1 & 2 & 4 \\ -2 & 0 & 1 \\ 4 & 3 & 2 \end{bmatrix}\begin{bmatrix} 4 & -1 \\ 5 & 3 \\ -2 & 0 \end{bmatrix}$

10. $\begin{bmatrix} 3 & 7 \\ 2 & 5 \\ -1 & -2 \end{bmatrix}\begin{bmatrix} 4 & 8 \\ 0 & -1 \end{bmatrix}$

11. $\begin{bmatrix} 6 & -1 & 3 \\ 4 & 8 & 7 \\ 2 & 4 & -1 \end{bmatrix}\begin{bmatrix} 1 & 0 & 0 \\ 0 & 1 & 0 \\ 0 & 0 & 1 \end{bmatrix}$

Mixed Review Exercises

Find the real solutions of each system.

1. $4x + 7y = 5$
   $x - 2y = 5$

2. $x^2 + 4y^2 = 16$
   $x^2 + y^2 = 4$

3. $x^2 - y^2 = 1$
   $y = -1 - x^2$

4. $5x + 2y = 3$
   $3x + 4y = -15$

5. $xy = 4$
   $y = -4x^2$

6. $x + y + z = 4$
   $2x + y - z = 8$
   $x - y + z = -2$

Solve.

7. $5x + 7 = 3 - 3x$

8. $\sqrt{3x - 2} = x$

9. $\dfrac{m}{m^2 - 1} + \dfrac{1}{m + 1} = 1$

NAME ______________________ DATE ______________

16–4 Applications of Matrices

Objective: To solve problems using matrices.

Vocabulary

Communication matrix A matrix representing routes by which data can be transmitted and received.

Example 1 Write the matrix that illustrates the network of computers shown at the right.

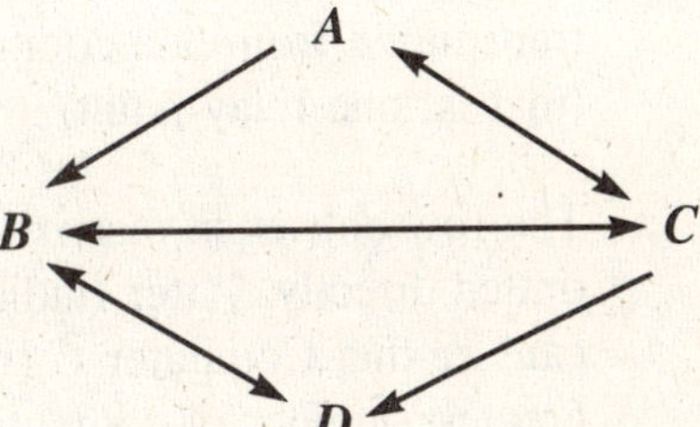

Solution The arrows in the drawing indicate the direction of communication in the network. When writing the matrix for this network, use a "1" to indicate that direct communication from one computer to another is possible and a "0" to indicate that direct communication is not possible. For example, Computer *A* cannot send data to itself directly, so a "0" goes in the first row, first column of the matrix. Likewise, Computer *A* can send data to Computer *B* directly, so a "1" goes in the first row, second column of the matrix.

$$\begin{array}{ccc} & & \text{To computer} \\ & & \begin{array}{cccc} A & B & C & D \end{array} \\ \text{From computer} & \begin{array}{c} A \\ B \\ C \\ D \end{array} & \begin{bmatrix} 0 & 1 & 1 & 0 \\ 0 & 0 & 1 & 1 \\ 1 & 1 & 0 & 1 \\ 0 & 1 & 0 & 0 \end{bmatrix} = X \end{array}$$

Matrix *X* is called a *communication matrix*.

Example 2 With regard to the network of computers given in Example 1, find the matrix that represents the number of ways that data can be sent from one computer to another using exactly one relay.

Solution The required matrix is the square of matrix *X* from the solution of Example 1:

$$X^2 = X \cdot X = \begin{bmatrix} 0 & 1 & 1 & 0 \\ 0 & 0 & 1 & 1 \\ 1 & 1 & 0 & 1 \\ 0 & 1 & 0 & 0 \end{bmatrix} \begin{bmatrix} 0 & 1 & 1 & 0 \\ 0 & 0 & 1 & 1 \\ 1 & 1 & 0 & 1 \\ 0 & 1 & 0 & 0 \end{bmatrix} = \begin{array}{c} \\ A \\ B \\ C \\ D \end{array} \begin{array}{c} \begin{array}{cccc} A & B & C & D \end{array} \\ \begin{bmatrix} 1 & 1 & 1 & 2 \\ 1 & 2 & 0 & 1 \\ 0 & 2 & 2 & 1 \\ 0 & 0 & 1 & 1 \end{bmatrix} \end{array}$$

The "1" in the first row, first column of X^2, for example, means that Computer *A* can send data to itself via Computer *C*. Likewise, the "2" in the first row, fourth column means that Computer *A* can send data to Computer *D* via either Computer *B* or Computer *C*.

NAME _______________ DATE _______________

16–4 Applications of Matrices *(continued)*

Example 3 Use Examples 1 and 2 to solve these problems.

a. Which computer can transmit directly to the greatest number of computers in the network?

b. Which computer can receive messages from the greatest number of computers in the network?

c. Find the matrix that represents the total number of routes that data can be transmitted from one computer to another using no more than 2 steps (that is, one relay point).

Solution

a. The row entries in matrix X represent routes on which data can be transmitted directly. After finding the sum of the numbers in each row, you can see that Computer C (with three routes) can send messages to the greatest number of computers.

b. The column entries in matrix X represent routes on which data can be received directly. After finding the sum of the numbers in each column, you can see that Computer B (with three routes) can receive messages from the greatest number of computers.

c. Matrix X represents the number of direct message routes, and matrix X^2 represents the number of routes using one relay station. So the sum $X + X^2$ represents the number of routes using *no more than* 2 steps.

$$X + X^2 = \begin{bmatrix} 0 & 1 & 1 & 0 \\ 0 & 0 & 1 & 1 \\ 1 & 1 & 0 & 1 \\ 0 & 1 & 0 & 0 \end{bmatrix} + \begin{bmatrix} 1 & 1 & 1 & 2 \\ 1 & 2 & 0 & 1 \\ 0 & 2 & 2 & 1 \\ 0 & 0 & 1 & 1 \end{bmatrix} = \begin{bmatrix} 1 & 2 & 2 & 2 \\ 1 & 2 & 1 & 2 \\ 1 & 3 & 2 & 2 \\ 0 & 1 & 1 & 1 \end{bmatrix}.$$

For Exercises 1–7, use the radio network shown at the right. (*Note*: A "hook" indicates that two line segments do *not* intersect.)

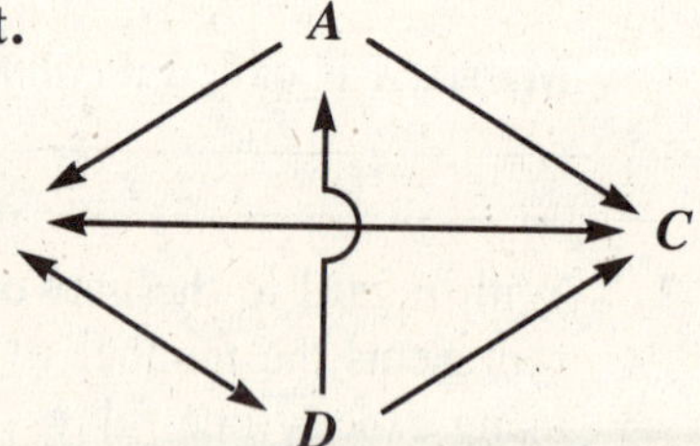

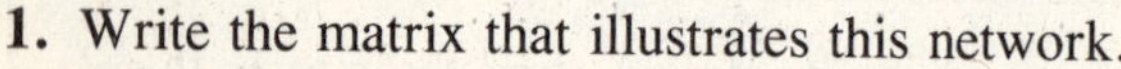

1. Write the matrix that illustrates this network.
2. Find the matrix that represents the number of routes a message can be sent from one station to another using exactly one relay.

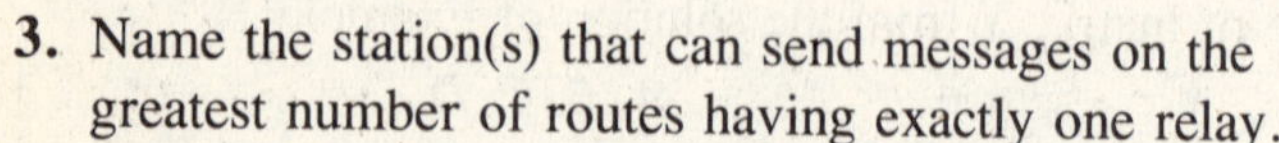

3. Name the station(s) that can send messages on the greatest number of routes having exactly one relay.
4. Name the station(s) that can receive messages on the greatest number of routes having exactly one relay.
5. Find the matrix that represents the number of routes a message can be sent from one station to another using at most one relay.
6. Name the station(s) that can send messages on the greatest number of routes having at most one relay.
7. Name the station(s) that can receive messages on the greatest number of routes having at most one relay.

NAME ______________________ DATE ____________

16–5 Determinants

Objective: To find the determinant of a 2×2 or 3×3 matrix.

Vocabulary

Determinant A real number associated with a *square* matrix. The determinant is usually displayed in the same form as the matrix, but with vertical bars rather than brackets enclosing the elements.

Order of a determinant The number of elements in any row or column of the determinant. Example: $A_{2\times 2}$ has order 2, and $B_{3\times 3}$ has order 3.

Determinant of a 2×2 matrix Let $A = \begin{bmatrix} a & b \\ c & d \end{bmatrix}$. Then the determinant of A, denoted by det A, is defined as:

$$\det A = \begin{vmatrix} a & b \\ c & d \end{vmatrix} = ad - bc$$

Determinant of a 3×3 matrix Let $B = \begin{bmatrix} a_1 & b_1 & c_1 \\ a_2 & b_2 & c_2 \\ a_3 & b_3 & c_3 \end{bmatrix}$. Then det B is defined as:

$$\det B = \begin{vmatrix} a_1 & b_1 & c_1 \\ a_2 & b_2 & c_2 \\ a_3 & b_3 & c_3 \end{vmatrix}$$

$$= a_1b_2c_3 + a_2b_3c_1 + a_3b_1c_2 - a_3b_2c_1 - a_2b_1c_3 - a_1b_3c_2$$

Diagonal matrix A square matrix whose only nonzero elements are on the *main diagonal* from upper left to lower right. Example: $\begin{bmatrix} -1 & 0 & 0 \\ 0 & 2 & 0 \\ 0 & 0 & 5 \end{bmatrix}$.

The determinant of a diagonal matrix is the product of the elements on the main diagonal.

Symbol

det A (the determinant of square matrix A)

Example 1 Evaluate: $\begin{vmatrix} 3 & -2 \\ 5 & -4 \end{vmatrix}$.

Solution $\begin{vmatrix} 3 & -2 \\ 5 & -4 \end{vmatrix} = 3(-4) - (-2)5 = -12 + 10 = -2$

Evaluate.

1. $\begin{vmatrix} 5 & 1 \\ 3 & 2 \end{vmatrix}$

2. $\begin{vmatrix} -1 & 3 \\ 5 & 2 \end{vmatrix}$

3. $\begin{vmatrix} 3 & -2 \\ 4 & -1 \end{vmatrix}$

4. $\begin{vmatrix} 15 & 2 \\ -1 & -3 \end{vmatrix}$

5. $\begin{vmatrix} 4 & 1 \\ 0 & 3 \end{vmatrix}$

6. $\begin{vmatrix} 3 & -4 \\ 5 & -6 \end{vmatrix}$

16–5 Determinants *(continued)*

Example 2 Evaluate: $\begin{vmatrix} 5 & -2 & 1 \\ 3 & 2 & 4 \\ -2 & 1 & -1 \end{vmatrix}$

Solution A convenient method for finding the six terms needed to evaluate a 3 × 3 determinant is shown below. This method works *only* for 3 × 3 determinants.

1. Copy the first two columns, in order, to the right of the third column.

$$\begin{array}{rrr|rr} 5 & -2 & 1 & 5 & -2 \\ 3 & 2 & 4 & 3 & 2 \\ -2 & 1 & -1 & -2 & 1 \end{array}$$

2. Multiply each element in the first row of the *original* matrix by the other two elements on the left-to-right downward diagonal. These products are the first three terms of the determinant.

$$\begin{array}{rrr|rr} 5 & -2 & 1 & 5 & -2 \\ 3 & 2 & 4 & 3 & 2 \\ -2 & 1 & -1 & -2 & 1 \\ & & & \mathbf{-10} & \mathbf{16} \quad \mathbf{3} \end{array}$$

3. Multiply each element in the last row of the *original* matrix by the other two elements on the left-to-right upward diagonal. The *opposites* of these products are the last three terms of the determinant.

$$\begin{array}{rrr|rr} & & & \mathbf{-4} & \mathbf{20} \quad \mathbf{6} \\ 5 & -2 & 1 & 5 & -2 \\ 3 & 2 & 4 & 3 & 2 \\ -2 & 1 & -1 & -2 & 1 \end{array}$$

$\therefore \det A = -10 + 16 + 3 - (-4) - 20 - 6 = -13$

Evaluate.

7. $\begin{vmatrix} 2 & 4 & 1 \\ -3 & -4 & 1 \\ 0 & 6 & 2 \end{vmatrix}$

8. $\begin{vmatrix} 5 & 4 & 0 \\ 0 & 2 & 1 \\ 3 & 0 & -1 \end{vmatrix}$

9. $\begin{vmatrix} 14 & 12 & 2 \\ 0 & -1 & 0 \\ 3 & -2 & -1 \end{vmatrix}$

10. $\begin{vmatrix} 5 & 0 & 0 \\ 0 & 7 & 0 \\ 0 & 0 & -2 \end{vmatrix}$

11. $\begin{vmatrix} 0 & 0 & 2 \\ 0 & -3 & 0 \\ 12 & 0 & 0 \end{vmatrix}$

12. $\begin{vmatrix} 5 & 1 & 3 \\ 6 & 4 & 2 \\ 0 & 1 & -1 \end{vmatrix}$

Mixed Review Exercises

Write in simplest form without zero or negative exponents.

1. $\dfrac{x^2 - 4}{x + 3x + 2}$

2. $\dfrac{x^5y^4}{x^4y^7}$

3. $\dfrac{2}{x + 2} + \dfrac{3}{3x - 3}$

4. $\sqrt[4]{a^5} \cdot \sqrt[4]{a^7}$

5. $\dfrac{1 + x^{-1}}{x - x^{-1}}$

6. $\dfrac{x + 3}{x - 2} \div \dfrac{x^2 + 7x + 12}{x^2 - 4}$

7. $(-7xy^2z^4)(2x^3y^3z^3)$

8. $\dfrac{x^3 - 1}{x^2 + x + 1}$

9. $(3^\pi)^{1/\pi}$

NAME ______________________ DATE ______________

16–6 Inverses of Matrices

Objective: To solve systems of equations using inverses of matrices.

Vocabulary

Inverse matrices Square matrices A and B are inverses if their product is the identity matrix (that is, $AB = BA = I$).

Inverse of a 2×2 matrix If $A = \begin{bmatrix} a & b \\ c & d \end{bmatrix}$ and $\det A \neq 0$, then the inverse of A, denoted by A^{-1}, is given by:

$$A^{-1} = \frac{1}{\det A}\begin{bmatrix} d & -b \\ -c & a \end{bmatrix} = \begin{bmatrix} \frac{d}{ad - bc} & \frac{-b}{ad - bc} \\ \frac{-c}{ad - bc} & \frac{a}{ad - bc} \end{bmatrix}$$

If $\det A = 0$, then A has *no* inverse.

Matrix equation An equation in which the variable stands for a matrix. The system of linear equations

$$\begin{aligned} ax + by &= e \\ cx + dy &= f \end{aligned}$$

can be written as a matrix equation by using the definitions of matrix multiplication and equality of matrices as follows:

$$\begin{bmatrix} a & b \\ c & d \end{bmatrix}\begin{bmatrix} x \\ y \end{bmatrix} = \begin{bmatrix} e \\ f \end{bmatrix}$$

This equation has the form $AX = B$, where $A = \begin{bmatrix} a & b \\ c & d \end{bmatrix}$, $X = \begin{bmatrix} x \\ y \end{bmatrix}$, and $B = \begin{bmatrix} e \\ f \end{bmatrix}$. A is called the *matrix of coefficients* of the system.

Symbol A^{-1} (the inverse of matrix A)

Example 1 Find A^{-1} if $A = \begin{bmatrix} 5 & -2 \\ 4 & -3 \end{bmatrix}$.

Solution First find $\det A$: $\begin{vmatrix} 5 & -2 \\ 4 & -3 \end{vmatrix} = 5(-3) - (-2)4 = -7$

Since $\det A \neq 0$, A has an inverse. To find A^{-1}, use the rule for the inverse of a 2×2 matrix given above: Interchange 5 and -3, replace both -2 and 4 with their opposites, and then multiply the resulting matrix by $\frac{1}{\det A}$.

$$\therefore A^{-1} = -\frac{1}{7}\begin{bmatrix} -3 & 2 \\ -4 & 5 \end{bmatrix} = \begin{bmatrix} \frac{3}{7} & -\frac{2}{7} \\ \frac{4}{7} & -\frac{5}{7} \end{bmatrix}$$

Check: Multiply A and A^{-1} to see if you get the 2×2 identity matrix:

$$AA^{-1} = \begin{bmatrix} 5 & -2 \\ 4 & -3 \end{bmatrix}\begin{bmatrix} \frac{3}{7} & -\frac{2}{7} \\ \frac{4}{7} & -\frac{5}{7} \end{bmatrix} = \begin{bmatrix} 1 & 0 \\ 0 & 1 \end{bmatrix} = I_{2\times 2} \quad \surd$$

NAME ______________________________ DATE ______________

16–6 Inverses of Matrices *(continued)*

Find the inverse of each matrix. If the matrix has no inverse, say so.

1. $\begin{bmatrix} 0 & 1 \\ 1 & 0 \end{bmatrix}$

2. $\begin{bmatrix} 5 & -7 \\ -3 & 4 \end{bmatrix}$

3. $\begin{bmatrix} 2 & 3 \\ 4 & 6 \end{bmatrix}$

4. $\begin{bmatrix} 7 & 2 \\ -1 & 0 \end{bmatrix}$

5. $\begin{bmatrix} 4 & 10 \\ 2 & 5 \end{bmatrix}$

6. $\begin{bmatrix} 1 & 1 \\ 2 & -2 \end{bmatrix}$

7. $\begin{bmatrix} 4 & 3 \\ -1 & 2 \end{bmatrix}$

8. $\begin{bmatrix} 3 & 2 \\ 4 & 6 \end{bmatrix}$

9. $\begin{bmatrix} 5 & 0 \\ 0 & -2 \end{bmatrix}$

Example 2 Use matrices to solve this system:
$$2x + 3y = 1$$
$$-x + 4y = 16$$

Solution Write the matrix equation $AX = B$: $\begin{bmatrix} 2 & 3 \\ -1 & 4 \end{bmatrix}\begin{bmatrix} x \\ y \end{bmatrix} = \begin{bmatrix} 1 \\ 16 \end{bmatrix}$

Then find det A: $\begin{vmatrix} 2 & 3 \\ -1 & 4 \end{vmatrix} = 2(4) - (-1)(3) = 11$

Since det $A \neq 0$, A^{-1} exists. The solution of the equation $AX = B$ is, therefore, $X = A^{-1}B$. To find $A^{-1}B$, you must first find A^{-1}, the inverse of the matrix of coefficients:

$$A^{-1} = \frac{1}{11}\begin{bmatrix} 4 & -3 \\ 1 & 2 \end{bmatrix}$$

Then $\quad X = A^{-1}B$:

$$\begin{bmatrix} x \\ y \end{bmatrix} = \frac{1}{11}\begin{bmatrix} 4 & -3 \\ 1 & 2 \end{bmatrix}\begin{bmatrix} 1 \\ 16 \end{bmatrix}$$
$$= \frac{1}{11}\begin{bmatrix} -44 \\ 33 \end{bmatrix}$$
$$= \begin{bmatrix} -4 \\ 3 \end{bmatrix}$$

∴ the solution of the system is $(-4, 3)$.

Check: $2(-4) + 3(3) = 1$ √
$-(-4) + 4(3) = 16$ √

Use matrices to find the solution of each system of equations. If a system has no unique solution, say so.

10. $x + 2y = 11$
$3x + y = 18$

11. $2x + 3y = 12$
$5x + 3y = 3$

12. $3x + 2y = 9$
$x + y = 3$

13. $4x - 2y = 5$
$2x - y = 3$

14. $3x + 7y = -4$
$2x + 5y = -3$

15. $3x + 2y = 14$
$4x - y = 15$

16. $x - y = 11$
$3x + 2y = 3$

17. $6x + 10y = 7$
$9x + 15y = 3$

NAME ______________________ DATE ____________

16–7 Expansion of Determinants by Minors

Objective: To evaluate third-order determinants using expansion by minors.

Vocabulary

Minor of an element in a determinant The determinant resulting from the deletion of the row and column containing the element. Example: For the determinant $\begin{vmatrix} 1 & 2 & 3 \\ 4 & 5 & 6 \\ 7 & 8 & 9 \end{vmatrix}$, the minor of the element 5 is $\begin{vmatrix} 1 & 2 & 3 \\ 4 & 5 & 6 \\ 7 & 8 & 9 \end{vmatrix}$, or $\begin{vmatrix} 1 & 3 \\ 7 & 9 \end{vmatrix}$.

Expanding a determinant by minors To expand a determinant by minors:

1. Multiply each element of a given row or column by its minor.
2. Add the number of the row and the number of the column for each element. If the sum is odd, multiply the product obtained in Step 1 by -1.
3. Add the products to obtain the value of the determinant.

(*Note*: A determinant may be expanded by minors about any row or column. To simplify your work, always choose a row or column having the greatest number of zeros as elements.)

Example 1 Expand $\begin{vmatrix} 4 & 2 & 1 \\ 3 & 5 & 6 \\ -2 & -1 & 2 \end{vmatrix}$ by the minors of the first row.

Solution Multiply each element of the first row by its minor. Also, multiply the product for the element in the *first* row, *second* column by -1, since the sum $1 + 2$, or 3, is odd. Then add the products.

$$\begin{vmatrix} 4 & 2 & 1 \\ 3 & 5 & 6 \\ -2 & -1 & 2 \end{vmatrix} = 4\begin{vmatrix} 4 & 2 & 1 \\ 3 & 5 & 6 \\ -2 & -1 & 2 \end{vmatrix} + (-1)(2)\begin{vmatrix} 4 & 2 & 1 \\ 3 & 5 & 6 \\ -2 & -1 & 2 \end{vmatrix} + 1\begin{vmatrix} 4 & 2 & 1 \\ 3 & 5 & 6 \\ -2 & -1 & 2 \end{vmatrix}$$

$$= 4\begin{vmatrix} 5 & 6 \\ -1 & 2 \end{vmatrix} - 2\begin{vmatrix} 3 & 6 \\ -2 & 2 \end{vmatrix} + 1\begin{vmatrix} 3 & 5 \\ -2 & -1 \end{vmatrix}$$

$$= 4[10 - (-6)] - 2[6 - (-12)] + 1[-3 - (-10)]$$

$$= 4(16) - 2(18) + 1(7)$$

$$= 64 - 36 + 7$$

$$= 35$$

Expand each determinant about the given row or column.

1. $\begin{vmatrix} 4 & 0 & 5 \\ -3 & -2 & -1 \\ 1 & 2 & 3 \end{vmatrix}$; row one

2. $\begin{vmatrix} 5 & 1 & -1 \\ 2 & 3 & 4 \\ -2 & 0 & -3 \end{vmatrix}$; row three

3. $\begin{vmatrix} 1 & -1 & 0 \\ 0 & 1 & 2 \\ 2 & 1 & -1 \end{vmatrix}$; column one

4. $\begin{vmatrix} 3 & 0 & 2 \\ 1 & -1 & 0 \\ 0 & 4 & 2 \end{vmatrix}$; column two

NAME ______________________________ DATE ______________

16–7 Expansion of Determinants by Minors *(continued)*

Example 2 Expand $\begin{vmatrix} 3 & 4 & 0 \\ 2 & 0 & 1 \\ -1 & 0 & 5 \end{vmatrix}$ about any row or column.

Solution Choose the second column, because it contains more zeros than any row and any other column.

$$\begin{vmatrix} 3 & 4 & 0 \\ 2 & 0 & 1 \\ -1 & 0 & 5 \end{vmatrix} = (-1)(4)\begin{vmatrix} 2 & 1 \\ -1 & 5 \end{vmatrix} + 0\begin{vmatrix} 3 & 0 \\ -1 & 5 \end{vmatrix} + (-1)(0)\begin{vmatrix} 3 & 0 \\ 2 & 1 \end{vmatrix}$$

$$= -4(11) + 0 + 0$$

$$= -44$$

Expand each determinant about any row or column.

5. $\begin{vmatrix} 2 & 4 & 3 \\ 0 & 1 & 0 \\ 5 & 6 & -2 \end{vmatrix}$

6. $\begin{vmatrix} 0 & 1 & 2 \\ 2 & 1 & 0 \\ 0 & 2 & 1 \end{vmatrix}$

7. $\begin{vmatrix} 1 & -1 & 0 \\ 2 & -2 & 1 \\ -1 & 1 & 0 \end{vmatrix}$

8. $\begin{vmatrix} 3 & 0 & 0 \\ 4 & 2 & 1 \\ -3 & -6 & -2 \end{vmatrix}$

Example 3 Solve for x: $\begin{vmatrix} x & 1 & 0 \\ 3 & -1 & 4 \\ 0 & 2 & -3 \end{vmatrix} = -1$

Solution Expand the determinant about the first row:

$$x\begin{vmatrix} -1 & 4 \\ 2 & -3 \end{vmatrix} + (-1)(1)\begin{vmatrix} 3 & 4 \\ 0 & -3 \end{vmatrix} + 0\begin{vmatrix} 3 & -1 \\ 0 & 2 \end{vmatrix} = -1$$

$$x(-5) + (-1)(-9) + 0 = -1$$

$$-5x + 9 = -1$$

$$-5x = -10$$

$$x = 2$$

Solve each equation for x.

9. $\begin{vmatrix} -3 & 0 & 0 \\ 0 & x & 0 \\ 0 & 0 & 2 \end{vmatrix} = 30$

10. $\begin{vmatrix} -1 & 2 & -2 \\ 0 & 4 & x \\ -3 & 1 & 0 \end{vmatrix} = -39$

Mixed Review Exercises

Simplify.

1. $27^{-2/3}$
2. $\sin \frac{5\pi}{4}$
3. $\frac{7!}{2!\,5!}$
4. $\sqrt{-3} \cdot \sqrt{-12}$
5. $\log_5 25\sqrt{5}$
6. $\csc^2 50° - \cot^2 50°$

NAME ______________________ DATE ______________

16–8 Properties of Determinants

Objective: To use the properties of determinants to simplify the expansion of determinants by minors.

Vocabulary

Properties of determinants The following properties will help you evaluate determinants of any order.

1. If each element in any row (or any column) is 0, then the determinant is equal to 0.
2. If two rows (or two columns) of a determinant have corresponding elements that are equal, then the determinant is equal to 0.
3. If two rows (or two columns) of a determinant are interchanged, then the resulting determinant is the opposite of the original determinant.
4. If each element in one row (or one column) of a determinant is multiplied by a real number k, then the determinant is multiplied by k.
5. If each element of one row (or one column) is multiplied by a real number k and if the resulting products are then added to the corresponding elements of another row (or another column), then the resulting determinant equals the original one.

Example 1 Evaluate: $\begin{vmatrix} 3 & 2 & 1 \\ 4 & 3 & -1 \\ 3 & 2 & 1 \end{vmatrix}$

Solution Since the first and the third rows have corresponding elements that are equal, Property 2 tells you that the determinant is equal to 0.

Example 2 Evaluate: $\begin{vmatrix} 4 & -1 & 0 \\ 11 & 22 & 33 \\ 2 & 5 & 6 \end{vmatrix}$

Solution Factor 11 from the elements of the second row (Property 4):

$$\begin{vmatrix} 4 & -1 & 0 \\ 11 & 22 & 33 \\ 2 & 5 & 6 \end{vmatrix} = \begin{vmatrix} 4 & -1 & 0 \\ 11(1) & 11(2) & 11(3) \\ 2 & 5 & 6 \end{vmatrix}$$

$$= 11\begin{vmatrix} 4 & -1 & 0 \\ 1 & 2 & 3 \\ 2 & 5 & 6 \end{vmatrix}$$

Expand by minors about the first row:

$$= 11\left(4\begin{vmatrix} 2 & 3 \\ 5 & 6 \end{vmatrix} + (-1)(-1)\begin{vmatrix} 1 & 3 \\ 2 & 6 \end{vmatrix} + 0\begin{vmatrix} 1 & 2 \\ 2 & 5 \end{vmatrix}\right)$$

$$= 11[4(-3) + 1(0) + 0]$$

$$= -132$$

NAME ______________________ DATE ______________

16–8 Properties of Determinants *(continued)*

Example 3 Evaluate: $\begin{vmatrix} 1 & 6 & 7 \\ -2 & 9 & -13 \\ 3 & 12 & 11 \end{vmatrix}$

Solution Factor 3 from the elements of the second column (Property 4):

$$\begin{vmatrix} 1 & 6 & 7 \\ -2 & 9 & -13 \\ 3 & 12 & 11 \end{vmatrix} = \begin{vmatrix} 1 & 3(2) & 7 \\ -2 & 3(3) & -13 \\ 3 & 3(4) & 11 \end{vmatrix} = 3\begin{vmatrix} 1 & 2 & 7 \\ -2 & 3 & -13 \\ 3 & 4 & 11 \end{vmatrix}$$

Multiply the first row by 2 and add the result to the second row (Property 5):

$$= 3\begin{vmatrix} 1 & 2 & 7 \\ -2+2 & 3+4 & -13+14 \\ 3 & 4 & 11 \end{vmatrix}$$

$$= 3\begin{vmatrix} 1 & 2 & 7 \\ 0 & 7 & 1 \\ 3 & 4 & 11 \end{vmatrix}$$ Notice that the first column now contains a zero.

Multiply the first row by −3 and add the result to the third row (Property 5):

$$= 3\begin{vmatrix} 1 & 2 & 7 \\ 0 & 7 & 1 \\ 3+(-3) & 4+(-6) & 11+(-21) \end{vmatrix}$$

$$= 3\begin{vmatrix} 1 & 2 & 7 \\ 0 & 7 & 1 \\ 0 & -2 & -10 \end{vmatrix}$$ Notice that the first column now contains two zeros.

Expand by minors about the first column:

$$= 3\left(1\begin{vmatrix} 7 & 1 \\ -2 & -10 \end{vmatrix} + 0 + 0\right)$$

$$= 3(-68) = -204$$

Evaluate each determinant. Use Properties 1–5 to simplify the work.

1. $\begin{vmatrix} 24 & 12 & 12 \\ 0 & 6 & 6 \\ 5 & 8 & 8 \end{vmatrix}$ **2.** $\begin{vmatrix} 3 & 0 & 12 \\ 0 & 0 & 0 \\ 4 & 17 & 21 \end{vmatrix}$ **3.** $\begin{vmatrix} 5 & 4 & -5 \\ -1 & 6 & 1 \\ 2 & 7 & -2 \end{vmatrix}$

4. $\begin{vmatrix} 6 & 0 & 0 \\ 0 & 6 & 0 \\ 0 & 0 & 6 \end{vmatrix}$ **5.** $\begin{vmatrix} 3 & -2 & 4 \\ 7 & 9 & 6 \\ -3 & 2 & -4 \end{vmatrix}$ **6.** $\begin{vmatrix} 3 & -1 & 14 \\ 2 & 5 & 21 \\ -3 & 4 & 7 \end{vmatrix}$

7. $\begin{vmatrix} 1 & 4 & 8 \\ 0 & -4 & -6 \\ 1 & 4 & -2 \end{vmatrix}$ **8.** $\begin{vmatrix} 3 & 1 & 1 \\ 1 & 3 & 1 \\ 1 & 1 & 3 \end{vmatrix}$ **9.** $\begin{vmatrix} 19 & 17 & 18 \\ 19 & 14 & 13 \\ 19 & 15 & 12 \end{vmatrix}$

NAME ______________________ DATE ____________

16–9 Cramer's Rule

Objective: To solve systems of equations using determinants.

Vocabulary

Cramer's rule The unique solution of a system of n linear equations in n variables is given by $x = \frac{D_x}{D}$, $y = \frac{D_y}{D}$, . . . where D is the determinant of the matrix of coefficients of the variables ($D \neq 0$) and D_x, D_y, . . . are derived from D by replacing the coefficients of x, y, . . . , respectively, by the constants.

CAUTION If $D = 0$, the system may have no solution or infinitely many solutions. To distinguish between the two cases, find the value of D_y. If $D = 0$ and $D_y \neq 0$, the system has no solution. If $D = 0$ and $D_y = 0$, the system has infinitely many solutions.

Example 1 Use Cramer's rule to solve this system:

$$3x - 2y = 8$$
$$5x + 7y = 3$$

Solution The determinant of the coefficient of matrices for this system is:

$$D = \begin{vmatrix} 3 & -2 \\ 5 & 7 \end{vmatrix} = 21 - (-10) = 31$$

To find D_x replace the x-column of D by the constants:

$$D_x = \begin{vmatrix} \mathbf{8} & -2 \\ \mathbf{3} & 7 \end{vmatrix} = 56 - (-6) = 62$$

To find D_y replace the y-column of D by the constants:

$$D_y = \begin{vmatrix} 3 & \mathbf{8} \\ 5 & \mathbf{3} \end{vmatrix} = 9 - 40 = -31$$

So $x = \frac{D_x}{D} = \frac{62}{31} = 2$ and $y = \frac{D_y}{D} = \frac{-31}{31} = -1.$

$\therefore$ the solution of the system is $(2, -1)$.

Example 2 Use Cramer's rule to solve this system:

$$x - y + 4z = 4$$
$$2x + 3y - 3z = 1$$
$$3x + 2y - 2z = -1$$

Solution

$$D = \begin{vmatrix} 1 & -1 & 4 \\ 2 & 3 & -3 \\ 3 & 2 & -2 \end{vmatrix} = 1\begin{vmatrix} 3 & -3 \\ 2 & -2 \end{vmatrix} + (-1)(-1)\begin{vmatrix} 2 & -3 \\ 3 & -2 \end{vmatrix} + 4\begin{vmatrix} 2 & 3 \\ 3 & 2 \end{vmatrix}$$

$$= 1(0) + 1(5) + 4(-5) = -15$$

$$D_x = \begin{vmatrix} \mathbf{4} & -1 & 4 \\ \mathbf{1} & 3 & -3 \\ \mathbf{-1} & 2 & -2 \end{vmatrix} = 4\begin{vmatrix} 3 & -3 \\ 2 & -2 \end{vmatrix} + (-1)(-1)\begin{vmatrix} 1 & -3 \\ -1 & -2 \end{vmatrix} + 4\begin{vmatrix} 1 & 3 \\ -1 & 2 \end{vmatrix}$$

$$= 4(0) + 1(-5) + 4(5) = 15$$

(Solution continues on the next page.)

16–9 Cramer's Rule (continued)

$$D_y = \begin{vmatrix} 1 & \mathbf{4} & 4 \\ 2 & \mathbf{1} & -3 \\ 3 & \mathbf{-1} & -2 \end{vmatrix} = 1\begin{vmatrix} 1 & -3 \\ -1 & -2 \end{vmatrix} + (-1)(4)\begin{vmatrix} 2 & -3 \\ 3 & -2 \end{vmatrix} + 4\begin{vmatrix} 2 & 1 \\ 3 & -1 \end{vmatrix}$$

$$= 1(-5) - 4(5) + 4(-5) = -45$$

$$D_z = \begin{vmatrix} 1 & -1 & \mathbf{4} \\ 2 & 3 & \mathbf{1} \\ 3 & 2 & \mathbf{-1} \end{vmatrix} = 1\begin{vmatrix} 3 & 1 \\ 2 & -1 \end{vmatrix} + (-1)(-1)\begin{vmatrix} 2 & 1 \\ 3 & -1 \end{vmatrix} + 4\begin{vmatrix} 2 & 3 \\ 3 & 2 \end{vmatrix}$$

$$= 1(-5) + 1(-5) + 4(-5) = -30$$

So $x = \dfrac{D_x}{D} = \dfrac{15}{-15} = -1$, $y = \dfrac{D_y}{D} = \dfrac{-45}{-15} = 3$, and $z = \dfrac{D_z}{D} = \dfrac{-30}{-15} = 2$.

$\therefore$ the solution of the system is $(-1, 3, 2)$.

Use Cramer's rule to solve each system.

1. $9x + 5y = -6$
   $4x + 3y = 2$

2. $3x - 4y - z = 0$
   $-x + 2y + 2z = 5$
   $2x - y + 3z = 11$

3. $x - 3y - 2z = 3$
   $2x + y + 3z = -1$
   $-3x - 2y + z = 8$

4. $3x + 10y - 5z = -6$
   $x + 2y - z = -2$
   $-3x + 5y - 2z = 5$

Example 3 Determine the number of solutions for each system.

a. $3x + 5y = 2$
$6x + 10y = 7$

b. $2x - 4y = 6$
$3x - 6y = 9$

Solution

a. $D = \begin{vmatrix} 3 & 5 \\ 6 & 10 \end{vmatrix} = 30 - 30 = 0$ and $D_y = \begin{vmatrix} 3 & 2 \\ 6 & 7 \end{vmatrix} = 21 - 12 = 9$

Since $D = 0$ and $D_y \neq 0$, the system has no solution.

b. $D = \begin{vmatrix} 2 & -4 \\ 3 & -6 \end{vmatrix} = -12 - (-12) = 0$ and $D_y = \begin{vmatrix} 2 & 6 \\ 3 & 9 \end{vmatrix} = 18 - 18 = 0$

Since $D = 0$ and $D_y = 0$, the system has infinitely many solutions.

Determine whether each system has no solution, one solution, or more than one solution.

5. $3x - 2y = 4$
   $-6x + 4y = 7$

6. $5x + 2y = 3$
   $-10x - 4y = -6$

7. $4x - y = 5$
   $2x + y = 5$

8. $5x + 2y = 3$
   $10x + 4y = 6$

Mixed Review Exercises

Perform the indicated matrix operations.

1. $\begin{bmatrix} 3 & 5 \\ 2 & -1 \end{bmatrix} + \begin{bmatrix} 4 & -2 \\ 3 & -2 \end{bmatrix}$

2. $3\begin{bmatrix} 1 & -5 \\ 2 & -4 \end{bmatrix}$

3. $\begin{bmatrix} -1 & 1 \\ 2 & -2 \end{bmatrix}\begin{bmatrix} 0 & 2 \\ -1 & 1 \end{bmatrix}$

Graph each function.

4. $y = \dfrac{1}{2}\cos 4x$

5. $y = \dfrac{4}{5}x - 1$

6. $y = -x^2 + 2x - 6$